Functional Organic Liquids

Functional Organic Liquids

Edited by
Takashi Nakanishi

Editor

Dr. Takashi Nakanishi
National Institute for Materials Science (NIMS)
International Center for Materials Nanoarchitectonics (WPI-MANA)
1-1 Namiki
305 Tsukuba
Japan

Cover Images: Background: © oxygen/Getty Images, © Takashi Nakanishi

All books published by **Wiley-VCH** are carefully produced. Nevertheless, authors, editors, and publisher do not warrant the information contained in these books, including this book, to be free of errors. Readers are advised to keep in mind that statements, data, illustrations, procedural details or other items may inadvertently be inaccurate.

Library of Congress Card No.: applied for

British Library Cataloguing-in-Publication Data
A catalogue record for this book is available from the British Library.

Bibliographic information published by the Deutsche Nationalbibliothek
The Deutsche Nationalbibliothek lists this publication in the Deutsche Nationalbibliografie; detailed bibliographic data are available on the Internet at <http://dnb.d-nb.de>.

© 2019 Wiley-VCH Verlag GmbH & Co. KGaA, Boschstr. 12, 69469 Weinheim, Germany

All rights reserved (including those of translation into other languages). No part of this book may be reproduced in any form – by photoprinting, microfilm, or any other means – nor transmitted or translated into a machine language without written permission from the publishers. Registered names, trademarks, etc. used in this book, even when not specifically marked as such, are not to be considered unprotected by law.

Print ISBN: 978-3-527-34190-0
ePDF ISBN: 978-3-527-80492-4
ePub ISBN: 978-3-527-80495-5
oBook ISBN: 978-3-527-80494-8

Typesetting SPi Global, Chennai, India
Printing and Binding CPI books GmbH, Germany

Printed on acid-free paper

Contents

Preface *xi*

1 Room-Temperature Liquid Dyes *1*
Bhawani Narayan and Takashi Nakanishi
1.1 Introduction *1*
1.2 Design Strategy: Alkyl Chain Engineering *2*
1.3 Alkylated π-Molecular Liquids *3*
1.3.1 Carbazoles *3*
1.3.2 Azobenzenes *5*
1.3.3 Naphthalenes *6*
1.3.4 Anthracenes *6*
1.3.5 Pyrenes *8*
1.3.6 π-Conjugated Oligomers *10*
1.3.6.1 Oligo-(*p*-phenylenevinylene)s (OPVs) *10*
1.3.6.2 Oligo-(*p*-phenyleneethylene)s (OPEs) *11*
1.3.6.3 Benzothiadiazoles (BTDs) *12*
1.3.7 Porphyrins *12*
1.3.8 Fullerenes *12*
1.4 Alkylsilane-Chain-Appended π-Molecular Liquids *13*
1.4.1 Triarylamines *14*
1.4.2 Phthalocyanines *15*
1.4.3 Oligofluorenes *15*
1.5 Analytical Tools for Functional Molecular Liquids *16*
1.5.1 Analytical Tools for Bulk Physical Properties *16*
1.5.1.1 Structural Analysis *16*
1.5.1.2 Microscopy Techniques *16*
1.5.1.3 Rheology *16*
1.5.1.4 Calorimetric Techniques *17*
1.5.2 Analytical Tools for Spectroscopic Properties *17*
1.5.2.1 UV–vis Analysis *17*
1.5.2.2 Fluorescence Measurements *17*
1.5.2.3 Fluorescence Lifetime Analysis *17*
1.5.2.4 FTIR Measurements *17*
1.6 Conclusion *18*
References *18*

2	**Low-Melting Porphyrins and Their Photophysical Properties** *21*
	Agnieszka Nowak-Król and Daniel T. Gryko
2.1	Introduction *21*
2.2	Liquid Porphyrins *22*
2.3	Low-Melting *trans*-A_2B_2-Arylethynyl Porphyrins *28*
2.4	Liquid Crystalline *trans*-A_2B_2-Arylethynyl Porphyrins *31*
2.5	Bis-porphyrins *31*
2.6	Low-Melting Corroles *34*
2.7	Summary and Outlook *34*
	References *35*

3	**Porous Liquids** *39*
	Stuart L. James and Ben Hutchings
3.1	Introduction *39*
3.2	Porosity in Solids *40*
3.3	Porosity in Liquids *41*
3.4	Porous Liquids Reported in the Literature *43*
3.4.1	Type 1 *43*
3.4.2	Type 2 *46*
3.4.3	Type 3 *48*
3.4.4	Other Types of Porous Liquids and Theoretical Studies *48*
3.5	Opportunities for Applications and Current Challenges *49*
3.6	Concluding Remarks *50*
	References *50*

4	**Cyclic Host Liquids for the Formation of Rotaxanes and Their Applications** *53*
	Tomoki Ogoshi, Takahiro Kakuta, and Tada-aki Yamagishi
4.1	Introduction *53*
4.2	Liquid Pillar[*n*]arenes at Room Temperature *54*
4.2.1	Synthesis and Structure of Pillar[*n*]arenes *54*
4.2.2	Versatile Functionality of Pillar[*n*]arenes *55*
4.2.3	Molecular Design to Produce Liquid-State Macrocyclic Hosts *56*
4.2.3.1	Pillar[*n*]arenes *56*
4.2.3.2	Cyclodextrins *58*
4.2.3.3	Crown Ethers *60*
4.2.3.4	Calix[*n*]arenes and Cucurbit[*n*]urils *60*
4.3	Complexation of Guest Molecules by Pillar[5]arenes *61*
4.3.1	Host Properties of Pillar[5]arenes *61*
4.3.2	Complexation of Guest Molecules in Liquid Pillar[5]arenes *62*
4.4	High Yield Synthesis of [2]Rotaxane and Polyrotaxane Using Liquid Pillar[5]arenes as Solvents *63*
4.5	Conclusion and Remarks *70*
	References *71*

5	**Photochemically Reversible Liquefaction/Solidification of Sugar-Alcohol Derivatives** *75*	
	Haruhisa Akiyama	
5.1	Introduction *75*	
5.2	Mechanism of the Phase Transition Between Liquid and Solid State *76*	
5.3	Effect of Molecular Structure *79*	
5.3.1	Number of Azobenzene Units *79*	
5.3.2	Alkyl Chain Length *80*	
5.3.3	Mixed Arms *82*	
5.3.4	Structure of Sugar Alcohol *83*	
5.4	Summary *85*	
	Acknowledgments *85*	
	References *85*	
6	**Functional Organic Supercooled Liquids** *87*	
	Kyeongwoon Chung, Da Seul Yang, and Jinsang Kim	
6.1	Organic Supercooled Liquids *87*	
6.2	Stimuli-Responsive Organic Supercooled Liquids *88*	
6.2.1	Shear-triggered Crystallization *88*	
6.2.2	Scratch-Induced Crystallization of Trifluoromethylquinoline Derivatives *89*	
6.2.3	Highly Sensitive Shear-Triggered Crystallization in Thermally Stable Organic Supercooled Liquid of a Diketopyrrolopyrrole Derivative *91*	
6.3	Highly Emissive Supercooled Liquids *95*	
6.4	Conclusion *97*	
	References *97*	
7	**Organic Liquids in Energy Systems** *101*	
	Pengfei Duan, Nobuhiro Yanai, and Nobuo Kimizuka	
7.1	Introduction *101*	
7.2	Photoresponsive π-Liquids for Molecular Solar Thermal Fuels *102*	
7.3	Azobenzene-Containing Ionic Liquids and the Phase Crossover Approach *107*	
7.4	Photon Upconversion and Condensed Molecular Systems *113*	
7.5	TTA-UC Based on the Amorphous π-Liquid Systems *114*	
7.6	Photon Upconversion Based on Bicontinuous Ionic Liquid Systems *118*	
7.7	Conclusion and Outlook *121*	
	References *122*	
8	**Organic Light Emitting Diodes with Liquid Emitters** *127*	
	Jean-Charles Ribierre, Jun Mizuno, Reiji Hattori, and Chihaya Adachi	
8.1	Introduction *127*	
8.2	Organic Light-emitting Diodes with a Solvent-Free Liquid Organic Light-emitting Layer *129*	

8.2.1	Basics of Conventional Solid-state OLEDs *129*
8.2.2	First Demonstration of a Fluidic OLED Based on a Liquid Carbazole Host *130*
8.2.3	Introduction of an Electrolyte to Improve the Liquid OLED Performance *132*
8.2.4	Liquid OLED Material Issues *134*
8.3	Microfluidic OLEDs *135*
8.3.1	Refreshable Liquid Electroluminescent Devices *135*
8.3.2	Fabrication of Microfluidic Organic Light-Emitting Devices *137*
8.3.3	Large-Area Flexible Microfluidic OLEDs *137*
8.3.4	Multicolor Microfluidic OLEDs *140*
8.3.5	Microfluidic White OLEDs *143*
8.4	Conclusions *147*
	References *148*

9   Liquids Based on Nanocarbons and Inorganic Nanoparticles *151*
Avijit Ghosh and Takashi Nakanishi

9.1	Liquid Nanocarbons *151*
9.1.1	Introduction *151*
9.1.2	General Synthetic Strategies *151*
9.1.3	Liquid Fullerenes *152*
9.1.4	Liquid-Like Carbon Nanotubes *154*
9.1.5	Fluidic Graphene/Graphene Oxide *156*
9.2	Liquids Based on Inorganic Nanoparticles *158*
9.2.1	Background *158*
9.2.2	Liquid-Like Silica Nanoparticles *159*
9.2.3	Functional Colloidal Fluids *160*
9.2.4	Fluidic Functional Quantum Dots *161*
9.3	Conclusions *162*
	References *164*

10   Solvent-Free Nanofluids and Reactive Nanofluids *169*
John Texter

10.1	Introduction *169*
10.1.1	Solvent-Free Nanofluids *170*
10.1.2	Simulation and Theoretical Modeling *180*
10.1.3	Reactive Solvent-Free Nanofluids *183*
10.2	Syntheses of Nanofluids *184*
10.2.1	Core–Corona–Cap Nanofluid *184*
10.2.2	Core-Free Corona–Cap Nanofluid *186*
10.2.3	Core–Corona Nanofluid *186*
10.3	UV Reactive Nanofluids *187*
10.3.1	Model Coatings and Thermomechanical Characterization *187*
10.3.2	UV Protective Coatings *191*
10.4	Polyurethane and Polyurea Coupling of Nanofluids *191*
10.4.1	Air-Cured Polyurethane Coupling with Isothiocyanate Nanofluid *192*

10.4.2	Air-Cured TDI Coupling with Amino Nanofluid	195
10.4.3	Polyurethane Shape-Memory Materials	196
10.4.4	PDMS-Amino Nanofluids Coupling with HMDI	197
10.4.5	Polyurethane Coupling with Hydroxyl Nanofluid	198
10.5	Epoxy Coupling with Amino Nanofluid	198
10.6	Using Nanofluids to Make Composites Tougher	199
10.6.1	Nanosilica Polyacrylate Nanocomposites	199
10.6.2	MWCNT Polyamide Nanocomposites	200
10.6.3	$MnSn(OH)_6$ Thread Epoxy Nanocomposites	201
10.6.4	Graphene Oxide Epoxy Nanocomposites	201
10.7	Summary and Future Prospects	201
	Acknowledgments	203
	References	203

11 Solvent-Free Liquids and Liquid Crystals from Biomacromolecules 211

Kai Liu, Chao Ma, and Andreas Herrmann

11.1	Introduction	211
11.2	Solvent-Free Nucleic Acid Liquids	212
11.2.1	Fabrication of Solvent-Free Nucleic Acid Liquids	212
11.2.2	Electrical Applications Based on Solvent-Free Nucleic Acid Liquids	215
11.3	Solvent-Free Protein Liquids	217
11.3.1	Fabrication of Solvent-Free Protein Liquids	217
11.3.2	Electrochemical Applications Based on Solvent-Free Protein Liquids	222
11.3.3	Catalysis of Solvent-Free Enzyme Liquids	224
11.4	Solvent-Free Virus Liquids	226
11.5	Mechanism for the Formation of Solvent-Free Bioliquids	228
11.6	Conclusions and Outlook	229
	References	230

12 Ionic Liquids 235

Hiroyuki Ohno

12.1	What Is Ionic Liquid?	235
12.2	Some Physicochemical Properties	236
12.3	Preparation	238
12.4	IL Derivatives	239
12.4.1	Zwitterions	239
12.4.2	Self-Assembled ILs	239
12.4.3	Polymers	241
12.5	IL/Water Functional Mixture	241
12.6	Application	243
12.6.1	Reaction Solvents	243
12.6.2	Electrolyte Solution	243
12.6.3	Biomass Treatment	244
12.6.4	Solvents for Proteins and Biofuel Cell	246

12.7	Summary *247*
	Acknowledgments *247*
	References *247*

13 Room-Temperature Liquid Metals as Functional Liquids *251*
Minyung Song and Michael D. Dickey

13.1	Introduction: Room-temperature Liquid Metals *251*
13.1.1	Mercury *251*
13.1.2	Gallium-Based Alloys *252*
13.1.3	Oxide Skin on Ga Alloys *252*
13.2	Removal of Oxide Skin *252*
13.3	Patterning Techniques for Liquid Metals *253*
13.3.1	Lithography-enabled Processes *254*
13.3.2	Injection *255*
13.3.3	Subtractive *256*
13.3.4	Additive *256*
13.4	Controlling Interfacial Tension *257*
13.4.1	Surface Activity of the Oxide on Liquid Metal Droplets *258*
13.5	Applications of Liquid Metals *261*
13.6	Conclusions and Outlook *263*
	References *263*

Index *273*

Preface

Liquids are matter participating deeply in our daily life. The most common and simplest one is water, which is something that we cannot live without. Similarly, organic solvents such as alcohol, ether, and aliphatic and aromatic solvents are essential media to produce new products from chemicals and polymers. As functional fluids, lubricants, fuel, inks, and even eatable honey are unintentionally and casually utilized as quite obvious items/materials for our support. These advantageous functions of liquids are ultimately their soft, flexible, and fluidic nature; and thus liquids can show an expedient deformability adaptable in the various shapes and geometries of surfaces and substrates. Liquids can, by themselves, flow into even very narrow spaces as a result of capillary action. They also exhibit self-repairing, self-healing characteristics independent of stimuli applied. Therefore, if one can develop a highly functional liquid, it can be a seminal fluidic material toward various practical applications. For instance, liquids possessing optoelectronic functions would be applicable for bendable and stretchable devices.

Apart from amorphous polymers, ionic liquids, functional organic liquids are a completely new concept among soft materials. The functional organic liquids are mainly composed of two molecular units. One is a functional core unit, e.g. luminescence dye, semiconducting π-conjugated molecule, cyclic host molecule as well as nanocarbons, and the other is a side chain which induces the material's fluidity even at room temperature. There are basically no cationic and anionic charges in the liquids; this can be categorized differently with ionic liquids whose weak ionic interaction decides their physical properties. Functional organic liquids having a well-wrapped functional core with bulky and soft side chains show great thermal and photostability. This durability is hardly achieved in organic dye molecules that decompose gradually under ambient conditions as well as in gels that degrade their performance with solvent evaporation or temperature variation. In addition, the prediction of their function in bulk state is available since the molecular intrinsic properties are rationally maintained in the core-isolated environment. Another noteworthy point in functional organic liquids is the accommodation of small dopants, e.g. acceptor dyes and nanoparticles, behaving like solvent and polymer matrix, and resulting in fluidic composites with tuned functions.

With the above-listed promising perspectives, functional organic liquids have been paid the most attention in the past few years in various research fields. By adapting the functional organic liquids' molecular concept, inorganic nanoparticles and biomolecules are also prepared as solvent-free liquid matter. Those liquid materials exhibit uncommon properties apart from their solid ones.

This book shows up contrasts to other liquid matter as well, in particular, ionic liquids and liquid metals. What is the difference between those liquid materials and functional organic liquids? Their advantages and disadvantages compared to functional organic liquids are commented on in the last two chapters.

I would like to emphasize that functional organic liquids are new-born liquid matter, and thus they have a tremendous potential for technological applications in various research directions. This field of research has just reached a point in which a book examining the current state of science and technology for functional organic liquids is necessary and will be more than welcome. I am sure this organized book would provide a summary of the currently existing knowledge in the area of functional organic liquids and would be a useful guide as well as stimulate researchers for future development and application.

6 September 2018

Dr. Takashi Nakanishi
Frontier Molecules Group
International Center for Materials
Nanoarchitectonics (WPI-MANA)
National Institute for Materials Science (NIMS)

1

Room-Temperature Liquid Dyes

Bhawani Narayan[1,2] and Takashi Nakanishi[1]

[1] *National Institute for Materials Science (NIMS), Frontier Molecules Group, International Center for Materials Nanoarchitectonics (WPI-MANA), 1-1 Namiki, Tsukuba 305-0044, Japan*
[2] *St. Joseph's College, Department of Chemistry, 36, Langford Road, Langford Gardens, Bangalore, 560027, Karnataka, India*

1.1 Introduction

Optoelectronically active functional molecules derived from π-conjugated chromophores [1] are becoming increasingly desirable. Their flexibility, light weight, processability, and low cost of fabrication provide them with an advantage over their inorganic counterparts, which are intrinsically hard and have no deformability [2]. The tailorable chemical functionalization on the π-conjugated unit, leading to tunable properties, makes them suitable for various applications in the field of optoelectronics [3]. Due to the strong π–π interactions between the π-conjugated units, these molecules generally exist as solids. As solids cannot be directly fabricated on the devices, they are either dissolved in organic solvents and coated onto a substrate, or used as gels [4] or as self-assembled phases [5]. They have been also applied as interfacial films. However, defect-free fabrications and homeotropic alignments are rather difficult due to the strong ordering and tendency to crystallize. The most common soft materials in this regard are liquid-crystalline (LC) materials. LCs are mesophases existing between the crystalline and isotropic phases [6]. As LC materials exhibit long-range order (micrometer scale), they are advantageous over locally ordered nanomaterials. However, LC phases only form on a special class of molecules and can be either thermotropic or lyotropic. They need the assistance of temperature/solvent to exhibit their properties different from that of the intrinsic chromophores in the monomeric form. Thus, simple prediction of material properties from LCs is almost impossible, and deep investigation is needed in holding the perfectly ordered state.

A clever strategy to retain the intrinsic characteristics of the chromophore, more often in the monomeric form, in a solvent-free paradigm is by designing functional molecular liquids (FMLs) [7]. FMLs are uncharged, solvent-free liquids at room temperature and can be categorized as one of the organic soft materials (Scheme 1.1). A close-knit member in the family is the class of "ionic

Functional Organic Liquids, First Edition. Edited by Takashi Nakanishi.
© 2019 Wiley-VCH Verlag GmbH & Co. KGaA. Published 2019 by Wiley-VCH Verlag GmbH & Co. KGaA.

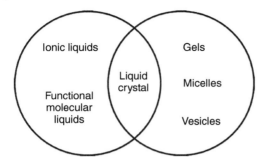

Scheme 1.1 Venn diagram representation of organic soft materials.

liquids" [8]. Ionic liquids, which are molten salts at a relatively low temperature (below 100 °C), result from weak ionic interaction and comprise bulky cationic and anionic counterparts [9]. They possess many useful properties, such as low volatility, low melting point, noninflammability, high thermal stability, and ionic conductivity. They have mainly been explored as solvents for carrying out reactions and as catalysts, and the recent trend of their utilization is in battery research.

FMLs are easily processable due to their fluidic nature and are environment friendly, as they do not require solvents in further fabrications and are nonvolatile. As most studied systems are based on chromophores, these liquids possess luminescent properties. Tunable emission and facile incorporation of other emissive materials into the liquid matrix have led to a deep understanding of phenomena such as excited-state energy transfer and white light emission. They are desired in the isotropic phase; however, interchromophoric (π–π) interactions may lead to local ordering, equipping them with properties such as liquid crystallinity at lower temperatures. This chapter discusses the strategy of designing room-temperature FMLs and gives detailed insights into the applications of these materials developed so far.

1.2 Design Strategy: Alkyl Chain Engineering

Organic molecules and polymers containing π-conjugated backbones are widely used in organic/polymeric optoelectronic devices by virtue of their suitable optical, semiconducting, and electronic properties. However, in general, these π-conjugated substances tend to exist as solids at room temperature due to the strong π–π interactions. Although conventional solution-processing techniques such as spin coating and solvent-assisted methods have been used for fabrication of ordered films, few molecules form poorly soluble random aggregates. This makes the rich semiconducting π-conjugated moieties unsuitable for showing proper optoelectronic performances such as charge transport. Furthermore, aerial oxidation and photooxidation have often been reported to cause photodimerization/polymerization, resulting in the loss of the semiconducting property of the π-conjugated backbone.

A simple synthetic technique to address the aforementioned challenges is to attach solubilizing linear or branched alkyl chains [10]. The introduction of flexible and bulky alkyl chains on the functional π-conjugated units gives us a tool for fine-tuning the balance between π–π interactions among neighboring chromophores and van der Waals (vdW) interactions governed by the attached alkyl chains. This technique not only softens the intrinsically rigid π-molecular-based material but also allows for unique phase transition ranging from solids to thermotropic LC to room-temperature liquids. Furthermore, the bulky alkyl chains can wrap/cover the π-conjugated units, preventing them from aerial oxidation/or photodimerization/polymerization, and thus can guarantee them a longer lifetime with their advanced functions.

The detailed discussions of the organic liquid dyes presented in the following sections have been organized in the order of the size of their functional cores (Scheme 1.2).

1.3 Alkylated π-Molecular Liquids

Substitution of multiple linear or branched alkyl chains has dramatically led to the phase transformation in many π-conjugated systems. This has made them useful for solvent-free direct applications in optoelectronics. In this section, the impact of the alkyl chain substitution on the phase behavior of several π-conjugated molecular systems has been briefly discussed. This section presents a detailed account of room-temperature FMLs. It covers the aspects of the synthetic strategies adopted to tune the physical properties of FMLs. An attempt is made to cover the development of these functional materials so far and the advances made in deriving functions from them.

1.3.1 Carbazoles

The first reported liquid carbazoles applied in an application was from the group of Peyghambarian and coworkers. 9-(2-ethylhexyl)carbazole (**1**, Figure 1.1a), the room-temperature liquid, was utilized as the solvent in ellipsometry measurements for determining the birefringence induced by electric field in photorefractive polymer composites [12]. Further insights into the charge-transport properties of **1** were provided by the research group of Wada and coworkers [13]. They demonstrated **1** to be a p-type semiconductor and determined the drift hole mobility by time of flight (TOF) method to be 4×10^{-6} cm^2 V^{-1} s^{-1}. Liquid carbazoles have more recently been employed as the liquid-emitting layer of organic light-emitting diodes (OLEDs) [14]. The liquid host **1** was used along with 6-(1)-naphthalene-2-carboxylic acid hexyl ester (BAPTNCE) as the guest emitter and tetrabutylammonium hexafluorophosphate (TBAHFP) as the electrolyte. Using a liquid emitter had two major advantages over conventional OLEDs. Firstly, the liquid-emitting layer and the electrodes do not lose contact upon significant bending of the device. Secondly, the liquid emitters that degraded in the OLED can be easily replenished by

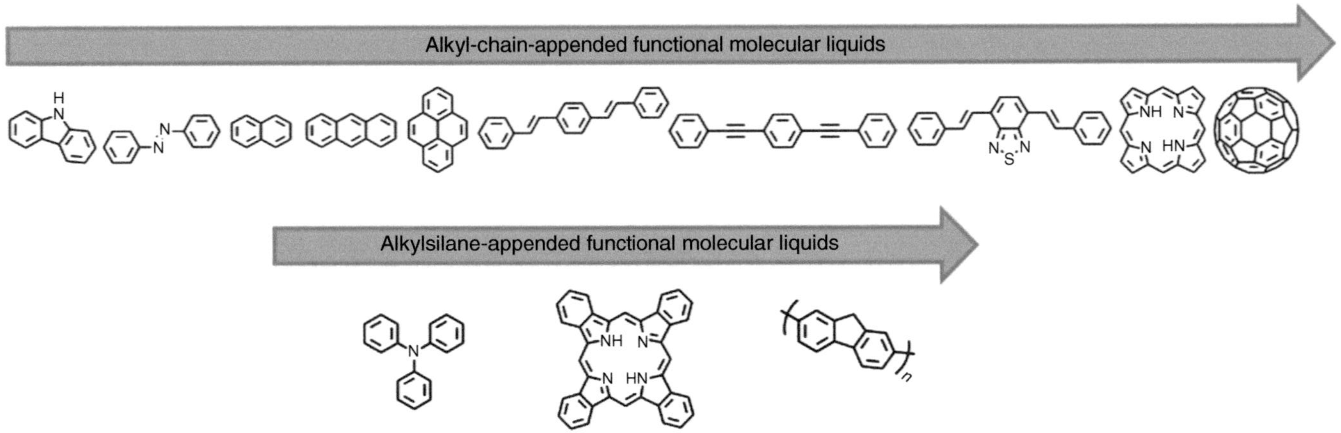

Scheme 1.2 Functional π-conjugated molecular units organized according to ascending size, as discussed in this chapter. These functional π-units have been appended with alkyl chains (from left to right: carbazole, azobenzene, naphthalene, anthracene, pyrene, oligo-(*p*-phenylenevinylene), oligo-(*p*-phenyleneethylene), benzothiadiazole, porphyrin and C$_{60}$ fullerene) or alkylsilane chains (from left to right: triarylamine, phthalocyanine, and oligofluorene) to develop functional liquid dyes at ambient temperature.

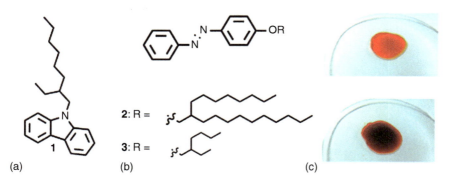

Figure 1.1 Chemical structures of (a) liquid carbazole **1** and (b) liquid azobenzenes **2** and **3**. (c) Photographs of the trans (orange, upper) and the cis state (deep red, bottom) conformations of **3** upon irradiation with UV at 365 nm. Source: Adapted with permission from Masutani et al. [11] Copyright 2014, Royal Society of Chemistry.

a fresh flow of the liquid, thus increasing the lifetime of the OLED over the existing ones in the device developed by Adachi and coworkers. The maximum external electroluminescence quantum yield (Φ_{EL}) achieved is 0.31 ± 0.07%, and the maximum luminance achieved is approximately 100 cd m^{-2}. The Φ_{EL} and the maximum luminance are 10 and 100 times, respectively, higher than the OLEDs reported so far. Thus, examples of this kind are advantageous over the conventionally applied materials in optoelectronic devices.

In addition, in the device developed by Adachi and coworkers, liquid carbazoles were employed to create liquid organic distributed feedback lasers built on flexible corrugated polymeric patterns [15]. **1** was used as a host matrix on which the blue-, green-, and red-emitting laser dyes were added. Cascade energy transfer between the chromophores efficiently channelized the lasing wavelength to cause a broad spectrum over the visible color range.

1.3.2 Azobenzenes

An interesting example of liquefaction of azobenzenes was demonstrated by Akiyama and Yoshida [16]. The design strategy was functionalizing a sugar alcohol scaffold with multiple azobenzene chromophores. When the azobenzenes were in the trans form, they were yellow crystalline solids. However, photoisomerization by UV light to the cis form resulted in the liquid state. Among a series of multiple azobenzene-arm-appended sugar scaffolds, the ones having an LC transition could be reversibly switched back to the trans state upon irradiation with visible light. However, the conversion to the cis form was a pure photochemical transformation, as the transition occurred even at room temperature. The cis-formed liquids were stable for at least half a day at room temperature. In this strategy, the sugar backbone acts as the flexible moiety. The liquids were cleverly employed as reusable adhesives and were demonstrated for sealing and detaching two quartz plates upon photoirradiation.

Our group had synthesized a liquid azobenzene by attaching a 2-octyldodecyl chain (**2**, Figure 1.1b) [17]. As we had observed directed self-assembly in the case of alkylated fullerenes (see Chapter 9), we desired to extend the concept of "hydrophobic amphiphile" with a smaller π-conjugated core. 2 served the purpose and showed a directed structural formation of micelles with an assist of alkane solvents.

An interesting application of a liquid azobenzene derivative has been demonstrated by Kimizuka and coworkers [11]. They employed an azobenzene carrying a 2-ethylhexyl chain (**3**, Figure 1.1b) as a solar thermal fuel. The trans-to-cis conversion has been extensively used for molecular switches and has been mainly performed in solid azobenzenes. However, employing trans-to-cis photoisomerization for storage of photon energy had been unsuccessful because such a phenomenon was suppressed in the condensed phase. The trans-to-cis conversion of compound **3** was performed upon photoirradiation with an ultrahigh-pressure mercury lamp at 365 nm (Figure 1.1c). High energy density was obtained in the process.

1.3.3 Naphthalenes

Preceeded by benzene, naphthalene is the smallest π-conjugated chromophore. As the absorption and emission features of naphthalenes can be conveniently probed using conventional techniques, this small dye unit was employed to create functional liquids with monomer-rich luminescence properties [18].

By a simple and effective alkylation strategy, 1- and 2-alkylated naphthalenes were synthesized. The regioisomeric alkylated naphthalenes were found to exist as liquids of very low complex viscosity (η^* in the range of 38.4–67.0 mPa s at 25 °C [$\omega = 10\,\text{rad}\,\text{s}^{-1}$]). Detailed optical investigations indicated the existence of excimeric state in 1-naphthalenes in the solvent-free state, whereas the 2-substituted regioisomers were found to be monomeric in nature. This simple design strategy serves as a guideline for synthesizing novel alkylated π-molecular liquids with predictable properties.

1.3.4 Anthracenes

Anthracenes, being the smallest member of the acene family, are well-sought out candidates for photosensitizer and organic laser dyes. Their blue emission and high quantum yield are the most attractive features. However, they are highly sensitive to light (dimerization) and air (oxidation) and highly crystalline. Therefore, fabricating uniform films from anthracenes for devices with long-term stability poses a grave challenge. Alkyl chain engineering on the anthracene unit has been able to address this issue effectively [19]. Functionalizing the anthracene unit with multiple branched alkyl chains has led to nonvolatile, blue-emitting, uncharged organic liquids that retain the intrinsic monomeric optical properties (Figure 1.2a,d). The wrapping of the alkyl chains on the anthracene unit can hamper the intermolecular π–π (anthracene–anthracene) interactions and provides better protection from atmospheric oxygen. This makes them more stable than their unsubstituted crystalline counterparts. While **4** was stable for 15 hours in a dilute dichloromethane solution and for 4 hours in the neat state under photoirradiation (365 nm, 100 W m^{-2}, Xe lamp), the unsubstituted

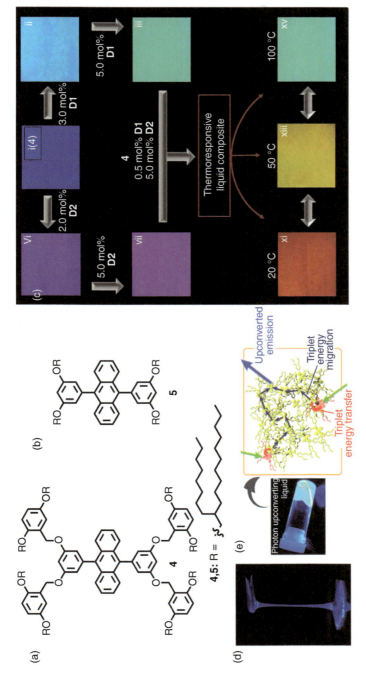

Figure 1.2 (a,b) Chemical structures of liquid anthracenes **4** and **5**; (c) photographs demonstrating the emission color tunability and thermoresponsiveness of the composites of **4**, **D1** (0.5 mol%) and **D2** (5 mol%); (d) photograph of solvent-free state of **4** under UV at 365-nm irradiation; (e) schematic of the photon upconversion system created by utilizing a liquid anthracene **5**. Source: Reproduced with permission from Babu et al. [19]. Copyright 2013, Macmillan Publishers Limited and adapted with permission from Duan et al. [20]. Copyright 2013, American Chemical Society.

anthracene started its decomposition after only 2 hours. Here, it is noted that there are no photodimerization reactions, but still its 9,10-endoperoxides are formed. Furthermore, the fluorescence quantum yields of the liquid anthracenes were around 55% in solvent-free neat state under ambient conditions. An alternative utilization of the highly blue emissive donor liquid was made by blending them with minute amounts of acceptor molecules and tuning their emission color through energy transfer (Figure 1.2c). Tunable luminescence was achieved by blending solid dopants 9,10-bis(phenylethynyl)anthracene (**D1**) and tris(1,3-diphenyl-1,3-propanedionato)(1,10-phenanthroline)europium (III) (**D2**) in minute amounts. The incorporation of incremental amounts of acceptor **D1** led to color tuning from blue to yellow at a single blue light excitation. This is because of the quenching of donor emission and the emergence of the emission of the acceptor and fluorescence resonance energy transfer (FRET) mechanism. The donor–acceptor composite was further made thermoresponsive by blending in **D2**. Incremental percentages of **D2** (2–5 mol%) in **4** tuned the color from violet to purple, due to weak energy transfer. A tricomponent mixture consisting of **4**, **D1** (0.5 mol%), and **D2** (5 mol%) resulted in a red emission. Due to the thermoresponsiveness of **D2**, heating the tricomponent mixture from room temperature to 50 °C resulted in yellow emission. A further increase in temperature led to quench the **D2**'s red emission and display emerald green emission, mainly contributed from **D1**'s emission. Therefore, a system of fluorescent thermometer was successfully constructed from composites of alkylated anthracenes.

Another interesting demonstration of the utilization of liquid anthracene **5** was made by Kimizuka and coworkers (Figure 1.2b,e). They successfully employed liquid **5** as an acceptor in an upconversion luminescent system [20]. Efficient upconversion systems known so far have been achieved in solution as the diffusion of the triplet molecules is a prime requisite. However, the volatility of organic solvents and deactivation of triplet state caused by molecular oxygen pose serious challenges. Harvesting the fluidic nature of **5** and its photostability due to shielding by multiple alkyl chains could address this issue. An alkyl-chain-substituted Pt(II) porphyrin derivative was chosen as the donor (sensitizer) which was highly miscible with liquid **5**, the blue emissive acceptor. An upconversion quantum efficiency of around 28% was achieved in solvent-free organic liquid system under ambient conditions, without matrix. It should be noted that the liquid **5** formed a liquid state, but in fact it was in a metastable supercooled liquid state [21]. Mixing with the alkyl-chain-substituted Pt(II) porphyrin derivative might secure its liquid form in their study.

1.3.5 Pyrenes

Adachi and coworkers synthesized a liquid pyrene **6** and used it as the emitting material in OLED (Figure 1.3a) [15]. The most important property of the OLED was its fast emission recovery because the liquid pyrene could be replenished as soon as it was consumed in the device. The design involved a mesh-structured cathode made of aluminum in which the liquid material could fill smoothly and uniformly, leading to the OLED having emission over a large area. A back reservoir enabled the injection of the liquid emitter through the cathode into the gap

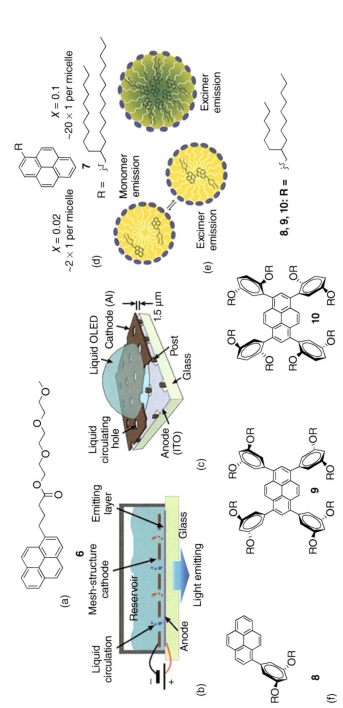

Figure 1.3 Liquid pyrenes: (a) **6** reported by Adachi and coworkers and employed in OLED; (b,c) show schematic representation and device structure made by microelectromechanical system (MEMS) processing; (d) an alkylated pyrene (**7**) reported by Hollamby and coworkers; (e) luminescent micellar structures formed from **7**; (f) alkylated pyrenes (**8–10**) with varied alkyl chain number and position of substitution that serve as a guide for molecular design of alkyl-π liquid dyes. Source: Adapted with permission from Shim et al. [15]. Copyright 2012, AIP Publishing LLC and reproduced with permission from Hollamby et al. [22]. Copyright 2016, Royal Society of Chemistry.

between the electrodes, thus causing convection circulation (Figure 1.3b,c). The external electroluminescence quantum efficiency (EQE) of the device was found to be 0.2%. This unique OLED structure has prospects in creating uniform and degradation-free lighting devices.

Hollamby and coworkers reported an excimeric-fluorescent liquid pyrene derivative **7** (Figure 1.3d) with a high fluorescence quantum yield in the solvent-free state (65%) [22]. The authors prepared stable oil-in-water emulsions with the liquid pyrene as the oil phase. Detailed structural investigations were carried out by small-angle neutron scattering (SANS). The SANS analysis revealed a cylindrical micellar structure with the microemulsions behaving as hard spheres at intermediate oil loading. Interestingly, the emission of the microemulsions could be tuned by the droplet size (Figure 1.3e). Taki et al. recently reported trialkyl-substituted pyrene liquids and their formation of color-tunable nanoparticles [23].

Although the field of FMLs is rapidly developing, there is no clear rationale for designing the molecule to extract the desired function. Our group has recently exploited the dual monomer–excimer fluorescence of liquid pyrenes to clearly lay down a guidance for molecular nanoarchitectonics [24]. We synthesized a series of alkylated pyrene liquids (**8–10**), which have branched alkyl chains of different substitution patterns and employed them as a rational FML model (Figure 1.3f). The number and substitution motif of the alkyl chain on the pyrene unit was modulated to create a series of derivatives that exhibit emission from the excimer to an uncommon intermediate to the monomer. The pyrene moiety was effectively isolated by wrapping of alkyl chains in the case of compound **10**, showing its monomer emission. While the pyrene moiety is only partially covered by the alkyl chains, i.e. **8**, its luminescence has a strong excimer feature. Thus, a direct correlation between the molecular structure and the π–π interactions was established by detailed investigation of liquid pyrenes.

1.3.6 π-Conjugated Oligomers

A few rod-shaped π-conjugated oligomers were substituted with branched alkyl chains and studied as luminescent solvent-free liquids at room temperature. A few of them were exploited as the donor matrix for acceptor dopant molecules and detailed investigations of energy transfer phenomenon were carried out. These are described here.

1.3.6.1 Oligo-(p-phenylenevinylene)s (OPVs)

Liquid oligo-(p-phenylenevinylene)s (OPV)s (**11–14**, Figure 1.4a) were synthesized by attaching branched alkyl chains to two different OPV cores [25]. OPVs have remarkable fluorescent blue emission characteristics, high stability, and self-assembling tendencies. It is, therefore, a good choice of chromophore for optoelectronics and is widely explored in this field. The liquid OPVs (**11–14**) were determined to have low viscosities ranging from 0.64 to 34.6 Pa s, at an angular frequency of $\omega = 10$ rad s^{-1}. Differential scanning calorimetry (DSC) showed glass transition temperature (T_g) of -43 to -55 °C. The liquid OPVs were blue emissive and had monomeric characteristics due to efficient

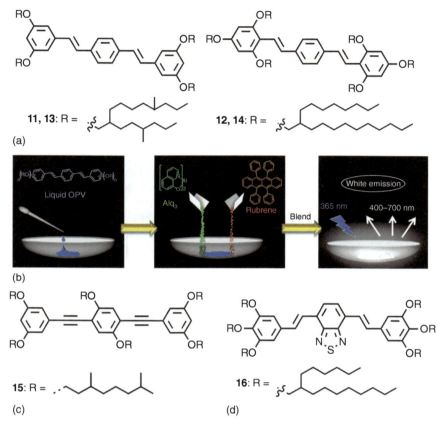

Figure 1.4 FMLs with π-conjugated oligomer unit: chemical structures of liquid (a) OPV and (b) schematic representation of using a liquid OPV as a matrix for making white-light-emitting liquid composite; chemical structures of liquid (c) OPE and (d) BTD. Source: Reproduced with permission from Babu et al. [25]. Copyright 2012, John Wiley & Sons.

isolation of the OPV unit by wrapping with the bulky branched alkyl chains. Furthermore, they were employed as a dispersion medium for various acceptor dopants, such as green-emitting tris(8-hydroxyquinolinato) aluminum (Alq_3) and orange-emitting 5,6,11,12-tetraphenylnaphthacene (rubrene) (Figure 1.4b); and the composite liquid showed white light emission with a chromaticity diagram coordinate value of (0.33, 0.34). The white-light-emitting liquid can be painted on substrates of various shapes and geometries as well as used as an ink for roller ball pens.

1.3.6.2 Oligo-(p-phenyleneethylene)s (OPEs)

In a follow-up work of the liquid OPVs, a liquid oligo-(p-phenyleneethylene)s (OPE) functionalized with multiple branched alkyl chains was reported [26]. The melting point of **15** was found to be 21 °C, proving that the compound was a liquid at above room temperature (Figure 1.4c). In this case, the substitution of multiple alkyl chains could not effectively isolate the OPE unit. This was suggested by a redshift in the absorption and fluorescence spectra of the solvent-free state compared to those in solution. The redshift in the absorbance and fluorescence spectra indicates aggregation of molecules governed by π–π interactions in the solvent-free state.

1.3.6.3 Benzothiadiazoles (BTDs)

Benzothiadiazoles (BTDs) are known as strongly electron-accepting dyes which are fluorescent in red color both in solution and in the bulk solid state. Recently, the group of Ishi-i reported BTDs containing electron-donating triphenylamines that emitted red light [27]. However, the emission was quenched in polar solvents. A recent effort from the same group was successful in appending BTD dyes with electron-rich tris(alkyloxy)phenylethene and creating BTD-based electron donors [28]. Interestingly, these produced efficient red light emission in solvents of all polarity and was also red emissive in the bulk solvent-free state. A branched alkyl chain derivative of BTD (substituted with C_2C_6 chain) was a red viscous liquid at room temperature (**16**, Figure 1.4d). By a strategy of donor–acceptor conjugation, the authors were successful in creating red-light-emitting materials with large Stokes shift. The development of red-emitting dyes is important because red is one of the three primary colors, and also for biological applications.

1.3.7 Porphyrins

Porphyrins have been widely used for various optical and electronic applications, such as in solar cells, transistors, electroluminescent materials, sensors, and nonlinear optics. Due to the strong π–π interactions among the large π-conjugated macrocyclic units, porphyrins have a strong tendency to form a self-assembled state. Many of the porphyrins have been reported to exist in solid state and to have poor solubility in organic solvents. This makes the optoelectronic applications of this class of molecules a formidable challenge. Easily processable analogs of porphyrins, such as in the LC state or in the liquid state for greater convenience, are therefore desirable. Linear alkyl-chain-substituted porphyrin derivatives were synthesized and investigated by the research groups of Maruyama [29] and Gryko [30] independently in 2010. Meso-substituted porphyrins bearing 3,4,5-trialkoxyphenyl substituents **17a–c** (Figure 1.5a) were obtained as liquids at room temperature and employed as dispersion media for electron acceptors such as C_{60} and carbon nanotubes (CNTs). On the other hand, investigation was carried out by Gryko and coworkers on almost the same series of porphyrins, but **17c** was described as a solid at room temperature. Furthermore, detailed investigations by this group revealed that alkyl chain lengths below C_9H_{19} and above $C_{12}H_{25}$ on these porphyrin derivatives yielded solids at room temperature. For further detailed discussions of the studies by Gryko on porphyrin derivatives, refer to Chapter 2.

1.3.8 Fullerenes

In contrast to pristine fullerene C_{60}, which is practically insoluble in any organic solvent except aromatic ones like chlorobenzene, substitution of alkyl chains

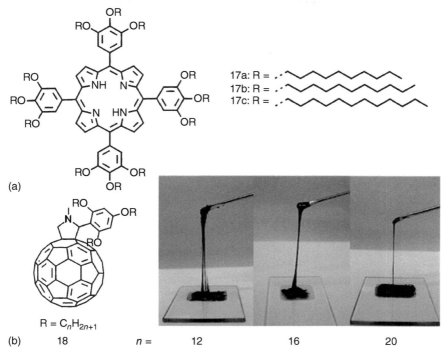

Figure 1.5 Chemical structures of series of (a) alkylated liquid porphyrins **17** and (b) liquid fullerene derivatives **18**; (c) photographs of the liquid fullerenes containing linear-alkyl chain lengths of 12, 16, and 20 (oligomethylene unit's number) respectively, showing their liquid nature. Source: Reproduced with permission from Michinobu et al. [31]. Copyright 2006, American Chemical Society.

not only increases the solubility but also tunes their phase from solid to liquid [31]. The chain length, nature of alkyl chain, and the substitution position have a great impact on the viscosity and the physical nature of the liquid fullerenes. C_{60} derivatives (**18**) with linear alkyl chains substituted on 2,4,6-tris(alkyloxy) phenyl group (Figure 1.5b) showed a clear decrease in the viscosity values with increasing chain lengths [31, 32]. While the derivative with $C_{12}H_{25}$ side chains had a complex viscosity value on the order 10^5 Pa s, the derivatives with $C_{16}H_{33}$ and $C_{20}H_{41}$ alkyl substitutions had viscosities on the order 10^4 and 10^3 Pa s, respectively, indicating a more fluid-like behavior (Figure 1.5b). Moreover, the derivative with shorter alkyl chains, C_8H_{14} was a solid, indicating stronger π–π interactions among C_{60} units. More details on liquid fullerenes, including branched alkyl-chain-substituted C_{60} and C_{70} derivatives and their studies on directed assembly using a concept of "hydrophobic amphiphile" can be found in Chapter 9.

1.4 Alkylsilane-Chain-Appended π-Molecular Liquids

Another strategy to create FMLs is by substituting alkylsilane chains on the π-conjugated dye unit. The commonly used moieties are polysiloxane chains, and their oligomeric counterparts. The siloxane chains effectively introduce amorphicity into the materials. Another advantage of this substitution could probably be the hydrophilicity provided by these functional groups. However, this aspect has not been explored. Although very few examples are available for this class of molecules, they deserve a mention because of their strategy and useful applications.

1.4.1 Triarylamines

Liquid triarylamines **19, 20** (Figure 1.6a,b) were synthesized by substituting silyl ethers onto a triarylamine unit [33]. This substitution considerably lowers the T_g and suppresses the crystallization tendency of these compounds (Figure 1.6c). Bender and coworkers carefully optimized the Piers–Rubinsztajn reaction for the synthesis of many liquid triarylamine derivatives [34]. They also performed detailed studies on the charge-transport properties of liquid triarylamines and proved that the molecular motions have negligible impact on the macroscopic charge-transport properties of a liquid semiconductor [35]. The hole mobility of liquid triarylamine **19** was found to be 1.92×10^{-3} cm^2 V^{-1} s^{-1}, well in accordance with the values reported for the nonliquid analogs.

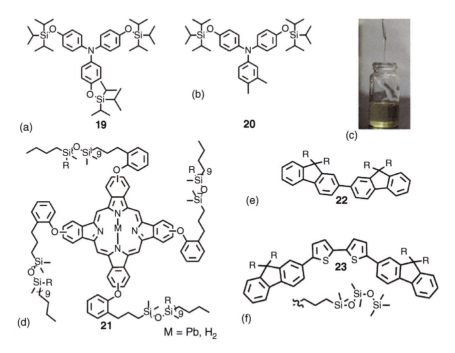

Figure 1.6 Alkylsilane-chain-appended FMLs: (a,b) triarylamines; (c) photograph of liquid triarylamine; (d) phthalocyanine; (e,f) oligofluorenes. Source: Adapted with permission from Kamino et al. [33]. Copyright 2011, American Chemical Society.

1.4.2 Phthalocyanines

A series of isotropic liquid phthalocyanines were successfully synthesized by appending poly(dimethylsiloxane) chains onto a phthalocyanine unit. The research on liquid phthalocyanines carried out by Maya et al. was aimed at thermorefractive properties of a silicone fluid and nonlinear optical properties of phthalocyanine chromophore [36]. Phthalocyanines are known to possess a large reverse saturable optical absorption by virtue of their larger absorption cross-section in the excited state compared to their ground state. The metal-free and Pb-substituted derivatives were synthesized and studied (**21**, Figure 1.6d). Those liquids were viscous, with the metal-free derivative having a T_g of 14 °C and the Pb-substituted derivative having a T_g of 3 °C.

The concept was extended to more polysiloxane derivatives [37]. The refractive index could be varied with temperature. A large negative thermorefractive coefficient was obtained, which was similar in magnitude to that observed from neat siloxanes. Detailed analysis through nonlinear transmission, transient absorption, z-scan, and degenerate four-wave mixing studies was performed. The studies proved that the liquid phthalocyanines substituted with Pb have a broad spectral region of induced absorption.

Linear or branched alkyl-chain-substituted liquid phthalocyanines have been also developed very recently in our group [38, 39], and showed complex rheology phenomena and redox-coupled spin switching as well as visible color tuning. Macrocyclic oligopyrroles, i.e. porphyrins and phthalocyanines, can form a metal complex and also its axial ligand coordinations; thus, these liquid substances could have many other possibilities in applications.

1.4.3 Oligofluorenes

Adachi and coworkers successfully synthesized room-temperature oligofluorene liquids [40]. The design strategy was based on functionalization of the fluorene core with alkylsiloxane side chains (Figure 1.6e,f). The fluidity conferred by the siloxane chains is effective in disrupting the π-stacking interactions of the fluorene units. Oligofluorenes are well known to have superior charge-transport and lasing properties. These properties make this chromophore a multifunctional moiety for application in optoelectronics. A major problem so far had been the defect-free fabrication of these materials in devices, which was made possible by this approach. The alkyl siloxane chains do not interfere with the properties of the π-functional core. The charge-transport properties of **22** and **23** were investigated by the TOF technique. A nondispersive ambipolar charge transport was observed. The hole and electron mobility of **22** were found to be around 7.5×10^{-5} and 1.5×10^{-5} cm^2 V^{-1} s^{-1}, respectively. **23** had a hole and electron mobility of 1×10^{-4} and 1.2×10^{-4} cm^2 V^{-1} s^{-1}, respectively, which are 1 order of magnitude higher than those of **22**. The charge-transport properties of **23** were found to improve upon increasing the electric field, in contrast to **22**. Furthermore, the lasing properties were investigated. Very low amplified stimulated emission (ASE) thresholds of 1.4 and 22 µJ cm^{-2} were obtained for both liquid oligofluorenes, respectively.

1.5 Analytical Tools for Functional Molecular Liquids

The most striking feature of FMLs is their liquid physical properties. The fluidity and amorphous features distinguish them from crystalline solids and ordered self-assembled phases. These properties limit the analytical tools available for characterization and investigation of this class of molecules. This section discusses the most commonly used techniques along with unconventional probes used to gain deep insights into the structure and properties of neat liquid substances.

1.5.1 Analytical Tools for Bulk Physical Properties

1.5.1.1 Structural Analysis

The order and disorder in FMLs is one of the highly disputed topics. Fluid matter is known to be generally disordered, but localized ordered domains may be present in alkylated π-molecular liquids, due to the interaction of the π-conjugated cores. Conventional wide-angle X-ray diffraction techniques are not suitable, as the central core units are separated in nanometer length scale regimes by attached bulky side chains. In this regard, small- and wide-angle X-ray scattering (SWAXS) in capillary mode is a better technique. The small-angle profiles obtained give a picture of the average distance of the π-conjugated units and the wide-angle profiles provide average alkyl chain distance as a common halo signal. In relatively ordered LC-like samples, SWAXS analysis would result in assignable peaks, resulting in well-defined molecular packing, like hexagonal and lamellar. SANS has also been used as a useful tool as neutrons scatter from the nuclei. Due to the multiple alkyl (hydrocarbon) chains, a good contrast is obtained by the scattering of neutrons by hydrogen and deuterium, when the sample is prepared with deuterated alkane solvents.

1.5.1.2 Microscopy Techniques

Optical microscopy (OM) in the bright field mode and polarized optical microscopy (POM) mode have been commonly employed for the characterization of FMLs. These techniques probe the microscale ordering/homogeneous isotropic nature of the liquid sample. Neat liquid samples sandwiched between cover slides are observed directly under the microscope with/without temperature control.

1.5.1.3 Rheology

Rheology is one of the highly informative tools with respect to the chemistry of fluids. Rheology measurements give us an estimate of the storage modulus (G') and the loss modulus (G''); for liquids, the G'' should be greater than the G'. It thus gives a quantitative definition to liquids. More importantly, rheology analysis gives us the complex viscosity of the liquid sample, which is a very important parameter in correlating the other physical properties observed. The angular frequency vs. strain curve helps us categorize whether the liquid has Newtonian or non-Newtonian behavior. Newtonian liquids exhibit a constant complex viscosity under frequency sweep with strain.

1.5.1.4 Calorimetric Techniques

Calorimetric analysis is very important in defining the phase transitions occurring in a liquid. While the decomposition temperature of the FMLs can be determined by thermogravimetric analysis (TGA), DSC can give very valuable information regarding the phases. It is commonly observed that liquids have a very low T_g. Special phenomena such as supercooling effect were observed in the case of a few FMLs (see details in Chapter 6) [21, 41]. In a few cases, external parameters like ramp rate, hold time at a particular temperature, and quenching at a desired temperature give us useful information regarding the liquid physical nature of the samples.

1.5.2 Analytical Tools for Spectroscopic Properties

Conventional spectroscopic tools used for the characterization of solutions, in general, can be applied to the same characterization of FMLs. However, a few modifications in the instrumental setup are sometimes necessary to make it suitable for FMLs. These details are discussed here.

1.5.2.1 UV–vis Analysis

The UV–vis instrument can be used for measuring the absorption of the samples. However, the neat sample should be sandwiched between glass/quartz plates, or coated onto one side of the plate as thin as possible and measured. The absorption spectra obtained in the neat state are broader than in the solution state, if there are intermolecular interactions. Completely disordered and chromophoric core isolated samples result in an almost identical absorption spectral feature as the monomeric species. In some neat liquids, the absorption spectra need to be recorded in the integrated sphere set up in the reflectance mode for better resolution.

1.5.2.2 Fluorescence Measurements

The fluorescence spectra can be easily recorded by sandwiching the neat sample between glass/quartz plates and by using the front-face geometry for recording the spectra. Another way is to place the sandwiched slide onto one face of a special prismatic cuvette and carry out the fluorescence measurements in normal geometry.

1.5.2.3 Fluorescence Lifetime Analysis

The fluorescence lifetime of neat samples can be recorded with a similar sample configuration as for the UV–vis and fluorescence measurements. Interesting phenomena such as phosphorescence and excimer formation have been confirmed by these measurements.

1.5.2.4 FTIR Measurements

The Fourier transform infrared (FTIR) of solvent-free, neat liquids can be carried out in the attenuated total reflectance (ATR) or kbr method.

1.6 Conclusion

The rational design of FMLs has increased new opportunities to create novel ultimately soft materials for applications that have never been explored before. The systematic softening of π-conjugated molecules from insoluble solids to liquid crystals to amorphous, neat liquids at room temperature proves a clever design of molecular architectonics. This field is a relatively new area, where the molecular design is simple, but handling liquid substances, characterizing them in depth and channelizing them to proper application are formidable challenges. Although many techniques have been explored for the characterization, the subtle balance of order and disorder in these amorphous paradigms poses a challenge for the detailed analysis of FMLs. A lot more needs to be done, and, thereby, this novel research field of FMLs opens the door to many unexplored regimes which may prove useful to all kinds of soft and fluidic materials in future.

References

1 Maggini, L. and Bonifazi, D. (2012). Hierarchised luminescent organic architectures: design, synthesis, self-assembly, self-organisation and functions. *Chem. Soc. Rev.* 41: 211–241.
2 Shirota, Y. (2000). Organic materials for electronic and optoelectronic devices. *J. Mater. Chem.* 10: 1–25.
3 Würthner, F. and Meerholz, K. (2010). Systems chemistry approach in organic photovoltaics. *Chem. Eur. J.* 16: 9366–9373.
4 Babu, S.S., Praveen, V.K., and Ajayaghosh, A. (2014). Functional π-gelators and their applications. *Chem. Rev.* 114: 1973–2129.
5 Aida, T., Meijer, E.W., and Stupp, S.I. (2012). Functional supramolecular polymers. *Science* 335: 813–817.
6 Laschat, S., Baro, A., Steinke, N. et al. (2007). Discotic liquid crystals: from tailor-made synthesis to plastic electronics. *Angew. Chem. Int. Ed.* 46: 4832–4887.
7 (a) Babu, S.S. and Nakanishi, T. (2013). Nonvolatile functional molecular liquids. *Chem. Commun.* 49: 9373–9382. (b) Ghosh, A. and Nakanishi, T. (2017). Frontiers of solvent-free functional molecular liquids. *Chem. Commun.* 53: 10344–10357.
8 Goossens, K., Lava, K., Bielawski, C.W., and Binnemans, K. (2016). Ionic liquid crystals: versatile materials. *Chem. Rev.* 116: 4643–4807.
9 Zhou, F., Liang, Y., and Liu, W. (2009). Ionic liquid lubricants: designed chemistry for engineering applications. *Chem. Soc. Rev.* 38: 2590–2599.
10 Lu, F. and Nakanishi, T. (2015). Alkyl- π engineering in state control toward versatile optoelectronic soft materials. *Sci. Technol. Adv. Mater.* 16: 014805.
11 Masutani, K., Morikawa, M.-a., and Kimizuka, N. (2014). A liquid azobenzene derivative as a solvent-free solar thermal fuel. *Chem. Commun.* 50: 15803–15806.

12 Hendrickx, E., Guenther, B.D., Zhang, Y. et al. (1999). Ellipsometric determination of the electric-field-induced birefringence of photorefractive dyes in a liquid carbazole derivative. *Chem. Phys.* 245: 407–415.
13 Ribierre, J.C., Aoyama, T., Muto, T. et al. (2008). Charge transport properties in liquid carbazole. *Org. Electron.* 9: 396–400.
14 Hirata, S., Kubota, K., Jung, H.H. et al. (2011). Improvement of electroluminescence performance of organic light-emitting diodes with a liquid-emitting layer by introduction of electrolyte and a hole-blocking layer. *Adv. Mater.* 23: 889–893.
15 Shim, C.-H., Hirata, S., Oshima, J. et al. (2012). Uniform and refreshable liquid electroluminescent device with a back side reservoir. *Appl. Phys. Lett.* 101: 113302.
16 Akiyama, H. and Yoshida, M. (2012). Photochemically reversible liquefaction and solidification of single compounds based on a sugar alcohol scaffold with multi azo-arms. *Adv. Mater.* 24: 2353–2356.
17 Hollamby, M.J., Karny, M., Bomans, P.H.H. et al. (2014). Directed assembly of optoelectronically active alkyl–π-conjugated molecules by adding n-alkanes or π-conjugated species. *Nat. Chem.* 6: 690–696.
18 Narayan, B., Nagura, K., Takaya, T. et al. (2018). The effect of regioisomerism on the photophysical properties of alkylated-naphthalene liquids. *Phys. Chem. Chem. Phys.* 20: 2970–2975.
19 Babu, S.S., Hollamby, M.J., Aimi, J. et al. (2013). Nonvolatile liquid anthracenes for facile full-colour luminescence tuning at single blue-light excitation. *Nat. Commun.* 4: 1969.
20 Duan, P., Yanai, N., and Kimizuka, N. (2013). Photon upconverting liquids: matrix-free molecular upconversion systems functioning in air. *J. Am. Chem. Soc.* 135: 19056–19059.
21 Lu, F., Jang, K., Osica, I. et al. (2018). Supercooling of functional alkyl-π molecular liquids. *Chem. Sci.* 9: 6774–6778.
22 Hollamby, M.J., Danks, A.E., Schnepp, Z. et al. (2016). Fluorescent liquid pyrene derivative-in-water microemulsions. *Chem. Commun.* 52: 7344–7347.
23 Taki, M., Azeyanagi, S., Hayashi, K., and Yamaguchi, S. (2017). Color-tunable fluorescent nanoparticles encapsulating trialkylsilyl-substituted pyrene liquids. *J. Mater. Chem. C* 5: 2142–2148.
24 Lu, F., Takaya, T., Iwata, K. et al. (2017). A guide to design functional molecular liquids with tailorable properties using pyrene-fluorescence as a probe. *Sci. Rep.* 7: 3416.
25 Babu, S.S., Aimi, J., Ozawa, H. et al. (2012). Solvent-free luminescent organic liquids. *Angew. Chem. Int. Ed.* 51: 3391–3395.
26 Naoya, A., Ryo, I., Masafumi, S., and Takayuki, N. (2014). Dispersion of fullerene in neat synthesized liquid-state oligo(*p*-phenyleneethynylene)s. *Chem. Lett.* 43: 1770–1772.
27 Kato, S.-i., Matsumoto, T., Shigeiwa, M. et al. (2006). Novel 2,1,3-benzothiadiazole-based red-fluorescent dyes with enhanced two-photon absorption cross-sections. *Chem. Eur. J.* 12: 2303–2317.
28 Ishi-i, T., Sakai, M., and Shinoda, C. (2013). Benzothiadiazole-based dyes that emit red light in solution, solid, and liquid state. *Tetrahedron* 69: 9475–9480.

29 Maruyama, S., Sato, K., and Iwahashi, H. (2010). Room temperature liquid porphyrins. *Chem. Lett.* 39: 714–716.
30 Nowak-Król, A., Gryko, D., and Gryko, D.T. (2010). Meso-substituted liquid porphyrins. *Chem. Asian J.* 5: 904–909.
31 Michinobu, T., Nakanishi, T., Hill, J.P. et al. (2006). Room temperature liquid fullerenes: an uncommon morphology of C_{60} derivatives. *J. Am. Chem. Soc.* 128: 10384–10385.
32 Li, H., Babu, S.S., Turner, S.T. et al. (2013). Alkylated-C_{60} based soft materials: regulation of self-assembly and optoelectronic properties by chain branching. *J. Mater. Chem. C* 1: 1943–1951.
33 Kamino, B.A., Grande, J.B., Brook, M.A., and Bender, T.P. (2011). Siloxane–triarylamine hybrids: discrete room temperature liquid triarylamines via the piers–rubinsztajn reaction. *Org. Lett.* 13: 154–157.
34 Kamino, B.A., Mills, B., Reali, C. et al. (2012). Liquid triarylamines: the scope limitations of piers–rubinsztajn conditions for obtaining triarylamine–siloxane hybrid materials. *J. Org. Chem.* 77: 1663–1674.
35 Kamino, B.A., Bender, T.P., and Klenkler, R.A. (2012). Hole mobility of a liquid organic semiconductor. *J. Phys. Chem. Lett.* 3: 1002–1006.
36 Maya, E.M., Shirk, J.S., Snow, A.W., and Roberts, G.L. (2001). Peripherally-substituted polydimethylsiloxane phthalocyanines: a novel class of liquid materials. *Chem. Commun.* 615–616.
37 Maya, E.M., Snow, A.W., Shirk, J.S. et al. (2003). Synthesis, aggregation behavior and nonlinear absorption properties of lead phthalocyanines substituted with siloxane chains. *J. Mater. Chem.* 13: 1603–1613.
38 Chino, Y., Ghosh, A., Nakanishi, T. et al. (2017). Stimuli-responsive rheological properties for liquid phthalocyanines. *Chem. Lett.* 46: 1539–1541.
39 Zielinska, A., Takai, A., Sakurai, H. et al. (2018). A spin-active, electrochromic, solvent-free molecular liquid based on double-decker lutetium phthalocyanine bearing long branched alkyl chains. *Chem. Asian J.* 13: 770–774.
40 Ribierre, J.-C., Zhao, L., Inoue, M. et al. (2016). Low threshold amplified spontaneous emission and ambipolar charge transport in non-volatile liquid fluorene derivatives. *Chem. Commun.* 52: 3103–3106.
41 (a) Machida, T., Taniguchi, R., Oura, T. et al. (2017). Liquefaction-induced emission enhancement of tetraphenylethene derivatives. *Chem. Commun.* 53: 2378–2381. (b) Chung, K., Kwon, M.S., Leung, B.M. et al. (2015). Shear-triggered crystallization and light emission of a thermally stable organic supercooled liquid. *ACS Cent. Sci.* 8: 94–102.

2

Low-Melting Porphyrins and Their Photophysical Properties

Agnieszka Nowak-Król and Daniel T. Gryko

Institute of Organic Chemistry, Polish Academy of Sciences, Kasprzaka 44-52, 01-224, Warsaw, Poland

2.1 Introduction

Porphyrins are ubiquitous in both nature and the scientific literature, but their omnipresence masks the unique character of each different porphyrin and often leads scientists to overlook their individual attributes. They remain the only group of functional dyes possessing the following combination of features: (i) ability to coordinate practically all metals from the periodic table; (ii) unusually large molar absorption coefficients (ε) reaching 400 000; and (iii) unprecedented easiness in fine-tuning optical, electrochemical, and other physicochemical properties via changing central metal and/or substituents (what typically can be achieved in straightforward manner) [1]. This combination of properties is responsible for extensive exploration of porphyrins in a large variety of fields spanning from applications in organic electronics, light-harvesting arrays, to complex supramolecular architectures [2]. More than a century after the pioneering work by Hans Fisher, porphyrins continue to reinvent themselves serving in new types of research and being modified in seemingly endless number of strategies. These so-called π-expanded porphyrins, π-extended porphyrins, and contracted porphyrinoids offer new possibilities for reaching unprecedented photophysical characteristics [2]. Their research applications in such fields as photodynamic therapy and dye-sensitized solar cells continue to blossom [2].

Porphyrins, however, owing to the large size of their π-scaffold and concomitant dispersion forces, as well as directing electrostatic interactions, willingly form stacks and aggregates [2]. This effect is even more pronounced for π-extended porphyrins possessing larger π-conjugated systems [2, 3]. As a result, melting points (m.p.) of these compounds are typically above 360 °C. Only a few examples of porphyrins display melting points below 200 °C. In view of the fact that for decades porphyrin chemistry was driven by such targets as modification of their electronic structure and utilizing their coordination chemistry to mimic functions of hemoglobin and chlorophyll, it was not until the 1980s when the first porphyrins, liquid crystals (LCs) at below 100 °C, were reported [4]. The latter investigations resulted in a plethora of porphyrin-based liquid crystals [5], albeit still with significant interplanar interactions. In spite of this relative

Functional Organic Liquids, First Edition. Edited by Takashi Nakanishi.
© 2019 Wiley-VCH Verlag GmbH & Co. KGaA. Published 2019 by Wiley-VCH Verlag GmbH & Co. KGaA.

abundance of liquid crystalline porphyrins, liquid porphyrins were until recently unknown [6].

In this chapter, we present the story of liquid porphyrins reported for the first time in 2010. Subsequently, we move toward closely related low-melting π-extended porphyrins, describing their synthesis and fundamental physicochemical and photophysical properties.

2.2 Liquid Porphyrins

Decoration of porphyrins with considerable number of aliphatic substituents is the most relevant starting point in a search toward liquid porphyrins. Both porphyrins' structure and the inherent logic of their synthesis suggest that these alkyl chains can be located either at meso or at β positions. In our endeavor to synthesize liquid porphyrins, we investigated several structural motives focusing mainly on meso-substituted A_4-porphyrins possessing trialkoxyphenyl substituents at all meso positions. We had particularly high expectations in two arrangements: 3,4,5- and 2,4,6-trialkoxyphenyl, since they were previously broadly exploited as effective building blocks to design liquid crystals of various aromatic scaffolds [7] and liquid fullerenes [8]. Along this line, we designed several A_4-porphyrins with different types of alkoxy side chains: linear and branched alkyl or possessing end-capping functional groups. The synthesis of designed porphyrins has been realized by means of condensation of trialkoxy-benzaldehydes with pyrrole under Lindsey's co-catalysis conditions [9]. The thermal behavior of these dyes was investigated by means of differential scanning calorimetry (DSC) and polarized optical microscopy (POM) (Figure 2.1B). The structures of A_4-porphyrins **1–4** and their isotropization temperatures are presented in Scheme 2.1 and Table 2.1, respectively. The comparison of porphyrins **1** and **2** gave the first insight into the influence of the substitution pattern on the melting points. In contrast to our expectations, porphyrin **1** displayed higher melting point than compound **2**, i.e. porphyrin **2** underwent isotropization at 152 °C, while its regioisomer at 184 °C [12]. Notably, **3** equipped with only four octyloxy substituents showed lower melting point (c. 119 °C) when compared to ester-functionalized porphyrins **1** and **2** [12]. These observations suggested that the combination of the alkyl chains with the 3,4,5-substitution pattern of the phenyl ring could bring us to the desired liquidity. n-Alkyloxy-substituted porphyrins **4d–h** and **4l** were isolated in high yields (36–53%), mainly as viscous fluids, and subjected to DSC measurements [10]. To our delight, among this homologous series we observed first room-temperature (RT) liquid porphyrins. DSC analyses showed a strong dependence of the crystal-isotropic liquid transition (Cry-Iso) on the number of carbon atoms in the side chain. Porphyrin **4d** substituted with octyl groups featured high crystallinity and melting point of 55.1 °C. Increase in the carbon chain length was accompanied by a dramatic decrease in the temperature of the Cry-Iso phase transition, reaching minimum for decyl chains. Accordingly, porphyrins **4f** and **4g** showed the lowest isotropization temperatures in the whole series, i.e. c. −55 and −24 °C, respectively (Figure 2.1A). Further elongation of the chains resulted in the increase of the isotropization temperature.

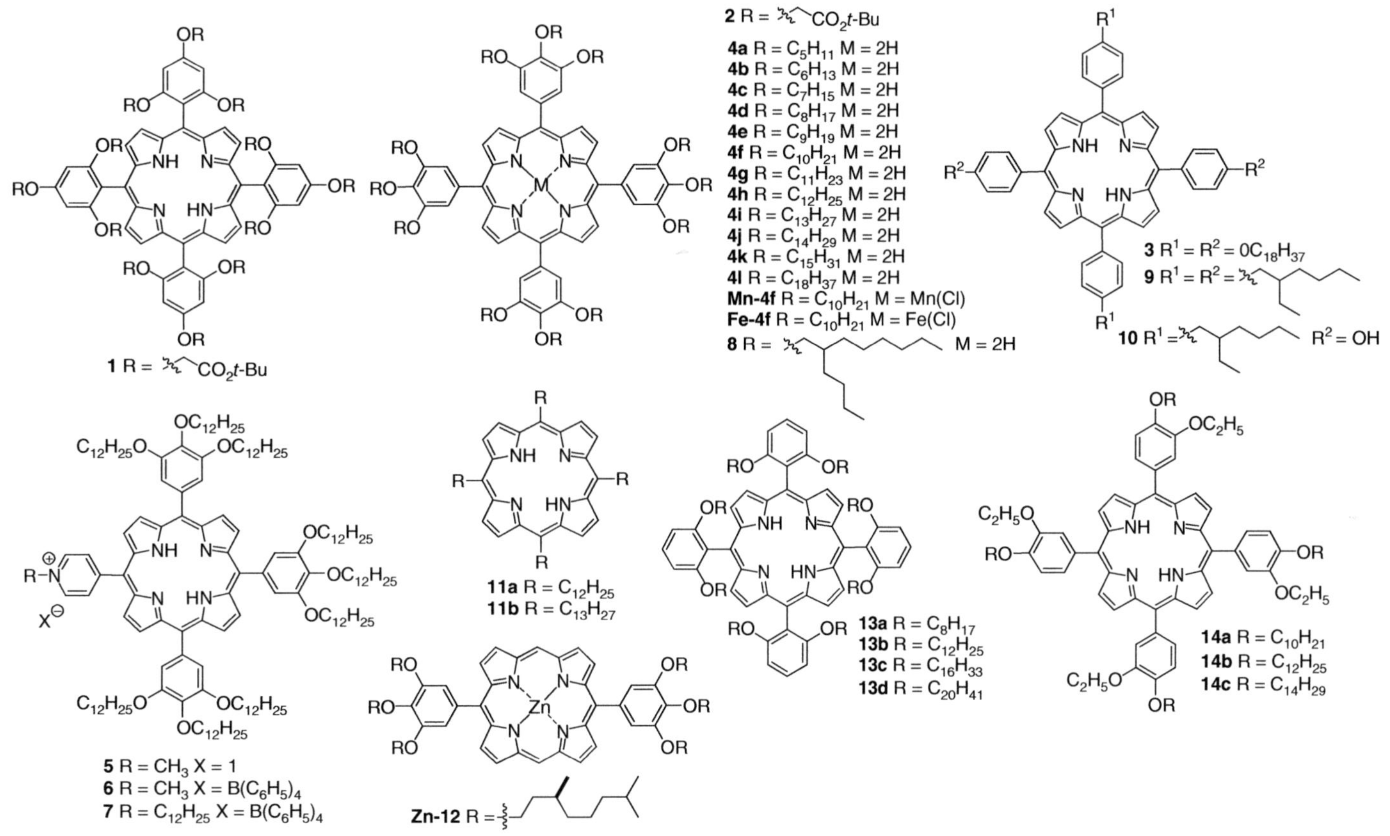

Scheme 2.1 Structures of meso-substituted arylporphyrins.

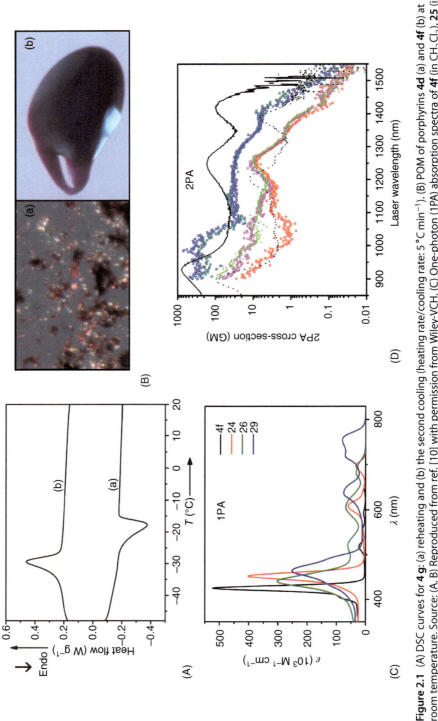

Figure 2.1 (A) DSC curves for **4g**: (a) reheating and (b) the second cooling (heating rate/cooling rate: 5 °C min^{-1}). (B) POM of porphyrins **4d** (a) and **4f** (b) at room temperature. Source: (A, B) Reproduced from ref. [10] with permission from Wiley-VCH. (C) One-photon (1PA) absorption spectra of **4f** (in CH$_2$Cl$_2$), **25** (in CH$_2$Cl$_2$), **27** (in CHCl$_3$), and **30** (in CCl$_4$). Source: Reproduced from ref. [11] with permission from The Royal Society of Chemistry. (D) Dependence of the 2PA cross-section (colored symbols) of **22** (red), **15** (pink), **18** (dark green), **19** (blue) and **23** (pale green) on laser wavelength in CCl$_4$ solution; the normalized 1PA absorption profile of **15** (dotted line) and **19** (solid line) is shown for comparison on the double transition wavelength scale.

Table 2.1 Melting points (m.p.) of selected porphyrins.[a]

Compound	Transition	m.p. (°C)	References
1	Cry[b] → Iso[c]	184.1	[12]
2	Cry → Iso	152.3	[12]
3	Cry(II) → Iso	118.5	[12]
4a	Cry → Iso	78.6	[13]
4b	Cry → Iso	83.9	[13]
4c	Cry → Iso	ND[d]	[13]
4d	Cry → Iso	55.1	[10]
4e	Cry → Iso	26.2	[10]
4f	Cry → Iso	−54.6	[10]
		−39.4	[13]
4g	Cry → Iso	−23.8	[10]
		−23.0	[13]
4h	Cry(II) → Iso	36.4[e]	[10]
		−6.0[f]	[13]
4i	Cry → Iso	11.2[f]	[13]
4j	Cry → Iso	42.0[e]	[13]
		18.4[f]	[13]
4k	Cry → Iso	34.3	[13]
4l	Cry → Iso	44.0	[10]
Mn-4c	Cry → Iso	15.3	[12]
6	ND	−23	[14]
7	ND	−22	[14]
8	Cry → Iso	37.5	[12]
9	ND	95–98[g]	[15]
10	ND	73–75[g]	[15]
11a	ND	73–76[g]	[15]
11b	ND	72–74[g]	[15]
14a	LC[h] → Iso	64.4	[16]
14b	LC → Iso	69.6	[16]
14c	LC → Iso	65.6	[16]
15	Cry → Iso	113.8	[17]
16	Cry(II) → Iso	188.5	[17]
17	Melts with decomposition	>300	[17]
18	Cry(II) → Iso	116.5	[17]
19	Cry(II) → Iso	122.5	[17]
20	Cry → Iso	107.4	[17]
24	SmA$_b$[i] → Iso	160.5	[18]

(Continued)

Table 2.1 (Continued)

Compound	Transition	m.p. (°C)	References
25	Col$_r$[j] → Iso	52.7	[18]
26	Col$_r$ → Iso	45.5	[18]
27	Cry → Iso	168	[19]
28	Cry → Iso	30.8	[19]
29	Cry(II) → Iso	156.9	[20]
31	Cry → Iso	64	[21]

a) Melting points determined from DSC measurements unless otherwise noted. In some cases, porphyrin underwent multiple phase transitions. For simplicity, only the phase transition to isotropic liquid is presented.
b) Crystal.
c) Isotropc liquid.
d) Not determined.
e) Measured for solidified compounds.
f) Measured for compounds as obtained.
g) Melting point determined by capillary method.
h) Liquid crystal; a specific phase was not assigned.
i) Biaxial smectic A phase.
j) Rectangular columnar phase.

DSC thermograms of **4d–h**, and **4l** showed usually the characteristics of one endothermic peak – corresponding to the Cry-Iso phase transition. Pronounced supercooling effect was distinctive of compounds **4d–f** and **4h**. Interestingly, lower melting solid porphyrins of this series remained in the form of a fluid over an extended period of time. It must therefore be recognized that the minimal isotropization temperature of the alkoxy-substituted porphyrins is determined by the subtle balance between π–π interactions of adjacent chromophores and attractive van der Waals interactions between side chains. For shorter chains, π–π stacking plays a dominant role, keeping the conjugated systems in a closer proximity and hence contributing to the increase in the melting points. The second extreme can be approached upon elongation of the alkyl chains, when van der Waals interactions start to prevail. The gradual increase in the carbon chain length magnifies the overall attractive interchain interactions, which also entails the increase in the isotropization temperature.

Independently, and also in 2010, a similar library of porphyrins was published by Maruyama and coworkers [13, 22]. Japanese authors screened compounds **4a–k** decorated with alkyl chains ranging from pentyl to pentadecyl for the same motivation, i.e. to obtain liquids at room temperature. Their DSC analysis showed results differing in some details from ours (Table 2.1). For example, for porphyrin **4e**, we observed clear melting point around 26 °C, while no transition in the range from −100 °C to 100 °C was noted by chemists from Dai Nippon Printing. Still, their most important conclusion corroborates ours – the lowest melting points are observed for porphyrins possessing $C_{10}H_{21}O$ and $C_{11}H_{23}O$ substituents (Table 2.1). The broader conclusion by Maruyama et al. is that the porphyrins are liquids below 25 °C, if alkyl chain length is in the range from

$C_{10}H_{21}$ to $C_{14}H_{29}$. Importantly, this statement is derived for samples as obtained, i.e. in a liquid state. Over time, some of the porphyrins undergo a very slow transformation into thermodynamically stable crystalline forms with different isotropization temperatures. For instance, the melting point registered by DSC for a crystallized sample of **4j**, increased to at around 42 °C [13]. Maruyama rightly points out that this behavior may be shared by other porphyrins, e.g. **4h** decorated with $C_{12}H_{25}$ chains, which could explain the difference in melting points observed by both groups for this material (see Table 2.1). The crystallization process of porphyrins bearing long alkyl chains can be very sluggish and take even months. For instance, in our case porphyrin **4h** solidified after half a year.

Needless to say, the insertion of metal into the porphyrin cavity increases the melting points of the resulting complexes vs. free bases. The question remained, however, as to what is the magnitude of this effect. Given that metal complexes of porphyrins display various interesting properties, we decided to obtain exemplary complexes **Mn-4f** and **Fe-4f** (Scheme 2.1). Both compounds proved highly soluble and, more importantly, liquid at room temperature. DCS curves of complex **Mn-4f** reveal one endothermic peak at 15 °C corresponding to the Cry-Iso phase transition [12]. Thus, the introduction of manganese into the cavity elevated the isotropization temperature of the free-base liquid porphyrin by around 70 °C.

In 2011, Barbe and coworkers revealed the first room-temperature ionic liquid porphyrins based on the same structural lead motif [14]. After thorough investigation of numerous meso-substituted porphyrins possessing both 3,4,5-tris(alkoxy)phenyl substituents and pyridinium quaternary substituents, they found three architectures **5**, **6** and **7** which have melting points below 25 °C (Scheme 2.1, Table 2.1). Thermogravimetric analysis clearly showed that salts possessing tetrakis(pentafluorophenyl)borate anion $(B(C_6F_5)_4^-)$ as a counterion are more thermally stable. The authors also discovered that porphyrins consisting of $B(C_6F_5)_4^-$ have significantly lower melting point than analogous bromides or iodides. Within this group, the lowest melting point (exothermic peak at −23 °C) belongs to salt **6**.

In the course of our studies on low-melting porphyrin materials, we also tested the impact of branching on the isotropization temperature by comparing compounds **4h** decorated with *n*-alkyl chains and **8** (Scheme 2.1) bearing branched alkyl chains of the same number of carbon atoms. Both porphyrins revealed comparable temperature [12]. Therefore, this approach toward molecular liquids, which proved effective for other types of chromophores, did not appear promising for meso-substituted A_4-aryl porphyrins with 3,4,5-tris(alkoxy)phenyl groups. A few years later, Calvete and coworkers tested the possibility of lowering isotropization temperature by introduction of 2-ethylhexyloxy groups. A_4-porphyrin **9** displayed m.p. of 95–98 °C [15]. Their experiments showed that lower temperatures of ~70 °C can be obtained for unsymmetrical porphyrin **10** lacking one alkyl chain or tetraalkylporphyrins **11a** and **11b** (Scheme 2.1, Table 2.1). Another branched alkyl chain i.e. (*S*)-3,7-dimethyloctyl was present in 5,15-di(3,4,5-trialkoxyphenyl)porphyrin **Zn-12** discovered by Morisue et al. (Scheme 2.1) [23]. This porphyrin turned out to be liquid at RT, although transition temperature was not determined. This structural motif was later utilized to design π-expanded porphyrin glasses exhibiting near-infrared region (NIR) luminescence [24].

Other approaches toward lowering melting points are illustrated by two groups of A_4-porphyrins substituted with bis(alkoxy)phenyl groups in different substitution patterns. The first represents sterically hindered porphyrins **13a–d** with equilength chains located at 2- and 6- position of the phenyl ring [25]. The isotropization temperatures of these compounds were in the range of 60–85 °C for C8, C12, C16, and C20 alkyl chains. Porphyrins **14a–c** decorated with eight *n*-alkoxy groups at 3- and 4- positions on the phenyl group constitute the second variant and exhibit melting temperatures as low as 64.4–69.6 °C for 10, 12, and 14 carbon atoms in one alkyl chain, while the length of the second chain was kept constant (OC_2H_5) [16]. It has to be added that attachment of 3,4-bis(alkoxy)phenyl and 2,6-bis(alkoxy)phenyl groups did not allow to obtain liquid porphyrins. In both cases, the authors obtained liquid crystalline materials with considerably higher isotropization temperatures than 3,4,5-tris(alkoxy)phenyl-porphyrins **4**.

2.3 Low-Melting *trans*-A_2B_2-Arylethynyl Porphyrins

Our key motivation behind the synthesis of liquid porphyrins was the perspective of development of liquid optical limiters, which would combine fast two-photon absorption (2PA) with somewhat slower excited-state absorption (reverse saturable absorption) [26]. Porphyrins are among the most promising molecular materials for use in such devices because they possess many important properties, including large excited-state absorption and long triplet lifetime [27]. However, *meso*-aryl porphyrin monomers display typically very small two-photon absorption cross-section (σ_2) of 1–10 GM (1 GM = 10^{-50} cm^4 s^{-1} $photon^{-1}$), which renders them unsuitable for any two-photon-oriented applications [11]. The key factors to increase the overall efficiency of the two-photon absorption processes are charge-transfer character and the elongation of π-conjugation [28]. The effective π-conjugation between a porphyrin core and meso- or β-substituents can be ensured by introducing a triple bond between the core and the aryl substituent. However, an addition of these groups entails a significant decrease in the solubility, which limits the processability of arylethynyl porphyrins and reduces their potential for a real-world application. These deficiencies could be eliminated in novel materials that would preserve optical properties of the more crystalline counterparts but display, at the same time, high solubility or liquidity. Consequently, we designed several groups of π-extended porphyrins possessing previously optimized 3,4,5-tris(alkoxy)phenyl groups. The series include *trans*-A_2B_2-diarylethynyl porphyrins bearing either two tris(decyloxy)phenyl or two tris(undecyloxy)phenyl meso-substituents (Scheme 2.2) [17, 18]. The thermal behavior of the target porphyrins was investigated by DSC. Melting points are gathered in Table 2.1. DSC thermograms of **15** and **20** depict only one sharp endothermic peak at 114 and 107 °C, respectively, corresponding to the Cry-Iso

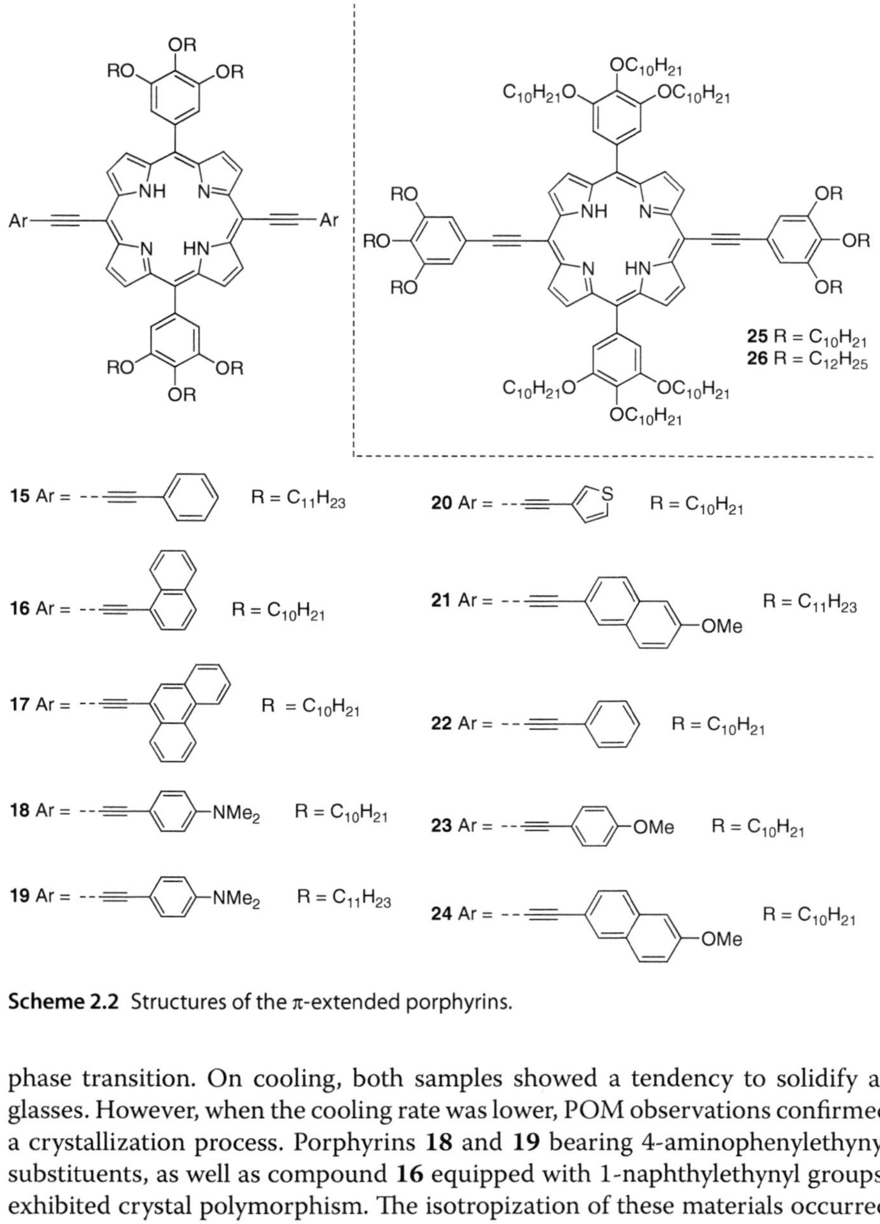

2.3 Low-Melting trans-A_2B_2-Arylethynyl Porphyrins

Scheme 2.2 Structures of the π-extended porphyrins.

phase transition. On cooling, both samples showed a tendency to solidify as glasses. However, when the cooling rate was lower, POM observations confirmed a crystallization process. Porphyrins **18** and **19** bearing 4-aminophenylethynyl substituents, as well as compound **16** equipped with 1-naphthylethynyl groups, exhibited crystal polymorphism. The isotropization of these materials occurred at 116.5, 122.5, and 188.5 °C. Given these results, it can be concluded that slight elongation of the pendant on the aryl substituents has a rather minor effect on the melting point. On the other hand, gradual extension of the aromatic π-system at lateral positions entailed a dramatic increase in melting points, which is illustrated by phenylethynyl, 1-naphthylethynyl, and phenantrenylethynylporphyrins **15**, **16**, and **17** (Table 2.1).

The effective conjugation granted by intervening triple bonds resulted in a bathochromic shift of absorption bands (1PA) in comparison to

Table 2.2 Optical properties of selected porphyrins and corrole **31**.

Compound	1PA[a] λ_{max} (nm)	ε[b] (10^3 cm^{-1} M^{-1})	2PA[c] λ_{max} (nm)	σ_2[d] (GM)	References
15	445	393.0	906	70	[17]
16	451	368.0	906	110	[17]
17	454	268.0	906	11	[17]
18	464	285.0	980	520	[17]
19	465	280.0	980	290	[17]
20	444	431.0	900	20	[17]
21	454	410	900	300	[17]
22	445	334	900	30	[17]
23	449	403.0	904	90	[17]
27	440	303	3000	940	[19]
29	455	310.0	915	3400	[20]
30	463	230.0	926	10000	[20]
Zn-29	464	240.0	920	6000	[20]
Zn-30	466	220.0	920	21000	[20]
31	413	131.0	ND	ND	[21]

a) One-photon absorption.
b) Molar absorption coefficient.
c) Two-photon absorption.
d) Two-photon absorption cross-section.

A_4-tetraarylporphyrins. The Soret band of **15** was located at 445 nm (Figure 2.1C). A further extension of the π-system upon introduction of electron-withdrawing, electron-donating groups and/or larger aromatic entities entailed the shift to lower energies (Table 2.2). In addition, dyes **18** and **21** displayed fluorescence in the NIR and moderately long lifetimes of the triplet-excited states ($\tau_T \sim 70$–130 μs). 2PA spectra of porphyrins **15–23** were measured by two-photon excited fluorescence (Figure 2.1D). Several factors, including electron effects of end-capping groups, replacement of a carboaromatic ring with a heteroaromatic substituent, the size of the aromatic unit and the variation of the alkyl chain length were studied. At longer wavelengths, this set of compounds presented low σ_2 (<10 GM). Cross-sections increased toward shorter wavelengths, which could be partially attributed to resonance enhancement. This phenomenon takes place when the 2PA excitation frequency is close to one-photon-allowed Q-transition. Porphyrins **15** and **22** equipped with neutral substituents showed low σ_2 values (70 and 30 GM, respectively). The enlargement of the π-conjugated system in **21** increased the cross-section around threefold as compared to **23** substituted with the same donor group (300 GM vs. 90 GM, respectively). Furthermore, the spectra of porphyrin **23** bearing methoxy groups and **15** bearing neutral substituents virtually overlapped, which was reflected by σ_2 in the same range (90 GM for **23**). In contrast, compounds **18** and **19** with NMe$_2$ end-capping groups displayed much higher σ_2, i.e. 520

and 290 GM, respectively. Moreover, strong electron-donating substituents enhanced peak 2PA values in the entire range of the wavelengths by at least an order of magnitude. This clearly demonstrates that substituents void of a strong donating or withdrawing properties have a reduced influence on 2PA properties.

2.4 Liquid Crystalline *trans*-A$_2$B$_2$-Arylethynyl Porphyrins

Attachment of aryl substituents via interfering triple bonds removes the steric barrier between the aryl groups and a porphyrin core and hence elongates the conjugated π-system of the porphyrin. Not only does the introduction of the ethynyl linkers modify the optical properties but it also has implications on the interchromophoric interactions. Since the steric congestion is released and a free rotation around the single bond is possible in arylethynyl porphyrins, aryl moieties do not play the role of steric stabilizers anymore and the overall tendency of porphyrins to undergo π–π stacking is increased. This propensity of the arylethynyl porphyrin π-system in combination with the flexible alkyl chains could induce liquid crystalline properties, if the balance between both incompatible parts is achieved. We approached this point by the design of *trans*-A$_2$B$_2$-arylethynyl porphyrins bearing decyl or dodecyl alkyl chains at each aryl substituents (Scheme 2.2). Molecules **25** and **26** possess four tris(alkoxy)phenyl substituents, two of which are directly attached to the porphyrin core, while the other two via ethynyl linkage [18]. Such an architecture afforded discotic liquid crystals exhibiting columnar phases. It is worth noting that free bases **25** and **26** formed liquid crystalline phases at room temperature with low isotropization temperatures of 52.7 and 45.5 °C, respectively (Table 2.1). Thus, introduction of interfering triple bonds increased melting points by around 100 °C vs. A$_4$-arylporphyrins. Porphyrin **25** exhibits two liquid crystalline phases. Based on X-ray powder diffraction (XRD) experiments, a lower-temperature and a higher-temperature phase was assigned to oblique columnar (Col$_o$) and rectangular columnar (Col$_r$) phases, respectively. According to DSC, a transition between Col$_o$ and Col$_r$ occurs at around 39 °C. An exchange of tris(alkoxy)phenylethynyl for 6-methoxy-2-naphthylethynyl substituents in molecule **24** afforded a liquid crystalline material with much higher (160.5 °C) temperature of Cry-Iso phase transition. More interestingly, this *trans*-A$_2$B$_2$-arylethynyl porphyrin formed smectic phase, which is very rare among porphyrinoids.

2.5 Bis-porphyrins

The rational approach toward two-photon active chromophores includes the utilization of conjugated centrosymmetric quadrupolar architectures of D-A-D or A-D-A types, composed of strong electron-donating and strong electron-withdrawing groups [28, 29]. From this perspective, bis-porphyrins with conjugated bridges serve as excellent building units for such systems owing to the electron-rich nature of porphyrin macrocycles. The enlargement of the structure, however, contributes to a substantial increase in the melting point. To minimize this effect and provide a reasonable solubility, bis-porphyrin **27** was decorated with both previously optimized 3,4,5-tris(decyloxy)phenyl substituents and with glycol chains (Scheme 2.3). For intervening unit, an electron-withdrawing diketopyrrolopyrrole moiety was selected, hence forming a D-A-D system. The combination of long alkyl and glycol chains effectively reduced the clearing temperature of this system. Bis-porphyrin **27** melted to isotropic liquid at 168 °C, with a low transition enthalpy implying weak interactions between adjacent bis-porphyrins. Apart from that, this molecule displayed other features that placed it at the border between crystal and liquid crystalline materials, such as easy orientation and low viscosity. Nevertheless, the detailed structural analysis, including XRD, POM, and DSC measurements, revealed that LC-like behavior was attributed to the formation of a plastic rather than a liquid crystal. Interestingly, *trans*-A_2B_2-porphyrin **28** substituted with the same tris(alkoxy)phenyl and glycol amine groups displayed isotropization temperature as low as ~31 °C (Scheme 2.3).

Two-photon transitions for **27** in the Q-band region (1100–1500 nm) coincided well with the 1PA transitions and corresponded to peak 2PA cross-section values lower than 1000 GM. Large intrinsic 2PA ($\sigma_2 = 3000$ GM) was recorded at 940 nm (in the NIR range of wavelengths).

A substantial perturbation of the ground-state properties of the porphyrin could be also realized in π-conjugated bis-porphyrins with butadiynyl bridging unit. Porphyrins **29** and **30** and their zinc complexes, bearing tris(decyloxy)phenyl substituents were prepared in a multistep linear synthesis (Scheme 2.3) [18]. The extension of the molecular structure entailed the rise of the melting points. DSC curves of compound **29** showed endothermic peak corresponding to isotropization at 156.9 °C, while the clearing temperature for porphyrin **30** was not observed up to 200 °C.

Linear absorption spectra of **29** and **30** and their zinc complexes show features typical of other butadiynyl-bridged bis-porphyrins (Figure 2.1C) [3, 30]. Strongly electron-donating amino groups in **30** and **Zn-30** exert a substantial effect on the absorption transitions observed as broadening of the Soret bands. Furthermore, all these bis-porphyrins featured large 2PA cross-sections with maxima located between 915 and 926 nm and represented 100-fold enhancement of two-photon absorption cross-section in comparison with typical monomeric porphyrins. A dramatic increase in σ_2 for porphyrin dimers is partially attributed to strong resonance enhancement. Other factors contributing to the large σ_2 values of bis-porphyrins are amplification of the lowest-energy one-photon Q transition, assigned to the intermediate state in the three-level model [31], and an increase in the transition dipole moment from the intermediate state to the two-photon allowed final state. Compound **29** displayed 2PA cross-section of 3400 GM. An increase of σ_2 was observed upon insertion of zinc to the cavity. **Zn-29** showed a 1.8-fold enhancement of σ_2 over the free-base counterpart with the cross-section of 6000 GM. This enhancement is associated with the increased electron density of the porphyrin core. In porphyrins **30** and **Zn-30**,

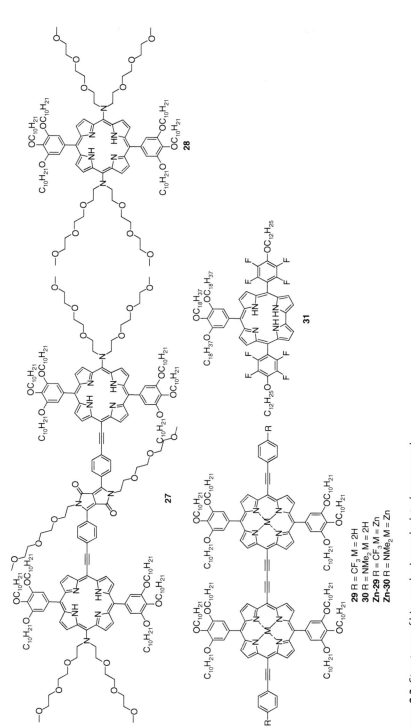

Scheme 2.3 Structures of bis-porphyrins and related compounds.

π-donor aniline groups induced considerable electronic perturbation on the bis-porphyrin π-core. Substitution with these groups enhanced σ_2 values to 10 000 and 21 000 for free-base **30** and zinc complex **Zn-30**, respectively. Thus, the effect of an end-capping group and a metal in the cavity clearly illustrate that strong electronic coupling is a prerequisite to achieve high σ_2.

2.6 Low-Melting Corroles

Corroles are tetrapyrrolic macrocyclic compounds related to porphyrins. Similar to porphyrins, they are aromatic and show characteristic absorption bands, as well as high fluorescent quantum yields. Interestingly, simple corroles proved more effective two-photon absorbers than did their porphyrin counterparts, which was of high importance for our studies. As a continuation of our work on two-photon active molecular liquids and low-melting materials, we designed corroles substituted with tris(alkoxy)phenyl moieties [21]. It is worth noting that higher electron density in the aromatic ring of corroles and lower oxidation potential reduces their stability. To compensate the influence of electron-donating tris(alkoksy)phenyl groups, and in turn to stabilize the structure of corroles, we introduced electron-withdrawing 4-alkoxy-2,3,5,6-tetrafluorophenyl substituents at two meso-positions. This approach proved effective and the synthesized compounds were stable over time and across the broad range of temperatures. Particularly interesting results were obtained for corrole **31** bearing long octadecyloxy chains (Scheme 2.3). DSC curves showed two endothermic events on the first heating cycle. The latter transition at 63.5–66 °C was assigned to the Cry-Iso phase transition. On cooling, the sample crystallized at 17.5–20.7 °C. The isotropization process on the second heating occurred at 27.6–32.4 °C with a heat effect similar to crystallization, which indicated reversibility of the events. These transitions were also reproducible during successive cycles. To get deeper insight into origin of thermal behavior of **31**, the melted sample (after first heating cycle) was allowed to age overnight at 4 °C. In this case, the second heating resulted in Cry-Iso transition at 63.5 °C, while the second cycle again showed lower temperature (27.6 °C). On this basis, it may be inferred that when a sample crystallizes from the melt, initially a kinetically favored species is formed, which melts at lower temperature. Over time, corrole **31** transforms in a slow kinetic process from the kinetically driven crystal packing to the thermodynamically stable crystalline state due to slow rearrangement of long alkyl chains. The annealing of the sample can accelerate the transition between both crystal packing states.

2.7 Summary and Outlook

Although unknown until 2010, liquid porphyrins underwent already a small evolution. It was discovered that the presence of four 3,4,5-tridecyloxyphenyl or 3,4,5-triundecyloxyphenyl substituents at meso-positions effectively prevented porphyrins aggregation and made it possible to lower their melting point from very high to below 0 °C. What has started from establishing this paradigm for design of liquid A_4-porphyrins evolved into studies oriented toward applications related mainly to two-photon absorption and optical limiting. The success of this design is visible in successful preparation of low-melting (typically 110–125 °C) π-extended porphyrins and conjugated bis-porphyrins. Given that preferred 3,4,5-tris(alkoxyphenyl) motive can be prepared from abundant gallic acid, these tetraaryl-porphyrins and arylethynyl-substituted porphyrins are easily accessible. Several future directions of research can be envisioned. Perhaps most importantly, a better understanding of the structural requirements that differentiate liquid porphyrins from liquid crystalline porphyrins should be targeted. It became clear that very small structural changes cause rather significant perturbations of various interactions involved. Yet, we are far from an in-depth understanding, which would allow making predictions before actual synthesis. One can also venture to hypothesize that the field of liquid porphyrins will develop further targeting various optoelectronic technologies such as bulk-heterojunction solar cells or organic field-effect transistors. This indeed may happen in the future since the discussion on the role of alkyl substituents in molecular packing (and of molecular packing on performance) is a hot and important subject of current research in materials chemistry. The analysis of research performed within the past 20 years in various aspects functional dyes' chemistry suggests that the future of liquid porphyrins will be only limited by our imagination.

References

1 Kadish, K.M., Smith, K.M., and Guilard, R. (eds.) (2000). *Porphyrin Handbook*, vol. 1–10. San Diego, CA: Academic Press.
2 Kadish, K.M., Smith, K.M., and Guilard, R. (eds.) (2010). *Handbook of Porphyrin Science*, vol. 1–35. Singapore: World Scientific.
3 Tanaka, T. and Osuka, A. (2015). Conjugated porphyrin arrays: synthesis, properties and applications for functional materials. *Chem. Soc. Rev.* 44: 943–969.
4 Goodby, J.W., Robinson, P.S., Teo, B.-K., and Cladi, P.E. (1980). The discotic phase of uro-porphyrin I octa-*n*-dodecyl ester. *Mol. Cryst. Liq. Cryst.* 56: 303–309.
5 Ohta, K., Nguyen-Tran, H.-D., Tauchi, L. et al. (2012). *Handbook of Porphyrin Sciences* (ed. K.M. Kadish, K.M. Smith and R. Guilard). World Scientific.
6 Henriques, C.A., Pinto, S.M.A., Canotilho, J. et al. (2016). Synthesis of low melting point porphyrins: a quest for new materials. *J. Porphyrins Phthalocyanines* 20: 843–854.
7 Percec, V., Ahn, C.-H., Bera, T.K. et al. (1999). Coassembly of a hexagonal columnar liquid crystalline superlattice from polymer(s) coated with a three-cylindrical bundle supramolecular dendrimer. *Chem. Eur. J.* 5: 1070–1083.

8 Michinobu, T., Nakanishi, T., Hill, J.P. et al. (2006). Room temperature liquid fullerenes: an uncommon morphology of C_{60} derivatives. *J. Am. Chem. Soc.* 128: 10384–10385.

9 Geier, G.W. III,, Riggs, J.A., and Lindsey, J.S. (2001). Investigation of acid cocatalysis in syntheses of tetraphenylporphyrin. *J. Porphyrins Phthalocyanines* 5: 681–690.

10 Nowak-Król, A., Gryko, D., and Gryko, D.T. (2010). *meso*-Substituted liquid porphyrins. *Chem. Asian J.* 5: 904–909.

11 Karotki, A., Drobizhev, M., Kruk, M. et al. (2003). Enhancement of two-photon absorption in tetrapyrrolic compounds. *J. Opt. Soc. Am. B* 20: 321–332.

12 Nowak-Król, A. (2013). Synthesis of liquid porphyrins and corroles with high two-photon absorption cross-section. Doctoral dissertation. Institute of Organic chemistry, Polish Academy of Sciences, Warsaw.

13 Maruyama, S., Sato, K., and Iwahashi, H. (2010). Room temperature liquid porphyrins. *Chem. Lett.* 39: 714–716.

14 Xu, H.-J., Gros, C.P., Brandès, S. et al. (2011). Room temperature ionic liquids based on cationic porphyrin derivatives and tetrakis(pentafluorophenyl)borate anion. *J. Porphyrins Phthalocyanines* 15: 560–574.

15 Morisue, M., Hoshino, Y., Shimizu, K. et al. (2015). Self-complementary double-stranded porphyrin arrays assembled from an alternating pyridyl–porphyrin sequence. *Chem. Sci.* 6: 6199–6206.

16 Zheng, W.Q., Liu, W.Y., and Yu, M. (2013). Liquid-crystalline properties of *meso*-tetra [(*p*-alkoxyl-*m*-ethyloxy)phenyl]porphyrins and their transition metal complexes. *Synth. React. Inorg. Met.-Org. Nano-Metal Chem.* 43: 175–177.

17 Koszelewski, D., Nowak-Król, A., Drobizhev, M. et al. (2013). Synthesis and linear and nonlinear optical properties of low-melting π-extended porphyrins. *J. Mater. Chem. C* 1: 2044–2053.

18 Salamończyk, M., Pociecha, D., Nowak-Król, A. et al. (2015). Liquid-crystalline properties of *trans*-A_2B_2-porphyrins with extended π-electron systems. *Chem. Eur. J.* 21: 7384–7388.

19 Nowak-Król, A., Grzybowski, M., Romiszewski, J. et al. (2013). Strong two-photon absorption enhancement in a unique bis-porphyrin bearing a diketopyrrolopyrrole unit. *Chem. Commun.* 49: 8368–8370.

20 Wilkinson, J.D., Wicks, G., Nowak-Król, A. et al. (2014). Two-photon absorption in butadiyne-linked porphyrin dimers: torsional and substituent effects. *J. Mater. Chem. C* 2: 6802–6809.

21 Nowak-Król, A., Fourie, E., Joubert, C.C. et al. (2016). Stable, low-melting *trans*-A_2B-corroles. *J. Porphyrins Phthalocyanines* 20: 1244–1255.

22 Maruyama, S., Saito, W. (2009). Liquid porphyrin derivative and method for producing the same. WO/2009/113511.

23 Morisue, M., Ueno, I., Nakanishi, T. et al. (2017). Amorphous porphyrin glasses exhibit near-infrared excimer luminescence. *RSC Adv.* 7: 22679–22683.

24 Henriques, C.A., Pinto, S.M.A., Pineiro, M. et al. (2015). Solventless metallation of low melting porphyrins synthesized by the water/microwave method. *RSC Adv.* 5: 64902–64910.

25 Arunkumar, C., Bhyrappa, P., and Varghese, B. (2006). Synthesis and axial ligation behaviour of sterically hindered Zn(II)–porphyrin liquid crystals. *Tetrahedron Lett.* 47: 8033–8037.
26 Calvete, M., Yang, G.Y., and Hanack, M. (2004). Porphyrins and phthalocyanines as materials for optical limiting. *Synth. Met.* 141: 231–243.
27 Hann, R.A. and Bloor, D. (1989). *Organic Materials for Nonlinear Optics*. London: The Royal Society of Chemistry.
28 Pawlicki, M., Collins, H.A., Denning, R.G., and Anderson, H.L. (2009). Two-photon absorption and the design of two-photon dyes. *Angew. Chem. Int. Ed.* 48: 3244–3266.
29 He, G.S., Tan, L.-S., Zheng, Q., and Prasad, P.N. (2008). Multiphoton absorbing materials: molecular designs, characterizations, and applications. *Chem. Rev.* 108: 1245–1330.
30 Taylor, P.N., Huuskonen, J., Rumbles, G. et al. (1998). Conjugated porphyrin oligomers from monomer to hexamer. *Chem. Commun.* 909–910.
31 Drobizhev, M., Stepanenko, Y., Dzenis, Y. et al. (2004). Understanding strong two-photon absorption in π-conjugated porphyrin dimers *via* double-resonance enhancement in a three-level model. *J. Am. Chem. Soc.* 126: 15352–15353.

3

Porous Liquids

Stuart L. James and Ben Hutchings

Queen's University Belfast, School of Chemistry and Chemical Engineering, David Keir Building Stranmillis Road, Belfast, Northern Ireland BT9 5GE, UK

3.1 Introduction

Until quite recently, the idea that liquids could support permanent empty cavities of molecular size was counterintuitive and unproved. While permanent porosity is a common feature of solid materials, it is enabled by their relatively fixed, rigid structures. The highly dynamic and fluid nature of liquids does not lend itself to forming large, regular, permanent voids. There is even an old Armenian proverb that attests to the apparent pointlessness of trying to make holes in liquids: "To get a favour from a miser you might as well try to make a hole in water" [1]. However, as explained more fully in the following sections, with careful design it is actually possible to form permanent cavities of molecular size in liquids, and the ability to do so opens up many new and interesting technological possibilities. A liquid containing permanent pores of the same size as typical molecules would be expected to be an exceptionally good solvent for molecules that fit well inside the pores, but not for those molecules that are too large. This presents the possibility of *size- and shape-selective dissolution*, which is not normal for conventional, nonporous liquids (where solubility depends more directly on factors such as polarity and/or specific interactions between the solvent and solute). This in turn suggests that porous liquids (PLs) could represent a generically new type of technology for chemical separations. Many chemical separations on which the modern world relies are technically difficult and energy-intensive. Therefore, any technology that enables a fundamentally new thinking and offers entirely new approaches to such processes is valuable.

Also, as stated in a recent article [2], porous liquids may prompt a rethink of any situation in which a solvent is currently used. For example, with the advent of porous liquids that can dissolve large amounts of gases, there is an opportunity to rethink how best to design and operate chemical processes that rely on gaseous reactants entering a liquid phase, in order to undergo a catalytic reaction, for example.

This chapter provides an overview of the current state of the art in the new field of porous liquids, covering background concepts such as the nature of porosity in solids and liquids and the historical context of porous materials which lead up to the invention of porous liquids. The examples of porous liquids that have been reported to date are described along with discussion on their properties. Recent theoretical papers on the topic are also discussed and the several prospects for eventual applications are considered.

3.2 Porosity in Solids

Persistent porosity is a fundamental property of materials, although until recently it was essentially restricted to materials in the solid state. Microporosity refers to materials with pores smaller than 2 nm (i.e. the size regime of typical molecules), whereas mesoporosity refers to materials with pores between 2 and 50 nm and macroporosity to those with pores of greater than 50 nm [3]. Some solids feature a hierarchy of pores in all three size ranges. Carbons are an important basic class of porous materials, obtained by carbonizing various natural or synthetic carbon-containing precursors at high temperatures [4]. Zeolites remain the archetypal examples of crystalline microporous materials [5]. Intensely developed now for several decades, they have been implemented in a range of commercial applications from petrochemical separations to ion-exchange additives in washing powders. Zeolites are composed mainly of aluminum, silicon, and oxygen; and the Si—O and Al—O bonds that form their frameworks are very strong (the Si—O bond enthalpy is typically 450 kJ mol^{-1} [6]). This strength imparts good thermal and mechanical stability, enabling applications at high temperatures, and/or where the materials are subjected to high mechanical stress. Their highly ordered, crystalline nature means that all the pores are essentially of uniform size and shape. This makes them suited to applications as size- and shape-selective molecular sieves.

Since the pioneering and visionary work of Richard Robson and coworkers [7], crystalline porous materials based on metal ions connected by bridging organic ligands, which have come to be known as porous coordination polymers (PCPs) [8] or metal organic frameworks (MOFs) [9], have been developed intensively and are now beginning to find commercial applications. These materials benefit from greater structural versatility than is found in zeolites, as a result of the wide variety of metal ions which may be used, but, more significantly, the essentially infinite variety of organic connectors that are available or can be synthesized. Thus, for example, the pore size may be increased through the use of longer organic linkers [10] and chirality may be introduced through chiral organic linkers [11]. PCPs/MOFs can also be quite structurally dynamic [12], changing form from closed to open forms on the incorporation of guests, for example. Neighboring adsorption sites may also bind guests cooperatively, leading to complex sorption dynamics and relatively exotic bulk sorption characteristics [13]. Although early examples showed low stability, an often-cited limitation of these materials in general, this is no longer the case with many water-stable and highly thermally stable MOFs based, for example, on aluminum or zirconium carboxylates [14, 15].

The explosion of interest in MOFs has been followed closely by inventions in the areas of purely organic porous materials. Framework organic materials, known variously as covalent organic frameworks (COFs) [16], porous organic polymers (POPs) [17], porous aromatic frameworks (PAFs) [18], etc., have become established classes of materials in recent years. Organic polymers with one-dimensional topologies (i.e. non–framework structures) known as polymers of intrinsic microporosity (PIMs) [19] have also been elegantly designed with rigid and highly tortuous structures such that they are incapable of packing efficiently and thus form microporous porous solid phases.

Isolated examples of molecular organic materials with permanent microporosity have been known for some time, such as Dianin's compound [20]. However, again in quite recent years, many more examples of porous molecular crystals have been discovered and designed, and these now represent a further distinct research area [2]. An important example of this class of compounds are porous organic cages (POCs) [21]. These are molecular host structures, often crystalline in the solid state, in which the porosity is not only interstitial (between the molecules) but also within the molecules themselves.

Among mesoporous materials, although less relevant to the focus of this chapter, it can be noted that advances have also been made in recent decades, such as in Mobil Crystalline Materials (MCMs), which are extended inorganic materials crystallized around large templating (pore-forming) liquid crystal structures [22].

Considering the brief summary, it is clear that research into new forms of porous solids has flourished in recent years. An interesting general trend can be identified as moving from the classical, purely inorganic, *extended* types of structures to include (mainly organic) *molecular* compositions. While all of the materials described are solids, it is interesting to consider that in making this move toward molecular materials, fundamentally new possibilities arise which are central to the topic of the chapter. In particular, while extended framework structures cannot be liquefied (i.e. melted or dissolved) without breaking down their structures, molecular materials can. Thus, in a sense, the general trend described, the scene was set for the advent of liquids with permanent porosity.

3.3 Porosity in Liquids

While the idea of having permanent microporosity in liquids seems counterintuitive at first, it is simultaneously true that all liquids possess a degree of porosity. It is therefore important to specify what we mean by this porosity in normal liquids. Since molecules in liquids are continually tumbling over each other in largely random ways, and because they do not pack with perfect efficiency, small pores continually appear and disappear in all liquids. As modeled by Pohorille et al. [23], these pores cover a range of sizes, but they are typically relatively small (around 1 Å in diameter), irregular in shape and, of course, transient. In each of these three regards, these pores are therefore very different from the pores that exist in microporous solids as described. Despite this, the presence of pores in liquids has long been recognized as being critical in influencing basic phenomena such as dissolution and diffusion. Specifically, scaled particle theory [24] has established that the main energy "penalty" to dissolution of a solute in a liquid is the energy required to form a cavity (pore) to accommodate the solute. In essence, this is the energy required to separate the molecules of the solvent from each other, to make a space (or pore) for the solute. Similarly, the rate at which a solute diffuses throughout the liquid can be related to its ability to move into neighboring cavities. Thus, the statistical likelihood of forming molecule-sized cavities in liquids has been an important concept in understanding the bulk properties of the liquid. Also, critically, if permanent pores of molecular size can be generated in the liquid state at high concentrations, then we can expect unusual bulk properties to be imparted to the liquid. This idea was the basis of a "concept" paper published in 2007, which suggested that liquids that have permanent well-defined porosity would be interesting materials, worthy of investigation, and that with careful thought it should be possible to design and synthesize such materials [25].

In essence, the concept paper referred to suggested that, although the voids *between* molecules of a liquid are difficult to control, the control over voids *within* molecules of the liquids should be achievable even with current knowledge. For ease of discussion, the voids between molecules are termed "extrinsic" and those within molecules of the liquids are termed "intrinsic" (see Figure 3.1). There is an inherent difficulty in controlling extrinsic porosity in liquids since it requires imposing order (to give a well-defined pore space) while retaining disorder to keep the material fluid. By contrast, if spaces can be generated within molecules of the liquids (intrinsic porosity), then the host species that define the pores can still potentially move freely within the liquid and fluidity is not compromised. In order to achieve states of "fluid hosts," however, significant challenges still remain. In particular, if we consider first the case of neat liquid hosts (known as "type 1" porous liquids, Figure 3.1), there is a significant challenge in achieving hosts with low melting points since, typically, hosts that can define molecule-sized pores are rather large species and they often have correspondingly high melting points. For example, unfunctionalized cyclodextrins normally decompose at several hundred Celsius without melting. One can imagine functionalizing their exterior hydroxyl groups with alkyl chains, for example, to lower the melting point, but then it is likely that the chains will occupy the host cavity so some degree, reducing or potentially removing the porosity. Also, it is essential that the host be sufficiently rigid not to collapse when empty. It rules out many archetypal well-known hosts such as typical calixarenes, etc.

An alternative approach is to allow the presence of a second component, which can be thought of as a solvent or plasticizer, to form a low-melting eutectic or a solution (such solutions are known as "type 2" porous liquids; Figure 3.1). In this approach, the melting point of the host is not so important, but it is essential that no part of the second component's structure can enter the hosts' pores, and extremely high concentrations of hosts will be needed to provide a substantial amount of open porosity. Therefore, the choice of solvents is limited, and hosts will need to have extremely high solubility and/or miscibility with the solvent, which may also present a challenge.

A third approach, to give "type 3" porous liquids, involves taking well-known types of framework porous solids (e.g. zeolites, MOFs, etc.) and dispersing them into appropriate liquid phases (Figure 3.1). Obtaining stable dispersions at high solid loadings may not be trivial, however, and, again, the liquid phase must not be able to enter the pores of the porous solid component. Despite the challenges

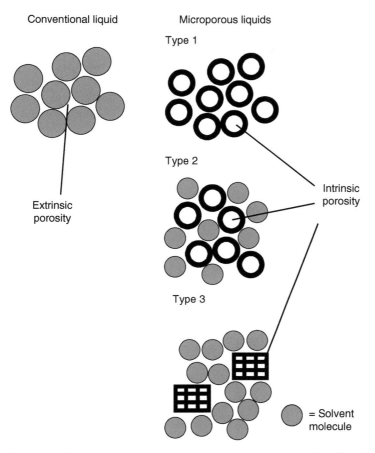

Figure 3.1 Cartoon diagram to represent molecules in a conventional liquid, which has only "extrinsic" porosity between the molecules (small, irregular, transient cavities), and microporous liquids that have "intrinsic" porosity within the molecules (molecule-sized, regular, permanent empty cavities). Type 1: neat liquid hosts that cannot collapse or interpenetrate. Type 2: rigid hosts dissolved in solvents that are too sterically hindered to occupy the cavities. Type 3: particles of microporous frameworks dispersed in sterically hindered solvents. Source: O'Reilly et al. 2007 [25]. Reproduced with permission from John Wiley & Sons.

presented in designing and synthesizing porous liquid phases, there have been some important recent successes that provide "proof-of-principle" for the existence of liquids with permanent porosity, as discussed in the next section.

3.4 Porous Liquids Reported in the Literature

3.4.1 Type 1

Organic cages based on the co-condensation of six 1,2-ethylene diamine molecules and four 1,3,5-benzenetricarboxaldehyde molecules have been explored as a platform for forming porous liquids [26–28]. The synthesis and solid-state properties of these POCs have been extensively investigated by Cooper and coworker [29]. They possess a number of advantageous features for forming porous liquids. In particular, they are sufficiently rigid not to collapse in the guest-free state. Also, they can be functionalized on the outside through the use of diamines containing various functional groups. A key structural feature is the central cavity of diameter approximately 5 Å which can be accessed through four tetrahedrally disposed windows. Such cages can host a variety of small molecules in the solid state.

In initial attempts to form type 1 porous liquids based on such cages, a range of alkyl chains of various lengths were added to the diamine groups in order to lower the melting cage point (on heating, the unfunctionalized cage does not melt before decomposition at around 300 °C; Figure 3.2) [26, 27]. This approach was successful in lowering the melting point to as low as around 40 °C by adding relatively long linear n-C_{12} chains. However, modeling of the liquid states of such cages indicated that the chains would quickly and completely occupy the cage cavities and remove the porosity. Modeling suggested that the addition of bulky

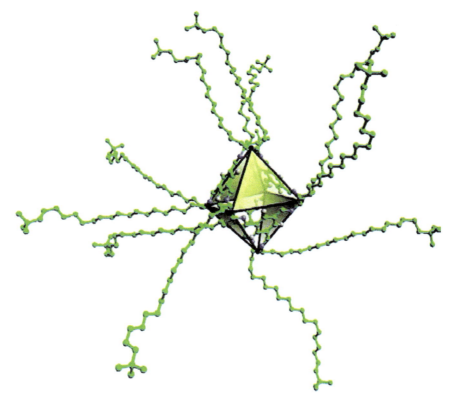

Figure 3.2 United atom representation of a cage functionalized with neo-C_{22} chains. The octahedron highlights the hollow cage core that defines an empty pore. The access windows are featured as transparent yellow planes. Source: Melaugh et al. 2014 [27]. Reproduced by permission from the PCCP Owners Societies.

groups to the ends of the chains (e.g. *t*-butyl) could slow down chain inclusion but could not prevent it indefinitely. However, interestingly, by modulating the *length* of the chain, complete occupation of the cages could be prevented. Specifically, in the liquid state of the cage bearing n-C_5 chains, around 30% of the cage cavities remained unoccupied at any given time. Thus, according to the modeling, this material was a type-1 porous liquid (denoted PL1–1). Modeling of gas inclusion into this phase was also undertaken. An important prediction of the modeling was that the presence of the pores in this material should increase the solubility of methane by a factor of 5 in comparison to nonporous analog materials. This provided an exciting hint that the properties of porous liquids could be remarkably different from those of normal solvents (in line with scaled particle theory, as discussed). However, in this case, by shortening the alkyl chains to n-C_5, the lowering of the melting point was compromised and the measured melting point of this material was 150 °C. This is prohibitively high for making experimental measurements of gas solubility.

Shortly afterward, a very different approach to type-1 porous liquids was reported by Dai and coworkers [30]. This approach made use of hollow silica nanospheres to define the pore space. These spheres have relatively large internal cavities of 14 nm, i.e. much greater than those in the organic cages described. Normally solids, they were rendered into liquid form by functionalization of the outside with ionic attachment of polyethylene glycol chains (Figure 3.3). This gave a liquid (denoted PL1–2) which was optically transparent and, although highly viscous, flowed under gravity. Gas solubility was not studied for PL1–2. In fact, gas solubility might not be expected to be increased much by the presence of such large pores since the effective "naked" surface area is probably not as great as in PL1–1, for example. However, the material was incorporated into a polymer film and gas permeability through the film was studied.

The permeability of CO_2 through the film was found to be higher (158 barrer) than when polyethylene glycol was used as an additive (75 barrer) or when nanoparticles without hollow cores were used. This general idea has also been extended to the use of hollow carbon nanospheres to form porous liquids. Because of the inert nature of their surfaces, functionlization was done by adsorbing polymerized ionic liquids onto the surface, then conducting anion exchange of bromide counterions for sulfonated polyethyleneglycol (PEG) chains. The resulting liquid phases (here denoted PL1–3) were found to dissolve

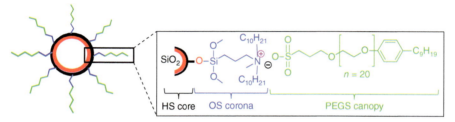

Figure 3.3 Cartoon representation of PL1–2 showing the hollow sphere (HS) core, organosilane (OS) corona, and polyethylene glycol (PEGs) canopy. Source: Zhang et al. 2015 [30]. Reproduced with permission from John Wiley & Sons.

more CO_2 at 1 atm pressure than analogous nonporous control materials, which was ascribed to the microporous nature of the carbon nanospheres. The N_2 solubility was found to remain low, however, which was ascribed to the PEG component acting as a barrier to N_2 diffusion. The application of such materials in CO_2/N_2 separation was suggested [31].

3.4.2 Type 2

To get around the problem of obtaining imine-type cages with low melting points as described, type 2 porous liquids based on these cages were targeted. The presence of a solvent component in type 2 porous liquids means that the cage component does not have to have a low melting point, removing a significant technical obstacle. However, the cage must still be rendered very highly soluble in the solvent used and the solvent must be sufficiently hindered not to enter the cage. Meeting these two criteria is also challenging. Nevertheless, success was achieved using 15-crown-5 as a bulky solvent with low surface curvature, and by functionalizing the outside of the cages with six crown ether groups (Figure 3.4) [28]. In this way, an empty organic cage could be dissolved in the bulky crown solvent at very high concentration, such that only 12 solvent molecules were needed per cage molecule to form a clear solution. Because of the high molecular weight of the cage, the resulting porous liquid (denoted PL2–1) was actually around 50% w/w cage to solvent. Extensive molecular modeling gave confidence that inclusion of the crown ether solvent molecules into the cages was energetically extremely unfavorable and therefore that the pores persisted in the bulk liquid state. Experimental support for the presence of empty pores was obtained through positron annihilation lifetime spectroscopy (PALS). In particular, positron lifetimes within the material were consistent with the presence of cavities of diameter 5 Å, as would be expected. Further,

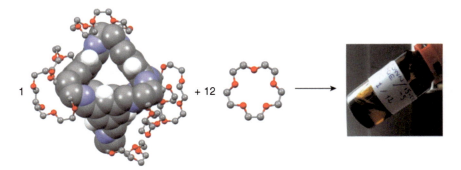

Figure 3.4 A type 2 porous liquid based on an organic cage. The empty, highly soluble cage molecule, left, defines the pore space; the 15-crown-5 solvent, middle, provides fluidity but cannot enter the cage cavities. The concentrated solution (porous liquid PL2–1), right, flows at room temperature, right. Key: C, gray; O, red; N, blue; H, white. Space-filling rendering highlights the core of the cage. Ball-and-stick rendering represents the crown-ether substituents on the cage and the 15-crown-5 solvent. All H atoms except those attached to aromatic rings of the cage compound have been omitted for clarity. Source: Giri et al. 2015 [28]. Reproduced with permission from Springer Nature.

the solubility of methane in the porous liquid was found to be around eight times that in the pure crown ether. This enhanced gas solubility was in line with predictions from molecular modeling, which also suggested that most methane molecules should occupy the cages rather than reside in the solvated regions between cages. This result was a critical experimental proof-of-principle that porous liquids could show unusual properties (such as very high gas solubility).

However, two drawbacks of the crown-ether-based porous liquid were that the synthesis of the cage component was lengthy (six steps) and the solvent component was highly toxic. An elegant alternative cage design with a much shorter synthesis was based on the use of mixtures of cages with subtly different external substituent. Specifically, using a mixture of 1,2-diaminocyclohexane and a dimethyl-substituted diamine in the cage synthesis resulted in a mixture of cages in which varying amounts of each diamine were randomly scrambled over the various cage vertices (Figure 3.5). Because the crystal packing of the mixture was frustrated by the structural disorder, the solubility of the cage mixture was found to be very high in several solvents. Perchloropropene was used as a bulky size-excluded solvent to produce a second type 2 porous liquid, denoted PL2–2 that also exhibited dramatically enhanced solubility for methane, as well as CO_2, Xe, and N_2 compared to neat perchloropropene. It was also demonstrated that release of the gases could be triggered by addition of very small amounts of competitive guests such as chloroform.

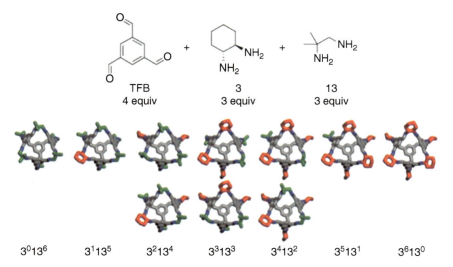

Figure 3.5 An alternative type 2 porous liquid. Reaction of a trialdehyde with two different diamines, 3 and 13, bearing cyclohexane and dimethyl groups, respectively, gives rise to a statistical mixture of cages which differ from each other in the number and positions of their dimethyl (green) and cyclohexane (red) groups. Nomenclature $3^x 13^y$ defines the number of cyclohexane groups (x) and number of dimethyl groups (y) for each cage. Three cages ($3^2 13^4$, $3^3 13^3$ and $3^4 13^2$) each exist as a pair of isomers, as indicated by the additional structures shown in the second row of products. The average structure of this scrambled mixture is $3^3 13^3$. Mixtures of "scrambled" cages were thus prepared in a scalable, one-pot synthesis. Hexachloropropene was used as a solvent which is size excluded from the cage cavities. Source: Giri et al. 2015 [28]. Reproduced with permission from Springer Nature.

The "scrambled" PL2–2 porous liquid was subsequently studied and modeled in further detail [32]. A number of interesting aspects were uncovered. In particular, the occupancy of cages in the presence of gases such as methane was found to be high (around 70%). Interestingly, this value is similar to the occupancy of cages in the solid state under similar conditions. In agreement with previous modeling of porous liquids [28], it was also shown experimentally that gas molecules did exchange between the cages and the solvent regions of the liquid. Also, ultrasonication was demonstrated as a method to accelerate the release of the dissolved gases.

3.4.3 Type 3

Type 3 porous liquids potentially offer a number of potential benefits over types 1 and 2, including more economical synthesis. Given the wide diversity of porous solids known, as well as the variety of liquid phases that may be used, it should be possible to formulate a range of stable dispersions which are likely to have gas solubilities that can be tailored to particular applications. At the time of writing, type 3 porous liquids remain largely unexplored, but this is doubtless an area of activity in future.

Slurries based on the porous MOF, Zn(methylimidazole)$_2$ (ZIF-8), in ethylene glycol or triethylene glycol including in the presence of large amounts of methylimidazole, have been studied for the uptake and release of CO_2 [33]. The study highlighted several interesting points, such as the ability to capture and release CO_2 with relatively low energy input. However, in the context of genuinely porous liquids, it remains to be conclusively shown that the liquid components of this system remain outside the MOF pores. ZIF-8 is known to be able to incorporate much larger guests than its window size would suggest, for example. Previously, dispersions of zeolite nanocrystals in liquid phases have also been described but without the aim of retaining the zeolite porosity [34].

3.4.4 Other Types of Porous Liquids and Theoretical Studies

A theoretical study was made of solvent-free liquid phases composed of rigid ring structures [35]. A range of ring shapes (circular, triangular, square, pentagonal, and hexagonal) were studied and liquid phases with a variety of degrees and types of structural order were modeled. Interestingly, with the circular rings, phases with packing fractions as low as 5% were observed, indicating high levels of porosity. Mimicking the perfect rigidity of these hypothetical ring structures and being able to modulate the attractive/repulsive forces between them to control their packing tendencies as well as fluidities in real systems would no doubt be highly challenging experimentally. Nevertheless, the work does suggest an interesting design strategy for forming type 1 PLs which should be considered further.

Recently there has been interest in the possibility of melting extended coordination polymers or MOFs. For example, Horike and Kitagawa showed that extended frameworks based on Zn^{2+}, HPO_4^{2-}, or $H_2PO_4^-$ and protonated nitrogen bases can melt at as low as 97 °C [36]. The solid phases of these materials are not porous, and porosity was not suggested to be present in the melt phases. However, importantly, this work demonstrates the possibility of melting coordination polymers at perhaps surprisingly low temperatures. The possibilities for preserving the porosity of MOFs in the melt state are intriguing. A study in this regard centred on the thermal behavior of $Zn(im)_2$ (im = imidazolate), ZIF-4 [37]. This material changes from a crystalline state to a glass state upon heating, and it is suggested that a fully liquid state is accessible upon further heating. Modeling of the liquid form (albeit at the prohibitively high temperature of 1200 K) suggested that significant interconnected porosity would be present in this state. Interesting behavior has also been reported for the purely hydrocarbon polymer isotactic poly(4-methyl-1-pentene) [38]. The pressure–volume behavior of the liquid state was studied at 270 °C, using either a gas (helium) or a solid piston to apply the pressure. Approximately 20% contraction in volume was observed when applying pressure via a piston compared to 10% when using He gas. The inference is that the difference between these two values can be accounted for by interstitial pores which become filled when He is used to apply pressure. The presence of pores presumably could, in principle, result from low-density packing caused by the bulky polymer side chains. In this sense, this work is reminiscent of PIMs [19].

3.5 Opportunities for Applications and Current Challenges

One obvious potential area of application for porous materials generally is in chemical separations, where the porous material acts as a selective sorbent. An interesting recent comment article summarizes seven chemical separations that are currently difficult and/or energy-intensive to perform [39]. If step changes in efficiency could be made in these example separations, then the consequences could be very significant. Although much research toward meeting these challenges focuses on the development of new porous solids, it is interesting to note that some current chemical separations are actually based on liquid solvents, such as the use of aqueous amines and/or oligoethers for the separation of CO_2 from mixed gas streams. The use of amines for CO_2 scrubbing is widely recognized as problematic because of the high regeneration costs, which result from the strong chemical binding of the CO_2 to the amine [39]. Porous liquids can be seen as hybrids of microporous solids and liquid solvents, combining the properties of high (and tunable) solubility with fluidity. Therefore, porous liquids with the right combination of properties could represent drop-in replacements with improved properties for these liquid-based separation systems, avoiding the need to retrofit plant that solid adsorbents would require. Also, being liquids, other aspects such as the ability to spray, or print porous liquids, or to form them into contiguous films and easily incorporate them into polymers etc. could represent advantages over crystalline porous solids. Along with these clear opportunities, it is important to recognize that current challenges faced by porous liquids include the cost of production. Type 3 porous liquids may be of particular interest in this regard if they can be formulated economically.

3.6 Concluding Remarks

Although initially the idea that permanent pores can be engineered into liquid states seems counterintuitive, several of the examples described in Sections 3.3–3.4 show that this can in fact be done through judicious design of the components that make up the liquid. Porous liquids now constitute an important class of functional molecular liquids [40]. Also, as alluded to in the introduction, the advent of porous liquids can be seen as a natural extension of the recent trend toward molecular porous solids to complement the classical types of extended inorganic solids. Even though to date only a small number of porous liquids have so far been prepared and definitively characterized, it is notable that remarkable properties such as very high gas solubility and fast gas diffusion have been observed. Given the pressing need for new platform technologies to address current and future global challenges such as sustainability, the field of porous liquids is set to undergo rapid development and there are several exciting prospects for new technologies based on these materials.

References

1 http://proverbicals.com/armenian-proverbs/ (accessed 20 November 2017).
2 Cooper, A.I. (2017). Porous molecular solids and liquids. *ACS Cent. Sci.* 3: 544.
3 Rouquerol, J., Avnir, D., Fairbridge, C.W. et al. (1994). Recommendations for the characterization of porous solids. *Pure Appl. Chem.* 66: 1739–1758.
4 Lee, J., Kim, J., and Hyeon, T. (2006). Recent progress in the synthesis of porous carbon materials. *Adv. Mater.* 18: 2073–2094.
5 Wright, P.A. (2007). *Microporous Framework Solids*. Cambridge: Royal Society of Chemistry.
6 Cottrell, T.L. (1958). *The Strengths of Chemical Bonds*, 2e. London: Butterworths.
7 (a) Hoskins, B.F. and Robson, R. (1989). Infinite polymeric frameworks consisting of three dimensionally linked rod-like segments. *J. Am. Chem. Soc.* 111: 5962. (b) Hoskins, B.F. and Robson, R. (1990). Design and construction of a new class of scaffolding-like materials comprising infinite polymeric frameworks of 3d-linked molecular rods. A reappraisal of the zinc cyanide and cadmium cyanide structures and the synthesis and structure of the diamond-related frameworks $[N(CH_3)_4][Cu^I Zn^{II}(CN)_4]$ and $Cu^I[4,4',4'',4'''$-tetracyanotetraphenylmethane$]BF_4 \cdot xC_6H_5NO_2$. *J. Am. Chem. Soc.* 112: 1546–1554. (c) Abrahams, B.F., Hoskins, B.F., Michail, D.M., and Robson, R. (1994). Assembly of porphyrin building blocks into network structures with large channels. *Nature* 369: 727–729.
8 Kitagawa, S., Kitaura, R., and Noro, S.I. (2004). Functional porous coordination polymers. *Angew. Chem. Int. Ed.* 43: 2334–2375.
9 Long, J.R. and Yaghi, O.M. (2009). The pervasive chemistry of metal-organic frameworks. *Chem. Soc. Rev.* 38: 1213–1214.

10 Eddaoudi, M., Kim, J., Rosi, N. et al. (2002). Systematic design of pore size and functionality in isoreticular MOFs and their application in methane storage. *Science* 295: 469–472.
11 Navarro-Sanchez, J., Argente-Garcia, A., Moliner-Martinez, Y. et al. (2017). Peptide metal-organic frameworks for enantioselective separation of chiral drugs. *J. Am. Chem. Soc.* 139: 4294–4297.
12 Chen, L., Mowat, J.P.S., Fairen-Jiminez, D. et al. (2013). Elucidating the breathing of the metal-organic framework MIL-53(Sc) with ab initio molecular dynamics simulations and in situ X-ray powder diffraction experiments. *J. Am. Chem. Soc.* 135: 15763–15773.
13 McDonald, T.M. et al. (2015). Cooperative insertion of CO_2 in Diamine-appended metal-organic frameworks. *Nature* 519: 303.
14 Kummer, H., Jeremias, F., Warlo, A. et al. (2017). A functional full-scale heat exchanger coated with aluminium fumarate metal-organic framework for adsorption heat transformation. *Ind. Eng. Chem. Res.* 56: 8393–8398.
15 Bai, Y., Dou, Y., Xie, L.H. et al. (2016). Zr-based metal-organic frameworks: design, synthesis, structure and applications. *Chem. Soc. Rev.* 45: 2327–2367.
16 Diercks, C.S. and Yaghi, O.M. (2017). The atom the molecule, and the covalent organic framework. *Science* 355: eaal1585.
17 Bildirir, H., Gregoriou, V.G., Avgeropoulos, A. et al. (2017). Porous organic polymers as emerging new materials of organic photovoltaic applications: current status and future challenges. *Mater. Horiz.* 4: 546–556.
18 Ben, T. and Qiu, S. (2013). Porous aromatic frameworks: synthesis, structure, and functions. *CrystEngComm* 15: 17–26.
19 McKeown, N.B. and Budd, P.M. (2006). Polymers of intrinsic microporosity (PIMs): organic materials for membrane separations, heterogeneous catalysis and hydrogen storage. *Chem. Soc. Rev.* 35: 675–683.
20 Barrer, R.M. and Shanson, V.H. (1976). Dianin's compound as a zeolitic sorbent. *J. Chem. Soc., Chem. Commun.* 333–334.
21 Hasell, T. and Cooper, A.I. (2016). Porous organic cages: soluble, modular and molecular pores. *Nat. Rev. Mater.* 1: 16053.
22 Zhao, X.S., Lu, G.Q., Millar, G.L., and Li, X.S. (1996). Synthesis and characterization of highly ordered MCM-41 in an alkali-free system and its catalytic activity. *Catal. Lett.* 38: 33–37.
23 Pohorille, A. and Pratt, L.R. (1990). Cavities in molecular liquids and the theory of hydrophobic solubilites. *J. Am. Chem. Soc.* 112: 5066–5074.
24 (a) Pierotti, R.A. (1963). The solubility of gases in liquids. *J. Phys. Chem.* 67: 1840–1845. (b) Pierotti, R.A. (1976). A scaled particle theory of aqueous and nonaqueous solutions. *Chem. Rev.* 76: 717–726.
25 O'Reilly, N., Giri, N., and James, S.L. (2007). Porous liquids. *Chem. Eur. J.* 13: 3020–3025.
26 Giri, N., Davidson, C.E., Melaugh, G. et al. (2012). Alkylated organic cages: from porous crystals to neat liquids. *Chem. Sci.* 3: 2153–2157.
27 Melaugh, G., Giri, N., Davidson, C.E. et al. (2014). Designing and understanding permanent microporosity in liquids. *Phys. Chem. Chem. Phys.* 16: 9422–9431.

28 Giri, N., Del Pópolo, M., Melaugh, G. et al. (2015). Liquids with permanent porosity. *Nature* 527: 216–220.

29 Slater, A.G. and Cooper, A.I. (2015). Function-led design of new porous materials. *Science* 348: 988–1000.

30 Zhang, J., Chai, S.-H., Qiao, S.-H. et al. (2015). Porous liquids: a promising class of media for gas separation. *Angew. Chem. Int. Ed.* 54: 932–936.

31 Li, P., Schott, J.A., Zhang, J. et al. (2017). Electrostatic-assisted liquefaction of porous carbons. *Angew. Chem. Int. Ed.* 56: 14958–14962.

32 Greenaway, R.L., Holden, D., Eden, E.G.B. et al. (2017). Understanding gas capacity, guest selectivity, and diffusion in porous liquids. *Chem. Sci.* 8: 2640–2651.

33 Liu, H., Liu, B., Lin, L.-C. et al. (2014). A hybrid absorption-adsorption methos to efficiently capture carbon. *Nat. Commun.* 5: 5147.

34 Devaux, A., Popovic, Z., Bossart, O. et al. (2006). Solubilisation of dye-loaded zeolite L nanocrystals. *Microporous Mesoporous Mater.* 90: 69–72.

35 Avendañoa, C., Jackson, G., Müller, E.A., and Escobedoc, F.A. (2016). Assembly of porous smectic structures formed from interlocking high-symmetry planar nanorings. *Proc. Natl. Acad. Sci. U.S.A.* 113: 9699–6703.

36 Umeyama, D., Horike, S., Inukai, M. et al. (2015). Reversible solid-to-liquid phase transition of coordination polymer crystals. *J. Am. Chem. Soc.* 137: 864–870.

37 Gaillac, R., Pullumbi, P., Beyer, K.A. et al. (2017). Liquid metal-organic frameworks. *Nat. Mater.* 16: 1149–1154.

38 Chiba, A., Inui, M., Kajihara, Y. et al. (2017). Isotactic poly(4-methyl-1-pentene) melt as a porous liquid: reduction of compressibility due to the penetration of pressure medium. *J. Chem. Phys.* 146: 194503.

39 Scholl, D.S. and Lively, R.P. (2016). Seven chemical separations to change the world. *Nature* 532: 435–437.

40 Ghosh, A. and Nakanishi, T. (2017). Frontiers of solvent-free functional molecular liquids. *Chem. Commun.* 53: 10344–10357.

4

Cyclic Host Liquids for the Formation of Rotaxanes and Their Applications

Tomoki Ogoshi[1,2,3]*, Takahiro Kakuta*[1,2]*, and Tada-aki Yamagishi*[1]

[1] *Kanazawa University, Graduate School of Natural Science and Technology, Kakuma-machi, Kanazawa, Ishikawa 920-1192, Japan*
[2] *Kanazawa University, WPI Nano Life Science Institute, Kakuma-machi, Kanazawa, Ishikawa 920-1192, Japan*
[3] *PRESTO, Japan Science and Technology Agency, 4-1-8 Honcho, Kawaguchi, Saitama 332-0012, Japan*

4.1 Introduction

Researchers have developed various types of supramolecular materials based on macrocyclic compounds since such compounds can capture guest molecules into their cavity by various physical interactions [1]. Cyclodextrins [2], calix[n]arenes [3], crown ethers [4], and cucurbit[n]urils [5] are well-known macrocyclic compounds that have long played a major role in supramolecular chemistry. Pillar[n]arenes, which were reported by our group in 2008 [6], are becoming new key players in supramolecular chemistry even though their history is only ten years old [7]. In order to exploit the properties of macrocyclic compounds for a range of different applications, their functionalization is a very important aspect. Among well-known host compounds, cyclodextrins and pillar[n]arenes possess reactive hydroxyl groups on both rims, which allow further installation of functional groups. Different functional groups introduced onto both rims can significantly affect physical properties such as solubility and crystallinity. However, the poor solubility of cyclodextrins in organic solvents prevents their functionalization. Reactions for the functionalization of cyclodextrins are mainly conducted in amphiphilic solvents such as dimethylformamide (DMF) and dimethyl sulfoxide (DMSO), as well as aqueous media [2g]. Therefore, many reactions taking place in organic solvents cannot be used for the functionalization of cyclodextrins. The limited availability of functionalization approaches generally causes low yields of cyclodextrins modified with functional groups. In contrast, due to the high solubility of pillar[n]arenes containing reactive functional groups such as phenols and bromide moieties in various organic solvents typically used in synthesis, pillar[n]arenes possess high functionality when compared with cyclodextrins [7]. One straightforward functionalization approach is based on an etherification reaction. Accordingly, various functionalized pillar[n]arenes have been synthesized by etherification [8]. Pillar[n]arenes with alkyne and azido groups are also useful key compounds

Functional Organic Liquids, First Edition. Edited by Takashi Nakanishi.
© 2019 Wiley-VCH Verlag GmbH & Co. KGaA. Published 2019 by Wiley-VCH Verlag GmbH & Co. KGaA.

because the copper(I)-catalyzed alkyne-azide cycloaddition (CuAAC) reaction is high yielding, functional group tolerant, and compatible with a wide range of substrates [9]. Pillar[*n*]arenes carrying bromide groups on both rims are also good key compounds due to the high reactivity of such groups [9f, 10]. Pillar[*n*]arenes are solids at room temperature in most cases, but we discovered that the modification of flexible and nonsymmetrical substituents gives rise to pillar[*n*]arenes in the liquid state at room temperature [10a, 11]. To the best of our knowledge, liquid-state macrocyclic compounds are little-known and very attractive materials. We investigated the bulk-state host–guest complexation and succeeded in the high-yield synthesis of mechanically interlocked molecules (MIMs) using a cyclic host liquid, tri(ethylene oxide)-modified pillar[5]arene, as a new type of solvent [11, 12]. In this chapter, first, we provide a brief description of the synthesis, structure, and functionality of pillar[*n*]arenes. The molecular design to produce liquid-state pillar[*n*]arenes based on their versatile functionality is also described. Secondly, we discuss the complexation of guest molecules into liquid pillar[*n*]arenes. Finally, we describe the high-yield synthesis of MIMs such as [2]rotaxanes and polyrotaxanes using liquid pillar[*n*]arenes as solvents.

4.2 Liquid Pillar[*n*]arenes at Room Temperature

4.2.1 Synthesis and Structure of Pillar[*n*]arenes

The first pillar[5]arene was prepared serendipitously while investigating the polymerization of featureless monomer, 1,4-dimethoxybenzene with paraformaldehyde in the presence of a Lewis acid (Scheme 4.1) [6].

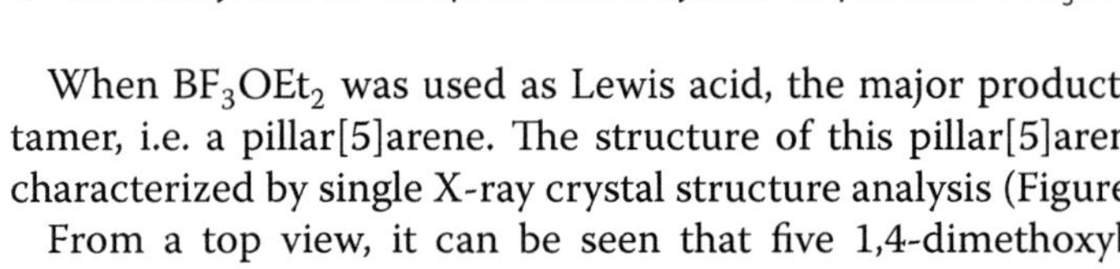

Scheme 4.1 Preparation of first pillar[5]arene, per-methylated pillar[5]arene by the reaction of 1,4-dimethoxybenzene with paraformaldehyde in the presence of BF_3OEt_2.

When BF_3OEt_2 was used as Lewis acid, the major product was a cyclic pentamer, i.e. a pillar[5]arene. The structure of this pillar[5]arene was completely characterized by single X-ray crystal structure analysis (Figure 4.1a).

From a top view, it can be seen that five 1,4-dimethoxybenzene units are connected by methylene bridges at their 2- and 5-positions (para position). The resulting shape corresponds to that of a regular pentagonal structure. From a side view, the shape is that of a highly symmetrical cylindrical structure. The regular pentagon and high symmetrical structure derive from the connection of the 1,4-dimethoxybenzene units at their 2- and 5-positions by methylene bridges. The chemical structure of this pillar[5]arene is very close to that of calix[5]arenes (Figure 4.1b). In contrast, in calix[5]arene, the phenolic units are linked by methylene bridges at their 2- and 6-positions (meta position). As a

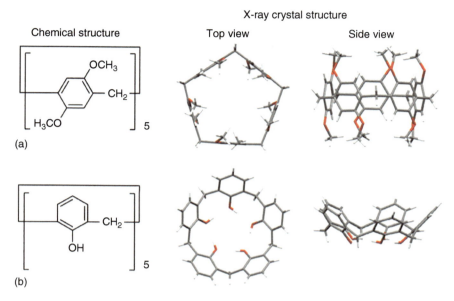

Figure 4.1 Chemical and X-ray crystal structures of (a) per-methylated pillar[5]arene and (b) calix[5]arene.

result of the different methylene bridge positions, the shapes of pillar[5]arenes are completely different from those of calix[5]arenes. Calix[n]arenes have open-ended, nonsymmetric calix-shaped structures, from which their name derives [3h]. In contrast, pillar[n]arenes have completely symmetric cylindrical structures as a result of the linkages at the 2- and 5-positions. Inspired by the highly symmetric pillars that constitute the Parthenon in Athens, we named the obtained highly symmetrical cyclic pentamer "pillar[5]arene." The yield of pillar[5]arene in our first report was 22% [6]. We later developed a facile synthetic route to provide pillar[5]arene, as we found that it could be formed even under air, and that its formation was completed within three minutes [13]. It could then be purified by recrystallization, which is a simple and easy purification method, and eventually isolated in over 70% yield. The starting compounds such as 1,4-dimethoxybenzene, paraformaldehyde, and BF_3OEt_2 were commercially available and inexpensive. The facile synthesis of pillar[5]arenes using commercially available inexpensive reagents is one of the reasons why many research groups became interested in pillar[n]arene chemistry.

4.2.2 Versatile Functionality of Pillar[n]arenes

The other important reason why the pillar[n]arene community is expanding is that pillar[n]arenes possess versatile functionality [7b, e]. In most cases, pillar[5]arene derivatives with various alkyl chains can be produced by macrocyclization of a corresponding monomer instead of 1,4-dimethoxybenzene. Pillar[5]arenes bearing linear alkyl chains and branched alkyl groups, such as ethoxy, propoxy, butoxy, pentyloxy, hexyloxy, dodecanoxy, and 2(S)-methylbutoxy groups, have been produced by macrocyclization (Scheme 4.2) [14].

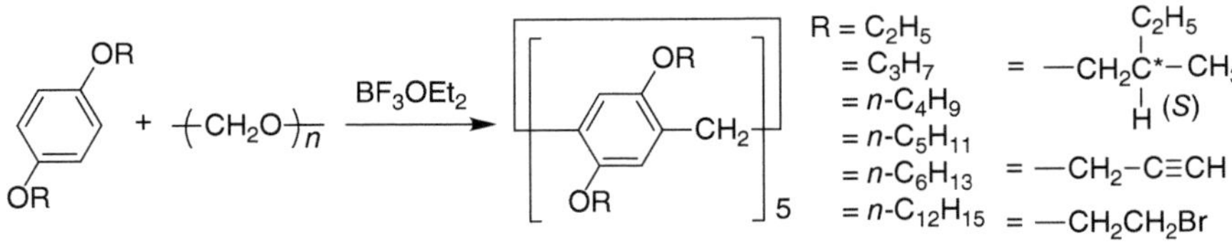

Scheme 4.2 Pillar[5]arenes with various lengths of linear and branched alkyl chains, and alkyne and bromide groups by macrocyclization of corresponding monomers.

Pillar[5]arenes with 10 alkyne and 10 bromide moieties were also produced by macrocyclization of 1,4-dialkoxybenzenes bearing alkyne and bromide groups, respectively [9b, c, 10a, 15]. These pillar[5]arenes are useful key compounds because bromide and alkyne moieties are highly reactive. Another important class of key compounds to produce various functionalized pillar[n]arenes are per-hydroxylated pillar[n]arenes. Per-hydroxylated pillar[n]arenes were synthesized by deprotection of the alkoxy moieties of preformed per-alkylated pillar[n]arenes with BBr$_3$ (Scheme 4.3a) [6, 16].

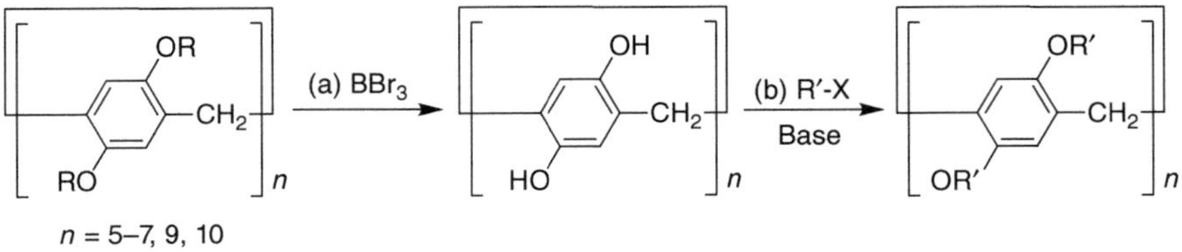

$n = 5$–$7, 9, 10$

Scheme 4.3 (a) Per-hydroxylated pillar[n]arenes by deprotection using BBr$_3$.
(b) Functionalization of pillar[n]arenes from per-hydroxylated pillar[n]arenes by etherification.

The etherification reaction is a straightforward approach to produce functionalized pillar[n]arenes from per-hydroxylated pillar[n]arenes (Scheme 4.3b). Pillar[n]arene derivatives carrying various functional groups are accessible by etherification of per-hydroxylated pillar[n]arenes, with the corresponding functional groups bearing good leaving groups in the presence of appropriate bases.

4.2.3 Molecular Design to Produce Liquid-State Macrocyclic Hosts

4.2.3.1 Pillar[n]arenes

The melting point of the first pillar[5]arene, per-methylated pillar[5]arene, was quite high (250 °C) due to its high crystallinity [6]. The melting points of simple per-alkylated pillar[5]arenes, which were prepared by cyclization of the corresponding monomers, decreased as the alkyl chain length increased up to five carbons (Figure 4.2, the melting point of per-pentylated pillar[5]arene was 96 °C), but increased again starting from a six carbon chain (the melting point of per-hexylated pillar[5]arene was 106 °C) [14a].

Since simple pillar[5]arenes bearing linear alkyl chains exist in the solid state, thus molecular design is required to obtain liquid pillar[5]arenes. The

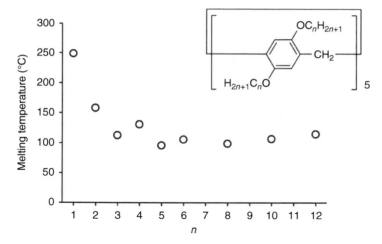

Figure 4.2 Melting points of per-alkylated pillar[5]arenes depending on the alkyl chain length.

first approach to produce liquid-state pillar[5]arenes was the introduction of branched alkanes onto both rims of pillar[5]arenes (Tomoki Ogoshi et al. Unpublished data). The introduction of branched alkane moieties into fullerene (C_{60}) and π-conjugated molecules is known to provide liquid-state C_{60} and π-conjugated molecules [17]. Therefore, a pillar[5]arene bearing branched alkyl chains was prepared by cyclization of hydroquinone diether with branched alkanes (Scheme 4.4).

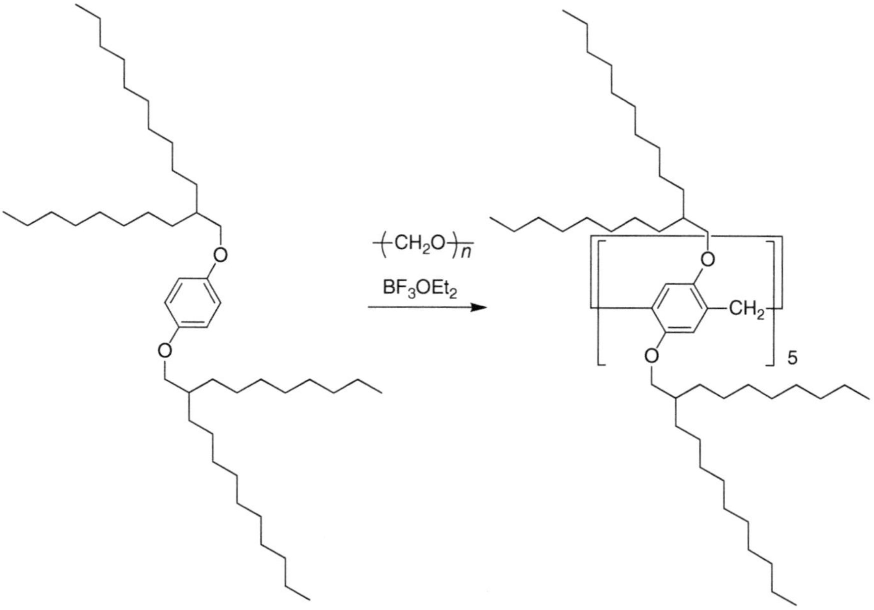

Scheme 4.4 Synthesis of liquid pillar[5]arene by macrocyclization of hydroquinone diether with branched alkane.

As expected, the obtained product was in the liquid state at room temperature, and even at −60 °C. Due to the branch-shaped structure, its crystallinity was quite low, contributing to its liquid state. The other strategy to obtain liquid pillar[5]arenes consists in the introduction of ionic liquid moieties on both rims of pillar[5]arenes. Ionic liquids, which are organic molten salts, have attracted much attention due to their high ionic conductivity, low volatility, and wide applications as reaction media. The introduction of ionic liquid moieties into organic compounds results in decreasing melting points. For example, Chujo and coworkers reported that room-temperature ionic liquids consisting of an octacarboxy polyhedral oligomeric silsesquioxane (POSS) core were synthesized by introduction of imidazolium moieties [18]. The rigid cubic structure of POSS was surrounded by imidazolium ionic liquid moieties, which contributed to the liquid state of POSS. Therefore, pillar[5]arenes with 10 imidazolium ionic liquid moieties were synthesized by three-step reactions (Scheme 4.5).

First, the cyclization of bis(1-bromopropoxy)benzene with paraformaldehyde afforded a pillar[5]arene carrying 10 bromide moieties. Upon reacting the decabromide with an excess of 1-methylimidazole, a pillar[5]arene containing 10 imidazolium bromides was formed. The decaimidazolium bromide was liquid, but it was eliminated due to its hygroscopic nature. The reaction of decaimidazolium bromide with an excess of sodium hexafluorophosphate and lithium bis(trifluoromethanesulfonyl)amide afforded decaimidazolium hexafluorophosphate and bis(trifluoromethanesulfonyl)amide, respectively, in high yields. Decaimidazolium hexafluorophosphate was a white solid, which melted at 119 °C. In contrast, decaimidazolium bis(trifluoromethanesulfonyl)amide was liquid at room temperature, and its melting point was observed at −7 °C. As with ionic liquids, the type of counter anions determined the state of the pillar[5]arene bearing 10 imidazolium moieties. The melting point of the unit model with two bis(trifluoromethanesulfonyl)amide anions was 88 °C, indicating a pillar-shaped structure, and showing that the modification of ionic liquid moieties into a pillar[5]arene core is very important to obtain liquid pillar[5]arenes.

The introduction of soft and flexible oligo(ethylene oxide) chains onto both rims of pillar[n]arenes is also a useful way to obtain liquid-state pillar[5]arenes. Pillar[n]arenes with mono- and tri(ethylene oxide) moieties were produced by etherification of per-hydroxylated pillar[n]arenes with mono- and tri(ethylene oxide)s bearing good leaving tosyl groups (Scheme 4.6) [11].

Both pillar[5]- and pillar[6]arenes with 10 and 12 tri(ethylene oxide) moieties were liquids, while pillar[5]arene with 10 mono(ethylene oxide) moieties was solid. An appropriate length of the oligo(ethylene oxide) chains is necessary to obtain liquid-state pillar[n]arenes.

4.2.3.2 Cyclodextrins

Liquid-state cyclodextrins were also reported. In many cases, the modification of flexible oligo(ethylene oxide) chains on both rims led to liquid-state cyclodextrins. However, in these reports, liquid cyclodextrins were produced by chance and no attention was paid to their preparation. Following the molecular design used to produce liquid-state pillar[n]arenes, tri(ethylene oxide) chains were introduced on both rims of cyclodextrins [11]. Liquid-state α-, β-, and

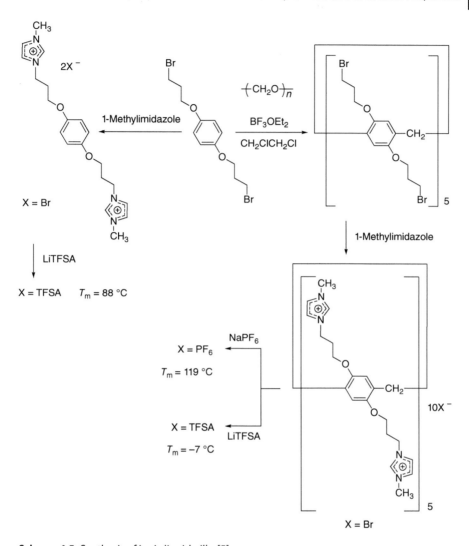

Scheme 4.5 Synthesis of ionic liquid pillar[5]arenes.

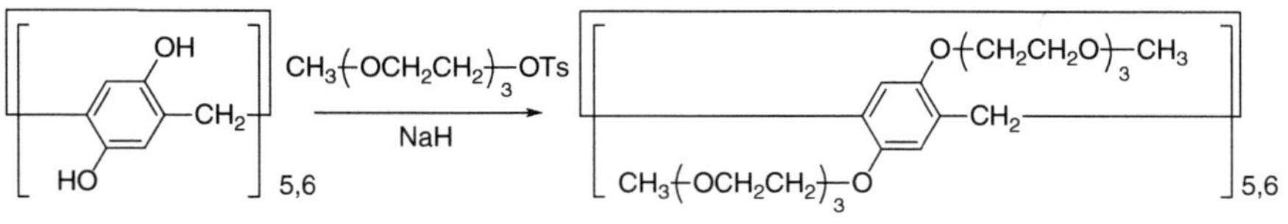

Scheme 4.6 Synthesis of liquid pillar[5,6]arenes from per-hydroxylated pillar[5,6]arenes by introduction of tri(ethylene oxide) chains by etherification.

γ-cyclodextrins were obtained by introducing tri(ethylene oxide) chains on both rims (Scheme 4.7), while the starting compounds, i.e. native α-, β-, and γ-cyclodextrins were in the solid state.

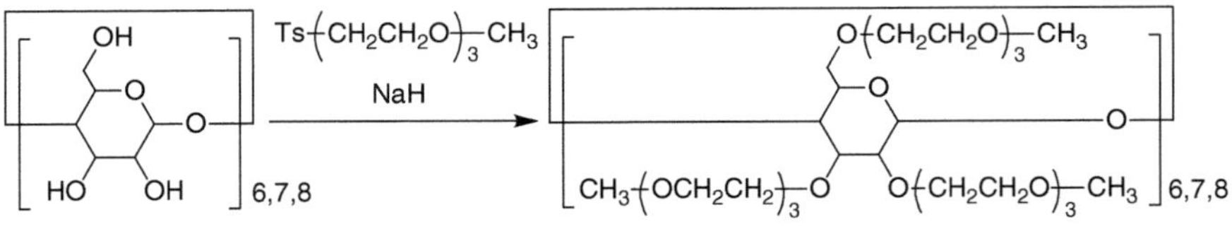

Scheme 4.7 Synthesis of liquid α-, β- and γ-cyclodextrins from native α-, β-, and γ-cyclodextrins by modification of tri(ethylene oxide) moieties by etherification.

These data indicate that the introduction of tri(ethylene oxide) chains is a useful approach to obtain liquid-state macrocyclic compounds.

4.2.3.3 Crown Ethers

Small ring-size crown ethers, such as 12-crown-4, 15-crown-5, and 18-crown-6 are in the liquid state at room temperature since their structures are constructed from oligo(ethylene oxide) moieties. The introduction of benzene moieties into crown ethers increased their melting points.

4.2.3.4 Calix[n]arenes and Cucurbit[n]urils

Calix[n]arenes possess highly reactive phenolic moieties on their lower rims. Functional groups can be introduced only onto lower rims; thus, the presence of functional groups at the reactive sites should change their physical properties. Regen and coworkers reported that the etherification between the phenolic groups of calix[n]arenes and oligo(ethylene oxide)s with good leaving groups afforded calix[n]arenes with oligo(ethylene oxide) chains on the lower rims (Figure 4.3) [19].

Calix[4]arenes with di- and tri(ethylene oxide) chains on the lower rims were oily liquids, while calix[6]arenes with oligo(ethylene oxide) chains were in the solid state at room temperature. To the best of our knowledge, there are no

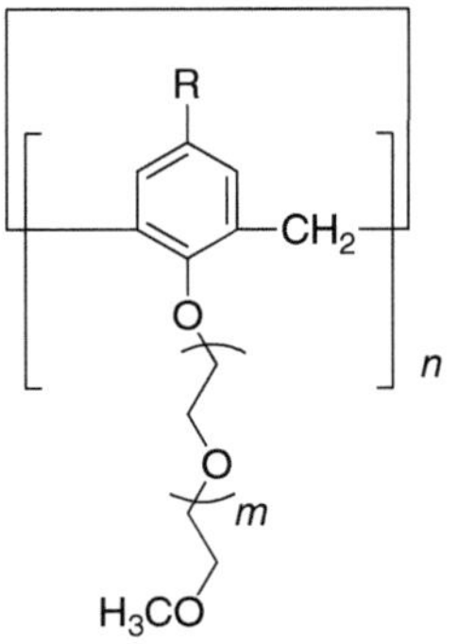

Liquid (oil), ($n = 4$, $m = 1$, R = H)
Liquid (oil), ($n = 4$, $m = 2$, R = H)
T_m = 97–99.5 °C, ($n = 4$, $m = 1$, R = C(CH$_3$)$_3$)
T_m = 65–67 °C, ($n = 4$, $m = 2$, R = C(CH$_3$)$_3$)
T_m = 216–217 °C, ($n = 6$, $m = 0$, R = H)
T_m = 131–131.5 °C, ($n = 6$, $m = 1$, R = H)
T_m = 87.5–88 °C, ($n = 6$, $m = 2$, R = H)
T_m = 40–42 °C, ($n = 6$, $m = 3$, R = H)
T_m = 175.5–176.5 °C, ($n = 6$, $m = 1$, R = C(CH$_3$)$_3$)
T_m = 129.5–132.5 °C, ($n = 6$, $m = 2$, R = C(CH$_3$)$_3$)

Figure 4.3 Liquid calix[n]arenes by modification of oligo(ethylene oxide) chains by etherification.

examples of liquid-state cucurbit[*n*]urils, probably due to their low functionality. The number of cucurbit[*n*]uril derivatives has been very low. However, Kim, Isaacs, and coworkers reported various functionalization approaches of cucurbit[*n*]urils, which hold the potential to afford liquid-state cucurbit[*n*]urils [20].

4.3 Complexation of Guest Molecules by Pillar[5]arenes

4.3.1 Host Properties of Pillar[5]arenes

Electrostatic effects play an important role in molecular recognition. The electron potential mapping of pillar[5]arene is shown in Figure 4.4a.

The inner cavity of pillar[5]arene is colored in red, to indicate that it has a negative charge. This is due to the fact that the cavities are constructed from π-electron-rich 1,4-dialkoxybenzene units. Furthermore, the tubular-shaped cavity of pillar[*n*]arenes contributes to increase the π-electron density. Therefore, pillar[5]arenes form complexes with cationic molecules and molecules with electron-withdrawing groups. The cavity size of pillar[5]arenes is ~4.7 Å based on its X-ray crystal structure. Therefore, pillar[5]arenes capture linear and simple aromatic molecules, fitting into their cavity size. Particularly, pillar[5]arenes form strong complexes with linear and simple aromatic molecules bearing cationic and electron-withdrawing groups. 4.4b shows some suitable guest molecules for pillar[5]arenes. Linear molecules with cationic groups such as ammonium, pyridinium, and viologen moieties are good guest molecules for pillar[5]arenes [6, 21]. Interestingly, neutral linear alkanes with

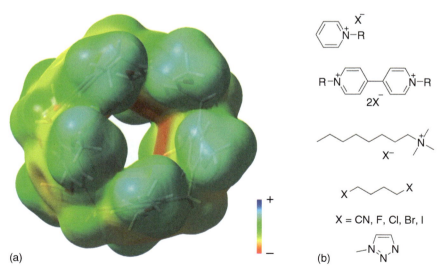

Figure 4.4 (a) Calculated electron potential profile (density functional theory (DFT) calculations, B3LYP/6-31G(d,p)) of per-methylated pillar[5]arene. (b) Guest molecules for pillar[5]arenes.

electron-withdrawing groups such as halogen, cyano, imidazole, and triazole groups fit into the cavity of pillar[5]arenes [22]. A four methylene linker between these electron-withdrawing groups is the best length for pillar[5]arenes, since both the increase and decrease in the methylene linker length resulted in unfavorable effects on complex formation. Pillar[5]arenes also form host–guest complexes with simple linear alkanes without electron-withdrawing groups. However, the association constants between n-alkanes and pillar[5]arenes are very small in $CDCl_3$ (K = c. $10\,M^{-1}$) due to weak CH/π interactions [11, 23].

4.3.2 Complexation of Guest Molecules in Liquid Pillar[5]arenes

In order to perform organic reactions, usually a liquid is used as solvent so that they can be carried out in homogeneous systems by dissolving the substrates in the liquid. Furthermore, it is also possible to control the reaction speed and remove the heat generated in the reaction using a liquid. However, a liquid might not be needed or might interfere with the reaction in some cases. In this event, bulk reaction systems are employed to make the reactions proceed more efficiently and quickly. For example, a bulk radical polymerization system can afford higher molecular weight polymers than a solution radical polymerization system. For the host–guest complexation reaction, solvents are also used in most cases, although they are not always required. Generally, macrocyclic hosts exist in the solid state at room temperature; thus, the host and guest molecules have to be dissolved in a liquid to form the corresponding host–guest complexes. However, on a molecular level (Figure 4.5), when host and guest are dissolved in a liquid, the liquid distances the guest molecules from the host.

This decreases the stability of the host–guest complexes and reduces the efficiency of these chemical reactions. Liquid pillar[5]arenes could be obtained by molecular design, and their host–guest complexation was investigated by ^{1}H NMR using liquid-state tri(ethylene oxide)-modified pillar[5]arene as solvent [11]. When 1,12-dibromododecane, which was used as guest, was mixed with an equimolar amount of liquid pillar[5]arene in the presence of $CDCl_3$, the proton signals of the methylene moieties of 1,12-dibromododecane were shifted (Figures 4.6a and b), although the peak shifts were very small.

The association constant of the complex between 1,12-dibromododecane and liquid pillar[5]arene determined by ^{1}H NMR titration was very small (K = $16.9 \pm 1.0\,M^{-1}$). The host–guest complexation in bulk state was also investigated by ^{1}H NMR using a double NMR tube. A 1:1 mixture of 1,12-dibromododecane and liquid pillar[5]arene was placed in the inner tube, while the external tube contained $CDCl_3$. In this case, the proton signals of 1,12-dibromododecane showed upfield shifts (Figures 4.6c and d), which were much larger than those of the sample in $CDCl_3$, indicating that the host–guest complexation in liquid pillar[5]arene was efficiently maintained compared with the host–guest complexation in $CDCl_3$. The same trends were observed when we investigated the complexation of other linear alkanes with and without $CDCl_3$. The solvation of host and guest molecules by $CDCl_3$ decreases the stability of the complexation, but no solvation effect occurs in bulk liquid pillar[5]arene systems. Thus, bulk liquid pillar[5]arene systems are excellent to maintain the

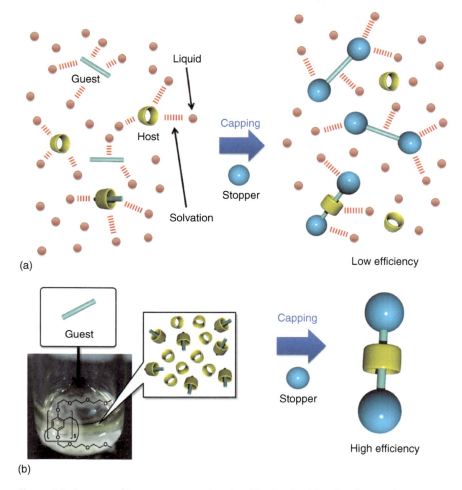

Figure 4.5 Concept of host–guest complexation (a) using liquid molecules as solvents and (b) in cyclic host liquids.

host–guest complexation even for unfavorable combinations of host–guest complexes.

4.4 High Yield Synthesis of [2]Rotaxane and Polyrotaxane Using Liquid Pillar[5]arenes as Solvents

Based on the efficient host–guest complexation observed in bulk liquid pillar[5]arene systems, we synthesized [2]rotaxanes by the end-capping method using liquid pillar[5]arene and 1,12-dibromododecane as solvent and axle, respectively [11]. Again, it was noted that pillar[5]arene formed a stable host–guest complex with 1,12-dibromododecane in bulk liquid pillar[5]arene,

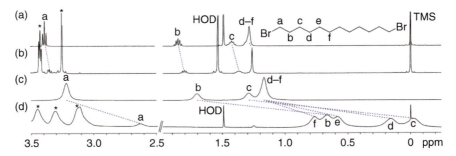

Figure 4.6 ¹H NMR spectra at 42 °C. (a) 1,12-dibromododecane (1 mM) in CDCl$_3$, (b) 1 : 1 mixture of 1,12-dibromododecane (1 mM) and tri(ethylene oxide)-modified pillar[5]arene (1 mM) in CDCl$_3$, (c) 1,12-dibromododecane without CDCl$_3$, and (d) equimolar mixture of 1,12-dibromododecane and tri(ethylene oxide)-modified pillar[5]arene without CDCl$_3$. Peaks with asterisk are proton resonances from tri(ethylene oxide)-modified pillar[5]arene. Source: Reproduced with permission from Ogoshi et al. [11]. Copyright 2012, American Chemical Society.

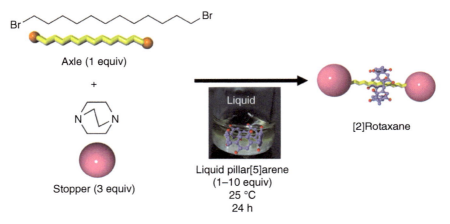

Figure 4.7 Synthesis of [2]rotaxane by end-capping reaction between 1,12-dibromododecane and 1,4-diazabicyclo[2.2.2]octane in liquid pillar[5]arene.

but not in CDCl$_3$. Bulky 1,4-diazabicyclo[2.2.2]octane was used as stopper (Figure 4.7).

The axle and stopper were dissolved in liquid pillar[5]arene, then stirred for 24 hours at 25 °C. Surprisingly, the conversion to [2]rotaxane was high when an excess of liquid pillar[5]arene was used as solvent. In particular, the conversion was 61% when an equimolar amount of liquid pillar[5]arene was used, and increased to 97% using 5 equiv of liquid pillar[5]arene. The isolated yield of [2]rotaxane was high (91%). In contrast, the conversion decreased to 8% even in the presence of a small amount of CDCl$_3$. The solvent interrupts the host–guest complexation, which results in the low conversion. Stoddart, Huang, and coworkers reported the synthesis of [2]rotaxanes using host–guest complexes between pillar[5]arene and linear alkanes [9e, 24]. However, the corresponding yields were too low owing to the use of weak host–guest complexes in CDCl$_3$.

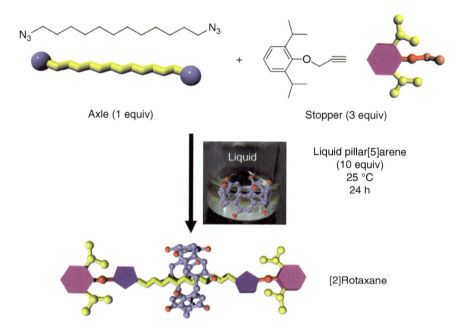

Figure 4.8 Synthesis of [2]rotaxane by CuAAC reaction between 1,12-diazidedodecane and a stopper with one alkyne moiety in liquid pillar[5]arene.

When we used a weak host–guest complex in CDCl$_3$, the yield of [2]rotaxane was also quite low. In contrast, [2]rotaxane was obtained in high yield in the liquid pillar[5]arene system. Thus, bulk complexation systems using liquid pillar[5]arene are powerful tools to produce [2]rotaxanes in high yields.

The CuAAC reaction is also used to synthesize [2]rotaxane in liquid pillar[5]arene (Figure 4.8).

1,12-Diazidedodecane and a stopper with one alkyne moiety were used as axle and stopper, respectively. The axle and stopper were mixed in an excess of liquid pillar[5]arene, then Cu(CH$_3$CN)$_4$PF$_6$ and tris[(1-benzyl-1H-1,2,3-triazol-4-yl)methyl]amine (TBTA) were added to the mixture. Next, the mixture was stirred for 24 hours at 25 °C. The conversion and isolated yield of the [2]rotaxane were 92% and 88%, respectively. The CuAAC reaction is widely used for the synthesis of MIMs; thus, the CuAAC reaction in liquid pillar[5]arenes can be used to synthesize various MIMs in high yields. We synthesized polyrotaxanes by the CuAAC reaction in liquid pillar[5]arenes (Figure 4.9) [12a].

We used polytetrahydrofuran (PolyTHF) having azide moieties at both ends and bulky compounds with an alkyne moiety as polymeric chain and a stopper to synthesize polyrotaxanes. PolyTHF and the stopper were dissolved in liquid pillar[5]arene, together with Cu(CH$_3$CN)$_4$PF$_6$ and TBTA. Polyrotaxanes were produced by stirring the mixture for 12 hours at 25 °C. Polyrotaxanes consisting of PolyTHF and many pillar[5]arene rings were obtained in moderate yield (44%) after purification. We also synthesized polyrotaxanes containing reactive pillar[5]arene rings using a new liquid pillar[5]arene. We synthesized a liquid pillar[5]arene with 10 reactive alkene ends. Using this reactive liquid pillar[5]arene

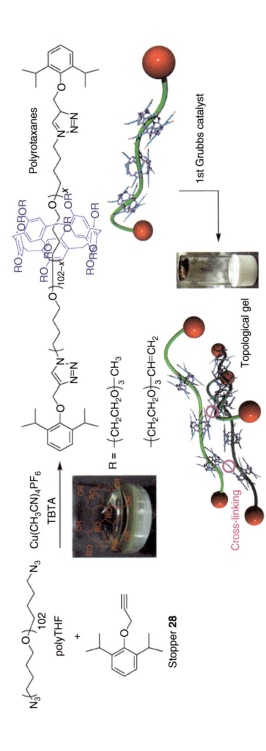

Figure 4.9 Synthesis of nonionic polyrotaxanes and a topological gel prepared from liquid pillar[5]arenes. Source: Reproduced with permission from Ogoshi et al. [12a]. Copyright 2014, The Royal Society of Chemistry.

as solvent, we also synthesized polyrotaxanes containing the pillar[5]arene with 10 alkene moieties as rings under the same reaction conditions used for the synthesis of polyrotaxanes. The intermolecular cross-linking between polyrotaxanes by olefin metathesis reactions afforded topological gels. The combination of CuAAC and olefin metathesis reactions in liquid pillar[5]arene resulted in the formation of topological gels, which is the first report of topological gels obtained using pillar[5]arene rings.

Pseudopolyrotaxanes were constructed using polyethylene as polymeric chain and per-ethylated pillar[5]arenes as rings since they can accommodate linear alkanes. However, polyethylene is not soluble in organic solvents at room temperature; thus, pseudopolyrotaxanes using polyethylene are quite difficult to obtain using a solvent. We then investigated bulk pseudopolyrotaxanation [25]. A pseudopolyrotaxane structure was formed by mixing per-pentylated pillar[5]arene with polyethylene in the melted state (Figure 4.10a). The molted polyethylene at 140 °C became solid upon addition of per-pentylated pillar[5]arene at 140 °C, indicating that pseudopolyrotaxanation has occurred.

From differential scanning calorimetry (DSC) measurements, the melting points of polyethylene and per-pentylated pillar[5]arene were 130 and 96 °C, respectively (Figure 4.10b). The pseudopolyrotaxanation by mixing melted polyethylene with melted per-pentylated pillar[5]arene at 140 °C led to the extension of the polyethylene chains, which dramatically increased the melting point of polyethylene from 130 to 152 °C. This process is very simple and industrially useful for enhancing the thermal properties of polyethylene.

[2]Rotaxanes were obtained by the end-capping method in the liquid pillar[5]arene system [12b]. We also obtained a pseudo[2]rotaxane in the liquid pillar[5]arene using the slippage method. If a dumbbell-shaped molecule containing an axle and bulky moieties at both ends can form a complex with a ring molecule at high temperature, but not at room temperature, pseudo[2]rotaxanes that are stable at room temperature can be produced by heating the mixture of the dumbbell and the ring components in a solvent. Thus, the formation of pseudo[2]rotaxanes by this slippage method is very easy: just mixing and heating the mixture of the dumbbell and ring components. However, the control of the reaction conditions is required to obtain pseudo[2]rotaxanes because the use of a liquid as solvent for the synthesis of pseudo[2]rotaxanes via the slippage method determines that a ring can slip over the bulky ends of the dumbbell at high temperature. Again, it should be noted that the bulk complexation system using liquid pillar[5]arene maximizes the complexation efficiency between host and guest. Thus, we speculated that the bulk complexation system using liquid pillar[5]arene is suitable to synthesize pseudo[2]rotaxanes by the slippage method. n-Dodecane with bulky valine derivative ends was used as dumbbell (Figure 4.11).

A pseudo[2]rotaxane that is stable at 25 °C was produced by heating the dumbbell in liquid pillar[5]arene at 100 °C. After the reaction was completed, the pseudo[2]rotaxane could be isolated by silica gel chromatography since it was stable at room temperature. Figure 4.12a shows the temperature effect on the conversion to pseudo[2]rotaxane.

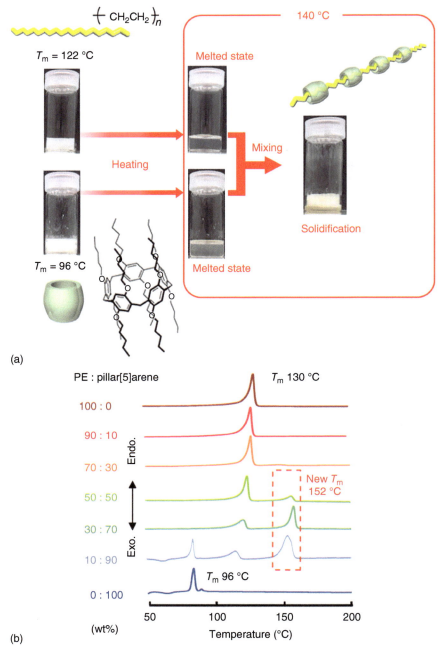

Figure 4.10 (a) The pseudopolyrotaxanation by mixing melted polyethylene with melted per-pentylated pillar[5]arene at 140 °C. (b) DSC traces of pristine polyethylene, per-pentylated pillar[5]arene and a mixture of polyethylene and per-pentylated pillar[5]arene with different feed ratios. Source: Reproduced with permission from Ogoshi et al. [25]. Copyright 2014, Nature Publishing Group.

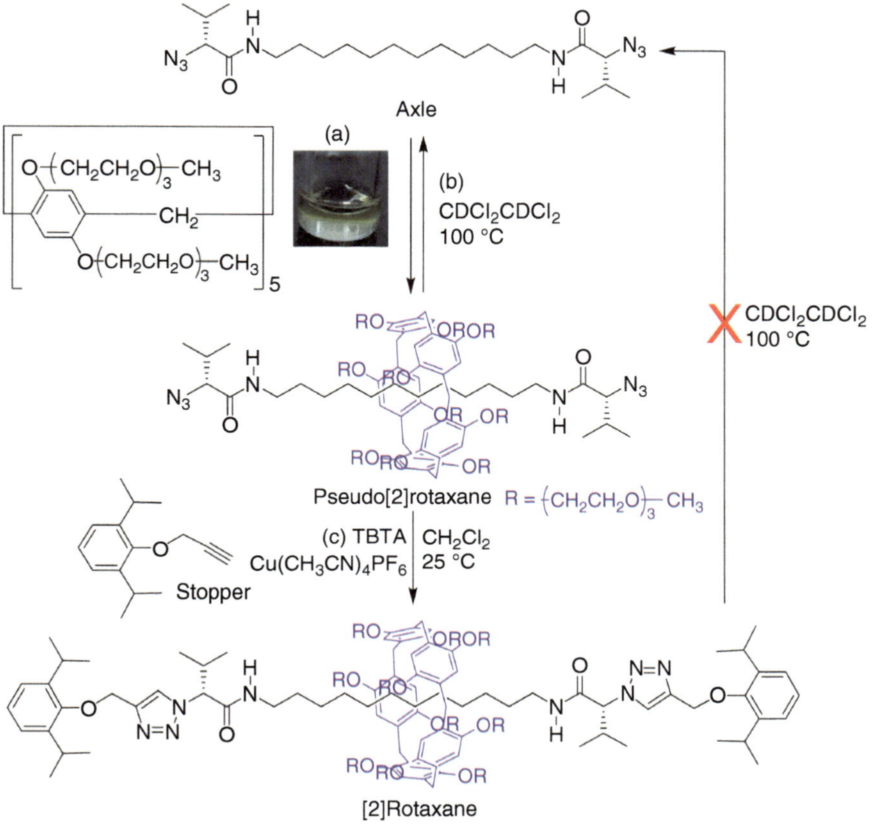

Figure 4.11 (a) Formation of pseudo[2]rotaxane in cyclic host liquid, tri(ethylene oxide)-modified pillar[5]arene, (b) dissociation of pseudo[2]rotaxane in $CDCl_2CDCl_2$ by heating at 100 °C, and (c) conversion of thermally dissociable pseudo[2]rotaxane to thermally stable [2]rotaxane by introducing bulky end-cap via CuAAC reaction. Source: Reproduced with permission from Ogoshi et al. [12]. Copyright 2016, The Royal Society of Chemistry.

The formation of pseudo[2]rotaxane did not occur at 25 °C. At 50 °C, the pseudo[2]rotaxane formation occurred, but did not reach the thermodynamic equilibrium even after 600 hours. At 70 °C, the conversion to pseudo[2]rotaxane at the equilibrium state was 51%. The conversion to pseudo[2]rotaxane decreased as the reaction temperature increased. Figure 4.12b shows the solvent effect on the conversion to pseudo[2]rotaxane. The pseudo[2]rotaxane was not found by heating the dumbbell in liquid pillar[5]arene with $CDCl_2CDCl_2$ as a solvent, while it was produced in the absence of $CDCl_2CDCl_2$. The complete dissociation of pseudo[2]rotaxane was observed upon heating in $CDCl_2CDCl_2$ at 100 °C. These results strongly suggest that bulk host–guest complexation using liquid pillar[5]arene is useful to synthesize pseudo[2]rotaxane by the slippage method. The recycling of the liquid pillar[5]arene and the axle could be accomplished by heating the pseudo[2]rotaxane in a solvent system. The combination of a cyclic host liquid and a normal solvent system enables efficient pseudo[2]rotaxane

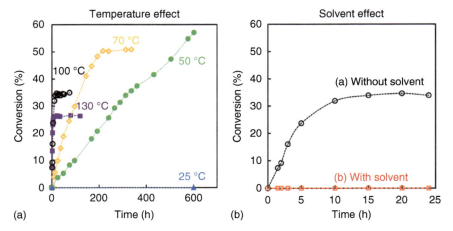

Figure 4.12 Percentage conversions to pseudo[2]rotaxane vs. reaction time (a) in cyclic host liquid, tri(ethylene oxide)-modified pillar[5]arene at 25 °C (blue triangles), 50 °C (green circles), 70 °C (yellow diamonds), 100 °C (black circles), 130 °C (purple squares) and (b) in cyclic host liquid, tri(ethylene oxide)-modified pillar[5]arene (without solvent, black circles) and in $CDCl_2CDCl_2$ (with solvent, red squares) at 100 °C. Source: Reproduced with permission from Ogoshi et al. [12b]. Copyright 2016, The Royal Society of Chemistry.

formation and dissociation. The dumbbell possesses reactive azido moieties at both ends; thus, a bulky stopper can be installed by the CuAAC reaction between the azido moieties in the dumbbell and the alkyne in the bulky stopper. Thermally stable [2]rotaxanes could be obtained by the slippage method followed by end-capping reactions via the CuAAC reaction in liquid pillar[5]arene.

4.5 Conclusion and Remarks

Cyclic host liquids are a new concept that we first mentioned in 2012 [11]. Using liquid pillar[5]arenes, we have constructed MIMs such as [2]rotaxanes, polyrotaxanes, and topological gels. The construction of other pillar[n]arene-based MIMs such as catenanes and polycatenanes could also be accessible using this concept. The use of liquid-state cyclodextrins, crown ethers, and calix[n]arenes for the synthesis of MIMs is also our next research target. Cyclic host liquids can be used as molecular separation systems owing to their molecular recognition ability. The application of cyclic host liquids as solvents for optic compounds is an interesting topic since optic compounds are efficiently located in the cavity of cyclic hosts, resulting in unique optical properties. Pillar[n]arenes with two or more types of functional groups have been synthesized. For example, mono-, di-, tri-, penta-, and per-functionalized pillar[5]arenes are accessible by various synthetic approaches [7b, e]. Thus, a precise molecular design of cyclic host liquids is possible in pillar[n]arene-based cyclic host liquids. The introduction of functional groups enables novel cyclic host liquids with new functions. In contrast, the following aspects need to be improved: (i) The prepared cyclic host liquids possess high viscosity. A decrease of viscosity is required to use cyclic host liquids in many reaction systems and improve handling. (ii) The prepared cyclic host liquids possess hydrophilic moieties; thus, they are good solvents for hydrophilic molecules, but not for hydrophobic molecules. The synthesis of hydrophobic cyclic host liquids is required for wider applications. The solution of these issues should lead to a further development of this concept. Although we are still in the early days of cyclic host liquid research, the creation of new applications of cyclic host liquids, which cannot be accomplished using typical solvents, is a very important aspect.

References

1 Lehn, J.M. (1997). *Supramolecular Chemistry*. Weinheim: Wiley.
2 (a) Rekharsky, M.V. and Inoue, Y. (1998). Complexation thermodynamics of cyclodextrins. *Chem. Rev.* 98: 1875–1918. (b) Uekama, K., Hirayama, F., and Irie, T. (1998). Cyclodextrin drug carrier systems. *Chem. Rev.* 98: 2045–2076. (c) Harada, A., Hashidzume, A., Yamaguchi, H., and Takashima, Y. (2009). Polymeric rotaxanes. *Chem. Rev.* 109: 5974–6023. (d) Crini, G. (2014). Review: a history of cyclodextrins. *Chem. Rev.* 114: 10940–10975. (e) Engeldinger, E., Armspach, D., and Matt, D. (2003). Capped cyclodextrins. *Chem. Rev.* 103: 4147–4174. (f) Harata, K. (1998). Structural aspects of stereodifferentiation in the solid state. *Chem. Rev.* 98: 1803–1828. (g) Khan, A.R., Forgo, P., Stine, K.J., and D'Souza, V.T. (1998). Methods for selective modifications of cyclodextrins. *Chem. Rev.* 98: 1977–1996. (h) Nepogodiev, S.A. and Stoddart, J.F. (1998). Cyclodextrin-based catenanes and rotaxanes. *Chem. Rev.* 98: 1959–1976. (i) Wenz, G., Han, B.H., and Müller, A. (2006). Cyclodextrin rotaxanes and polyrotaxanes. *Chem. Rev.* 106: 782–817.
3 (a) Ikeda, A. and Shinkai, S. (1997). Novel cavity design using calix[n]arene skeletons: Toward molecular recognition and metal binding. *Chem. Rev.* 97: 1713–1734. (b) Stewart, D.R. and Gutsche, C.D. (1999). Isolation, characterization, and conformational characteristics of p-tert-butylcalix[9–20]arenes. *J. Am. Chem. Soc.* 121: 4136–4146. (c) Morohashi, N., Narumi, F., Iki, N. et al. (2006). Thiacalixarenes. *Chem. Rev.* 106: 5291–5316. (d) Ajami, D. and Rebek, J. (2012). More chemistry in small spaces. *Acc. Chem. Res.* 46: 990–999. (e) Guo, D.-S. and Liu, Y. (2014). Supramolecular chemistry of p-sulfonatocalix[n]arenes and its biological applications. *Acc. Chem. Res.* 47: 1925–1934. (f) Brotin, T. and Dutasta, J.-P. (2009). Cryptophanes and their complexes – present and future. *Chem. Rev.* 109: 88–130. (g) Homden, D.M. and Redshaw, C. (2008). The use of calixarenes in metal-based catalysis. *Chem. Rev.* 108: 5086–5130. (h) Gutsche, C.D. (1989). *Calixarenes*. Cambridge: The Royal Society of Chemistry. (i) Vicens, V.B.J. (1991). *Calixarenes: A Versatile Class of Macrocyclic Compounds*. Dordrecht: Kluwer Academic.
4 Pedersen, C.J. (1967). Cyclic polyethers and their complexes with metal salts. *J. Am. Chem. Soc.* 89: 7017–7036.
5 (a) Kim, J., Jung, I.-S., Kim, S.-Y. et al. (2000). New cucurbituril homologues: Syntheses, isolation, characterization, and X-ray crystal structures of cucurbit[n]uril ($n = 5, 7$, and 8). *J. Am. Chem. Soc.* 122: 540–541.

(b) Baek, K., Hwang, I., Roy, I. et al. (2015). Self-assembly of nanostructured materials through irreversible covalent bond formation. *Acc. Chem. Res.* 48: 2221–2229. (c) Isaacs, L. (2014). Stimuli responsive systems constructed using cucurbit[n]uril-type molecular containers. *Acc. Chem. Res.* 47: 2052–2062. (d) Kaifer, A.E. (2014). Toward reversible control of cucurbit[n]uril complexes. *Acc. Chem. Res.* 47: 2160–2167. (e) Lee, J.W., Samal, S., Selvapalam, N. et al. (2003). Cucurbituril homologues and derivatives: new opportunities in supramolecular chemistry. *Acc. Chem. Res.* 36: 621–630.

6 Ogoshi, T., Kanai, S., Fujinami, S. et al. (2008). *para*-Bridged symmetrical pillar[5]arenes: their Lewis acid catalyzed synthesis and host-guest property. *J. Am. Chem. Soc.* 130: 5022–5023.

7 (a) Si, W., Xin, P., Li, Z.T., and Hou, J.L. (2015). Tubular unimolecular transmembrane channels: construction strategy and transport activities. *Acc. Chem. Res.* 48: 1612–1619. (b) Strutt, N.L., Zhang, H.C., Schneebeli, S.T., and Stoddart, J.F. (2014). Functionalizing pillar[n]arenes. *Acc. Chem. Res.* 47: 2631–2642. (c) Xue, M., Yang, Y., Chi, X.D. et al. (2012). Pillararenes, a new class of macrocycles for supramolecular chemistry. *Acc. Chem. Res.* 45: 1294–1308. (d) Ogoshi, T. (2016). *Pillararene*. Cambridge: The Royal Society of Chemistry. (e) Ogoshi, T. and Yamagishi, T. (2013). Pillararenes: versatile synthetic receptors for supramolecular chemistry. *Eur. J. Org. Chem.* 2961–2975. (f) Ogoshi, T. and Yamagishi, T. (2013). New synthetic host pillararenes: their synthesis and application to supramolecular materials. *Bull. Chem. Soc. Jpn.* 86: 312–332. (g) Ogoshi, T. (2012). Synthesis of novel pillar-shaped cavitands "Pillar[5]arenes" and their application for supramolecular materials. *J. Inclusion Phenom. Macrocyclic Chem.* 72: 247–262. (h) Ogoshi, T. and Yamagishi, T. (2014). Pillar[5]- and pillar[6]arene-based supramolecular assemblies built by using their cavity-size-dependent host-guest interactions. *Chem. Commun.* 50: 4776–4787. (i) Ogoshi, T., Yamagishi, T., and Nakamoto, Y. (2016). Pillar-shaped macrocyclic hosts pillar[n]arenes: new key players for supramolecular chemistry. *Chem. Rev.* 116: 7937–8002.

8 (a) Ogoshi, T., Hashizume, M., Yamagishi, T., and Nakamoto, Y. (2010). Synthesis, conformational and host-guest properties of water-soluble pillar[5]arene. *Chem. Commun.* 46: 3708–3710. (b) Ogoshi, T., Masaki, K., Shiga, R. et al. (2011). Planar-chiral macrocyclic host pillar[5]arene: no rotation of units and isolation of enantiomers by introducing bulky substituents. *Org. Lett.* 13: 1264–1266.

9 (a) Ogoshi, T., Shiga, R., Hashizume, M., and Yamagishi, T. (2011). "Clickable" pillar[5]arenes. *Chem. Commun.* 47: 6927–6929. (b) Deng, H.M., Shu, X.Y., Hu, X.S. et al. (2012). Synthesis of a fully functionalized pillar[5]arene by 'click chemistry' and its effective binding toward neutral alkanediamines. *Tetrahedron Lett.* 53: 4609–4612. (c) Niergarten, I., Guerra, S., Holler, M. et al. (2012). Building liquid crystals from the 5-fold symmetrical pillar[5]arene core. *Chem. Commun.* 48: 8072–8074. (d) Ogoshi, T., Yamafuji, D., Kotera, D. et al. (2012). Clickable di- and tetrafunctionalized pillar[n]arenes ($n = 5, 6$) by oxidation-reduction of pillar[n]arene units. *J. Org. Chem.* 77: 11146–11152. (e) Strutt, N.L., Forgan, R.S., Spruell, J.M. et al. (2011).

Monofunctionalized pillar[5]arene as a host for alkanediamines. *J. Am. Chem. Soc.* 133: 5668–5671. (f) Nierengarten, I., Guerra, S., Holler, M. et al. (2013). Macrocyclic effects in the mesomorphic properties of liquid-crystalline pillar[5]- and pillar[6]arenes. *Eur. J. Org. Chem.* 3675–3684.

10 (a) Ogoshi, T., Ueshima, N., Yamagishi, T. et al. (2012). Ionic liquid pillar[5]arene: its ionic conductivity and solvent-free complexation with a guest. *Chem. Commun.* 48: 3536–3538. (b) Ogoshi, T., Kayama, H., Yamafuji, D. et al. (2012). Supramolecular polymers with alternating pillar[5]arene and pillar[6]arene units from a highly selective multiple host guest complexation system and monofunctionalized pillar[6]arene. *Chem. Sci.* 3: 3221–3226. (c) Ogoshi, T., Ueshima, N., and Yamagishi, T. (2013). An amphiphilic pillar[5]arene as efficient and substrate-selective phase-transfer catalyst. *Org. Lett.* 15: 3742–3745.

11 Ogoshi, T., Aoki, T., Shiga, R. et al. (2012). Cyclic host liquids for facile and high-yield synthesis of [2]rotaxanes. *J. Am. Chem. Soc.* 134: 20322–20325.

12 (a) Ogoshi, T., Aoki, T., Ueda, S. et al. (2014). Pillar[5]arene-based nonionic polyrotaxanes and a topological gel prepared from cyclic host liquids. *Chem. Commun.* 50: 6607–6609. (b) Ogoshi, T., Tamura, Y., Yamafuji, D. et al. (2016). Facile and efficient formation and dissociation of a pseudo[2]rotaxane by a slippage approach using pillar[5]arene-based cyclic host liquid and solvent. *Chem. Commun.* 52: 10297–10300.

13 Ogoshi, T., Aoki, T., Kitajima, K. et al. (2011). Facile, rapid, and high-yield synthesis of pillar[5]arene from commercially available reagents and its X-ray crystal structure. *J. Org. Chem.* 76: 328–331.

14 (a) Ogoshi, T., Kitajima, K., Aoki, T. et al. (2010). Synthesis and conformational characteristics of alkyl-substituted pillar[5]arenes. *J. Org. Chem.* 75: 3268–3273. (b) Ogoshi, T., Shiga, R., Yamagishi, T., and Nakamoto, Y. (2011). Planar-chiral pillar[5]arene: chiral switches induced by multiexternal stimulus of temperature, solvents, and addition of achiral guest molecule. *J. Org. Chem.* 76: 618–622.

15 Yao, Y., Xue, M., Chi, X.D. et al. (2012). A new water-soluble pillar[5]arene: synthesis and application in the preparation of gold nanoparticles. *Chem. Commun.* 48: 6505–6507.

16 (a) Ma, Y., Chi, X., Yan, X. et al. (2012). *per*-Hydroxylated pillar[6]arene: synthesis, X-ray crystal structure, and host–guest complexation. *Org. Lett.* 14: 1532–1535. (b) Li, Z.T., Yang, J., Yu, G.C. et al. (2014). Water-soluble pillar[7]arene: synthesis, pH-controlled complexation with paraquat, and application in constructing supramolecular vesicles. *Org. Lett.* 16: 2066–2069. (c) Li, Z.T., Yang, J., Yu, G.C. et al. (2014). Synthesis of a water-soluble pillar[9]arene and its pH-responsive binding to paraquat. *Chem. Commun.* 50: 2841–2843. (d) Chi, X.D. and Xue, M. (2014). Chemical-responsive control of lower critical solution temperature behavior by pillar[10]arene-based host–guest interactions. *Chem. Commun.* 50: 13754–13756.

17 (a) Santhosh Babu, S., Aimi, J., Ozawa, H. et al. (2012). Solvent-free luminescent organic liquids. *Angew. Chem. Int. Ed.* 51: 3391–3395. (b) Michinobu, T., Nakanishi, T., Hill, J.P. et al. (2006). Room temperature liquid fullerenes:

an uncommon morphology of C_{60} derivatives. *J. Am. Chem. Soc.* 128: 10384–10385.
18 Tanaka, K., Ishiguro, F., and Chujo, Y. (2010). POSS ionic liquid. *J. Am. Chem. Soc.* 132: 17649–17651.
19 Conner, M., Kudelka, I., and Regen, S.L. (1991). Octopus molecules at the air-water interface. Mechanical control over tentacle orientation. *Langmuir* 7: 982–987.
20 (a) Jon, S.Y., Selvapalam, N., Oh, D.H. et al. (2003). Facile synthesis of cucurbit[n]uril derivatives via direct functionalization: expanding utilization of cucurbit[n]uril. *J. Am. Chem. Soc.* 125: 10186–10187. (b) Lucas, D., Minami, T., Iannuzzi, G. et al. (2011). Templated synthesis of glycoluril hexamer and monofunctionalized cucurbit[6]uril derivatives. *J. Am. Chem. Soc.* 133: 17966–17976. (c) Vinciguerra, B., Cao, L., Cannon, J.R. et al. (2012). Synthesis and self-assembly processes of monofunctionalized cucurbit[7]uril. *J. Am. Chem. Soc.* 134: 13133–13140.
21 (a) Li, C.J., Xu, Q.Q., Li, J. et al. (2010). Complex interactions of pillar[5]arene with paraquats and bis(pyridinium) derivatives. *Org. Biomol. Chem.* 8: 1568–1576. (b) Zhang, Z.B., Xia, B.Y., Han, C.Y. et al. (2010). Syntheses of copillar[5]arenes by co-oligomerization of different monomers. *Org. Lett.* 12: 3285–3287. (c) Li, C.J., Zhao, L., Li, J.A. et al. (2010). Self-assembly of [2]pseudorotaxanes based on pillar[5]arene and bis(imidazolium) cations. *Chem. Commun.* 46: 9016–9018.
22 (a) Li, C., Chen, S., Li, J. et al. (2011). Novel neutral guest recognition and interpenetrated complex formation from pillar[5]arenes. *Chem. Commun.* 47: 11294–11296. (b) Shu, X.Y., Fan, J.Z., Li, J. et al. (2012). Complexation of neutral 1,4-dihalobutanes with simple pillar[5]arenes that is dominated by dispersion forces. *Org. Biomol. Chem.* 10: 3393–3397. (c) Shu, X., Chen, S., Li, J. et al. (2012). Highly effective binding of neutral dinitriles by simple pillar[5]arenes. *Chem. Commun.* 48: 2967–2969.
23 Ogoshi, T., Demachi, K., Kitajima, K., and Yamagishi, T. (2011). Selective complexation of *n*-alkanes with pillar[5]arene dimers in organic media. *Chem. Commun.* 47: 10290–10292.
24 (a) Dong, S.Y., Han, C.Y., Zheng, B. et al. (2012). Preparation of two new [2]rotaxanes based on the pillar[5]arene/alkane recognition motif. *Tetrahedron Lett.* 53: 3668–3671. (b) Wei, P.F., Yan, X.Z., Li, J.Y. et al. (2012). Novel [2]rotaxanes based on the recognition of pillar[5]arenes to an alkane functionalized with triazole moieties. *Tetrahedron* 68: 9179–9185.
25 Ogoshi, T., Kayama, H., Aoki, T. et al. (2014). Extension of polyethylene chains by formation of polypseudorotaxane structures with perpentylated pillar[5]arenes. *Polym. J.* 46: 77–81.

5

Photochemically Reversible Liquefaction/Solidification of Sugar-Alcohol Derivatives

Haruhisa Akiyama

Research Institute for Sustainable Chemistry, National Institute of Advanced Industrial Science and Technology (AIST), 1-1-1 Higashi, Tsukuba, Ibaraki 305-8565, Japan

5.1 Introduction

The transition of a material between liquid and solid phase is generally induced by a temperature change. For instance, a solid material converts to a liquid or rubber state when heated above its melting point or glass transition temperature. Such a phase transition can also be induced by light exposure, a strategy that has been practically used in recording systems, such as CD or DVD media, in which a writing light is used as a thermal source to achieve local area heating. In these cases, a change in reflection occurs due to a solid–solid phase transition; this is a photophysical process. Reversible photochemical reactions of photochromic dyes have been also used as a trigger for phase transition of materials, liquid crystals/isotropic liquid transitions [1], and sol/gel transitions [2]. Recently, a photochemical solid–liquid phase transition was reported. In 1995, two research groups reported formation of a protrusion pattern on azobenzene polymer films in response to interference light exposure, also known as photoinduced surface relief grating (SRG) [3]. The phenomenon attracted the interest of many researchers because of its uniqueness and expected application in holographic recording [4]. In the early stages of research on SRG, fluidization of the polymers was expected because the formation of the protrusion pattern was considered to originate from the migration of the polymers under light illumination. In 2005, Karageorgiev et al. directly observed fluidization on the surface of an azobenzene polymer film via atomic force microscopy (AFM) [5]. Following this study, Uchida et al. reported that the melting point of diarylethene crystals changed to 30 °C in the vicinity of the surface [6]. Diarylethene has two photoisomerizable forms; while in one, the ring is opened, it is closed in the other. Photoisomerization of one isomer generated a mixture of the two isomers, followed by liquefaction of the low-melting-point mixture near the surface. Further irradiation was with light-induced crystallization of the other isomer. In 2011, Norikane et al. reported "photo-melting" of cyclic azobenzene dimer in which a photoinduced crystalline-to-isotropic phase transition occurred at ambient temperature [7]. An extension of this study revealed that further photoirradiation of the cyclic

Functional Organic Liquids, First Edition. Edited by Takashi Nakanishi.
© 2019 Wiley-VCH Verlag GmbH & Co. KGaA. Published 2019 by Wiley-VCH Verlag GmbH & Co. KGaA.

compound resulted in crystallization [8]. Unlike the diarylethene crystals, the cyclic azobenzene dimer has an intermediate form in which one of two azo dyes isomerizes. The melting point of this intermediate form is expected to be lower than ambient temperature. Further photoirradiation generated a crystalline isomer in which both the azobenzenes in the cyclic compound isomerized. Photochemical liquefaction of a single azobenzene compound has also been reported in hydrophobic rigid azobenzene moieties with a flexible tetra(ethylene glycol) chain [9] and azobenzene ionic crystals [10]. Simpler single azobenzene compounds, such as 3-methyl-4,4'-dialkoxyazobenzenes, have demonstrated reversible photoinduced solid-to-liquid phase transition and their use in simplification of the photolithography process [11]. A similar phase transition was observed in 4-alkyl-4'-alkoxyazobenzene [12]. Directional migration of materials was controlled via a photoinduced phase transition in single crystals of 3,3'-dimethylazobenzene on a substrate [13] and in 4-methoxybenzene on the water surface [14]. In 2012, we reported reversible photoinduced liquefaction and solidification of azobenzene derivatives with a multiarm structure composed of a sugar alcohol scaffold, and demonstrated their application as photochemically reversible adhesives [15].

5.2 Mechanism of the Phase Transition Between Liquid and Solid State

Figures 5.1 and 5.2 display the chemical structures of sugar alcohol derivatives of azobenzene and images of the photoinduced phase transition of the representative compound **1**, respectively [15]. Azobenzene is well known as a photoisomerizable compound (Figure 5.3). It has two stable isomers, trans and cis forms, which are thermally stable and metastable, respectively, at ambient temperature. Under light illumination, the two states are interchangeable. The trans isomer has a large absorption in the ultraviolet (UV) region, such that on irradiation with

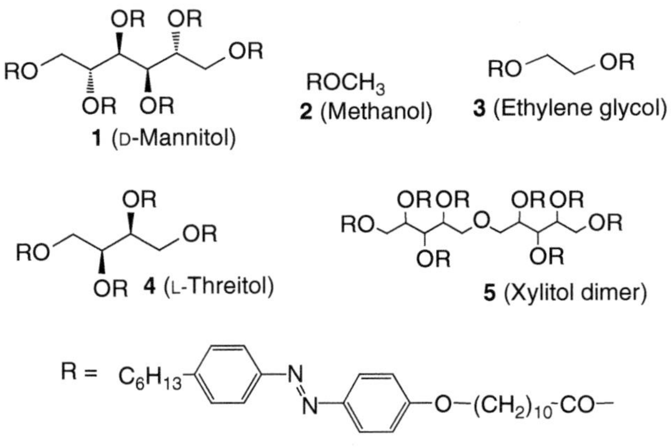

Figure 5.1 Chemical structures of azobenzene derivatives with different number of arms, where the sugar alcohols used as the starting materials are shown in parentheses.

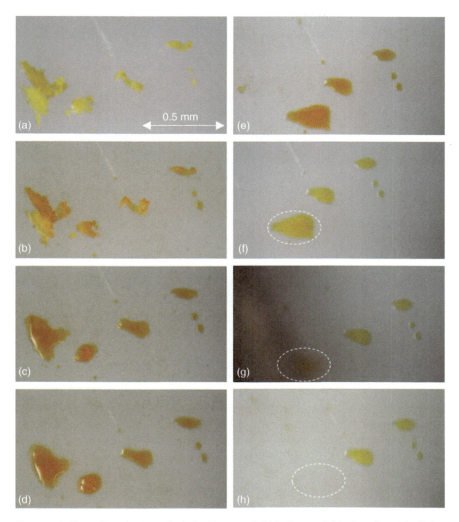

Figure 5.2 Photo-liquefaction of **1** (refer Figure 5.1): (a) before and (b) after ultraviolet light irradiation (λ_{max} = 365 nm, 40 mW cm^{-2}) for 3 minutes, (c) 30 minutes, and (d) 60 minutes at room temperature. Recovered solid state: (e) after ultraviolet light irradiation, (f) after visible light irradiation for 10 minutes (λ_{max} = 510 nm, 20 mW cm^{-2}), and (g) during and (h) after peeling off a piece of the solid (indicated by the white dashed circles). Source: Akiyama and Yoshida 2012 [15a]. Reproduced with permission from John Wiley & Sons.

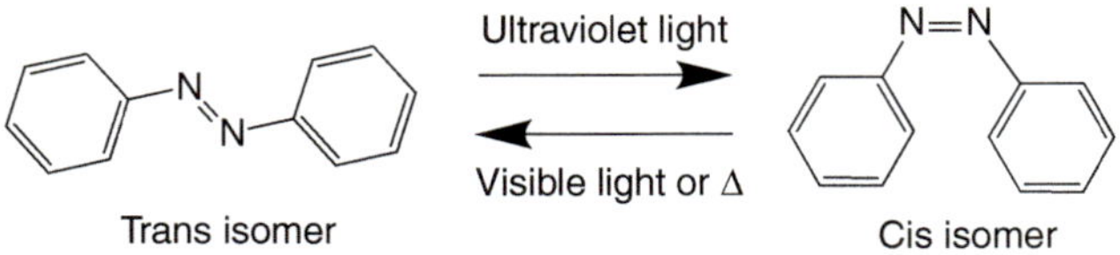

Figure 5.3 Isomerization of unsubstituted azobenzene.

UV light, it predominantly absorbs UV light and then isomerizes to a cis state. On the other hand, the cis isomer exhibits a relatively large absorption in the visible region (400–600 nm), and thus it is predominantly excited on irradiation with visible light, which induces its isomerization to the trans isomer. Lifetime of the metastable cis isomer is approximately one week at ambient temperature, depending on the substituent of azobenzene. A multi-azobenzene compound is obtained as a yellow powdered solid just after synthesis, in which the azobenzenes are in the thermally stable trans state. On irradiation with monochromatic UV light (365 nm), it gradually isomerizes to the cis state. As the isomerization progresses, the compound changes to an orange liquid. The liquid is stable for two days in the dark, even after stopping irradiation. The liquid solidifies, similar to the formation of a solid wax from a hot liquid on cooling. For solidification to occur, visible light is required to include blue or green light, although white light works as well. The rate of phase transition increases with light intensity; however, intensities of both the UV and visible light have to be lower than several tens of milliwatts per square centimeter because a high-intensity irradiation of more than 100 mW cm^{-2} increases the sample temperature.

As mentioned in the introduction, azobenzenes are widely used as photoreactive units in the photochemical phase transition of materials. The phase transition originates from the molecular shape change accompanied by the trans-to-cis isomerization, which shifts the melting point or glass transition temperature of the materials above and below ambient temperature. As a result, the stable state of the materials changes from solid to liquid at ambient temperature. There are two possibilities associated with the phase transition. One is that the melting points or glass transition temperatures of both the pure isomers are sufficiently different from that of the alternate stable state of the materials. Another possibility is that the mixed state of the two isomers, generated with isomerization, inhibits formation of an ordered structure such as a crystalline state and then decreases the melting point of the crystals. It has been already reported in the literature that the phase transition of diarylethene follows the second path. At the same time, however, it is not easy to prove the former mechanism. In case of an azobenzene system, a 100% cis state is impossible even when the photoisomerization proceeds toward a photostationary state. In order to directly prove the former assumption, we require to isolate the cis isomer from the mixture; however, thermal analysis of the metastable cis isomer of azobenzene derivatives is not easy because of the gradual transition to the trans isomer. Furthermore, it is known that the melting point of a bent structure is relatively lower in comparison to that of a linear structure. For example, the melting point of 9-(Z)-octadecenoic acid, i.e. oleic acid is 16 °C, while that of 9-(E)-octadecenoic acid, i.e. elaidic acid is 45 °C. These compounds are in a liquid and solid state, respectively, at ambient temperature. This fact indirectly supports the former mechanism. In addition, the isomerization ratio of sugar alcohol derivatives reaches up to 95% [15b]. Thus, we considered the former mechanism to be true for our case. Another important point to note is that the abovementioned photoinduced phase transition arises from the difference between the solid states. In low-molecular-weight materials, the solid states are basically crystals. For polymeric materials, a glassy solid state has to be considered. The sugar alcohol derivatives are intermediate compounds. The molecular weight of **1** is ~3000 g mol^{-1}, but without polydispersity. Compound **1** transitions to a glassy state at 80 °C from the liquid crystalline phase on cooling with an isotropic liquid (120 °C). The glassy sample crystallized on holding it at 100 °C. The compound has two solid phases, namely, glass and crystals, at ambient temperature. At the glass transition temperature, molecular motion is frozen before the appearance of the closed pack structure. Thus, the glassy state is more suitable for solid-state reactions because it can include a free volume inside to allow photoisomerization. On the other hand, the advantage of crystalline materials is the possibility of a cooperative effect that enables immediate liquefaction when the ordered structure is broken on the way to isomerization. On solidification, a crystalline compound requires a reverse process to form a crystalline order from an isotropic phase. In a glassy solid, the process is simplified since solidification occurs just at the moment when the molecular motion stops without ordering. Actually, the sugar alcohol derivative solidifies from a photochemically liquefied state just after completing photoisomerization to the trans form.

5.3 Effect of Molecular Structure

Herein, we discuss molecular design and effective structures to control photoinduced liquid–solid phase-transition properties of sugar alcohol derivatives.

5.3.1 Number of Azobenzene Units

We have found that the number of azobenzenes is very important in the phase transitions of sugar alcohol derivatives. Among the compounds bearing the azobenzene moieties labeled **1–5** in Figure 5.1 [15a], photoinduced phase transition was observed only in those compounds with more than four azobenzene units. Interestingly, the monomer (**2**) and dimer (**3**) did not exhibit any reactivity on exposure to light. In this series, only the monomer and dimer possess a crystalline solid phase (Table 5.1). We, therefore, considered that lack of space for

Table 5.1 Thermal phase behaviors of multi-azobenzene arm compounds.

	Alcohol	Rate (°C min^{-1})	Transition temperature(°C)
1	D-Mannitol	−5	Iso 107 SmC* 84 SmF/I Tg 83
		−2	Iso 109 SmC* 86 SmF/I
2	Methanol		Iso 65 Cr (Cr 72 Iso)
3	Ethylene glycol	−2	Iso 104 Cr
4	L-Threitol	−5	Iso 98 SmC* 87 SmB 79 SmX
		−2	Iso 100 SmC* 96 Cr
5	Xylitol dimer	−2	Iso 113 SmA–C* 85 SmF/I Tg
		+2	70 SmF/I 87 SmA–C* 114 Iso

isomerization in the crystal packing results in no exhibition of photoreactivity. In other words, it can be said that the melting point is affected by the molecular weight. The very low-molecular-weight compounds are generally in gas or liquid states, while in many cases the large molecular weight compounds are solids. It is to be noted that the high symmetry of materials and intermolecular interaction stabilize the crystal packing even for low-molecular-weight compounds. In general, polymeric materials are hard to crystalize perfectly. Taking into account these facts, we believe that photoisomerization of the sugar alcohol derivatives will be achieved with increasing number of azobenzene units because the polymer-like effect in the solid state makes closed packing difficult. Interestingly, the compounds with more than six azobenzene units exhibit glass transitions. In other words, isomerization is possible in high-molecular-weight compounds. Specifically, the isomerization to cis state depresses molecular alignment, and then decreases the glass transition temperature to fluidize the materials. Herein, we note that if the increase in molecular weight is large, entanglement would reduce the fluidity above glass transition temperature. Consequently, we have to pay attention to the limitation of molecular weight of the materials for realization of photoinduced solid–liquid transition.

5.3.2 Alkyl Chain Length

The alkyl chain length of multi-azobenzene compounds is also an important factor. We have prepared different lengths of azobenzene hexamers in order to compare them [15b]. There are two types of azobenzene tails, namely, no tail and hexyl tail, and two types of alkyl spacers, namely, penta methylene and deca methylene, between azobenzene and the sugar alcohol scaffold. In total, four types of materials, namely, **1, 6, 7,** and **8**, were compared (see Figures 5.1 and 5.4). The two compounds with hexyl tails, i.e. **1** and **6**, exhibited a liquid–solid phase transition regardless of the spacer length. The short spacer compound **6**, in the trans state, has only a highly ordered liquid crystalline phase SmF/I at ambient temperature (Table 5.2). Since the SmF/I phase is a quasi-solid phase without fluidity, it is ambiguous whether it is a solid phase or not. To clarify this point, we measured the dynamic viscoelastic properties. The storage modulus was much higher than the loss modulus at ambient temperature, which confirmed that it was a solid, but the difference in moduli is smaller than that of the long chain compound **1**.

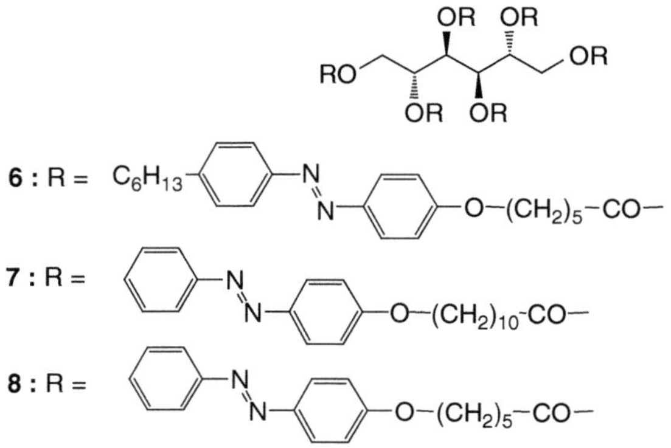

Figure 5.4 Chemical structures of azobenzene hexamers with different alkyl chains.

Table 5.2 Thermal phase behaviors of azobenzene hexamer.

	Tail, spacer of azobenzene	Rate (°C min^{-1})	Transition temperature (°C)
6	C6[a)], C5[b)]	−2	Iso 118 SmC* 59 SmF/I
		+2	SmF/I 121 SmC* 61 Iso
7	Non[c)], C10[d)]	−2	Iso 96 SmA 55 SmB Tg 50
8	Non, C5	−2	Iso 80 SmA 52 SmB
		+2	SmB 54 SmA 85 Iso
9	CN, C10	−2	Iso 153 SmA 63 SmB 48 Cr
		+2	Cr 50 SmB 68 SmA 157 Iso

a) C6, hexyl.
b) C5, pentamethylene.
c) Non, no substituent.
d) C10, decamethylene.

The two compounds without hexyl tails have rigid phenyl rings at molecular terminals (**7**, **8** in Figure 5.3). At first, we were doubtful whether such materials without alkyl terminals can undergo transition to a liquid phase, although in the long spacer compound **7**, the photoinduced solid–liquid transition was observed. Liquefaction by introduction of flexible tails into the rigid core has been reported [16]. In this case, there are no flexible terminals. The unexpected photoinduced liquefaction of the rigid terminal compound indicates that it is the difference in the direction of the rigid azophenyl ring that largely changes the transition temperature of the material. The shortest compound with no tail and short spacer **8** exhibits color change based on photoisomerization in response to light irradiation, but it always retains the solid state without any shape change. Absorption spectroscopy for the thin films revealed the reason behind the difference of the phase transition behaviors. In the long alkyl chain compounds, the absorption band of a trans isomer corresponding to π–π* transition at 365 nm largely shifts to shorter wavelength in comparison to that in chloroform solutions. It indicates formation of *H* aggregation among azobenzene moieties in a solid state. The solid state is stabilized by aggregation of azobenzene arms. As the isomerization progresses, the cis isomer increases in the solid. The bent structured cis isomers probably destabilize aggregation, and the material is completely liquefied. Furthermore, the shortest alkyl chain compound exhibits only a small spectral shift. The small contribution of aggregation indicates that it is near an amorphous solid. Thus, a solid state is stabilized by less mobility of all the molecules including the sugar alcohol scaffold. Under this condition, formation of bent cis isomers cannot affect mobility of the molecules. Consequently, the solid state becomes stable even in the cis state of the short compound. To answer the question, what would happen if the interaction among azobenzenes increased more, we prepared the cyanoazobenzene hexamer **9** (see Figure 5.4), which has a large dipole–dipole interaction between its azobenzenes. The material was obtained as a solid. Photoirradiation induced a color change, but phase transition to a liquid was not observed. We inferred that the large dipole moment due to introduction of the cyano group enhances molecular interaction even in the cis state and stabilizes the solid state [17].

5.3.3 Mixed Arms

One of the structural features of the sugar alcohol derivatives is the presence of multi-azobenzene arms in one molecule. The mixed chemical structure was used in the arms, which amplifies a variety of molecular design. At first, we describe the incompletely substituted compound [18]. On the synthesis of the azobenzene hexamer, we obtained a partially substituted compound **10** (see Figure 5.5) as a by-product. The nuclear magnetic resonance (NMR) spectra indicate the compound is composed of four azobenzene arms and two residual hydroxyl units on an average. The NMR spectra for proton of methine and methylene of D-mannitol scaffold is very complicate, which means that the compound is a mixture regardless of the substituent number and position. The mixture is an enantiotropic liquid crystal, which means that the liquid crystalline phase appears both on heating and cooling. The differential scanning calorimetry (DSC) curve did not show a clear glass transition, although a liquid crystalline phase and highly ordered phase were observed at 95 °C and 70 °C, respectively, on cooling from the isotropic phase (Table 5.3). At ambient temperature, the material had no fluidity. The corresponding azobenzene hexamer **1** had monotropic liquid crystals, which appeared only on cooling. The stabilization of liquid crystallinity is probably contributed by the mixed arm structures. Practically mixed liquid crystals are used for display to increased liquid crystalline temperature range. The mixed arm material, as well as the laser, were reversibly liquefied and solidified in response to light exposure.

The simple alkyl chain arm, octadecanoic acid, and the same azobenzene arm were used for synthesis of the other mixtures. The feed ratio of the azobenzene arm and alkyl arm was changed to 1/1, 1/3, and 1/9. As a result, we obtained mixed arm hexamers **11**, **12**, **13** (see Figure 5.5) at the ratio of 1/1, 1/3, and 1/9 which were estimated from the corresponding NMR spectra. The reaction was quantitative. In each mixed arm compound, several spots or a tailing spot on a thin layer chromatograph were observed. It indicates that mixed arm

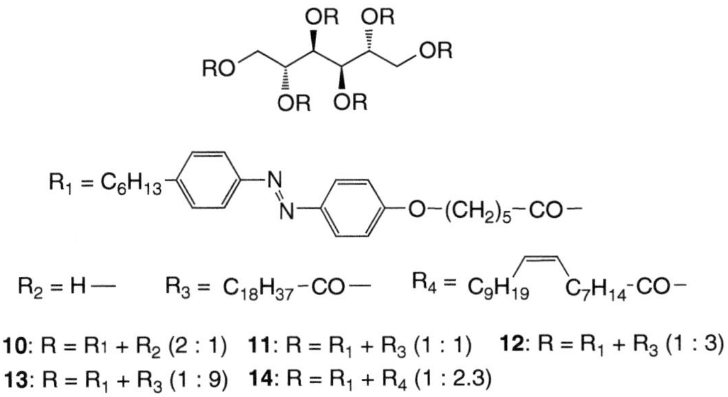

Figure 5.5 Chemical structures of mixed arm compounds.

Table 5.3 Thermal phase behaviors of mixed arm hexamers.

	Arms (molar ratio)	Rate (°C min^{-1})	Transition temperature (°C)
10	Azo[a], OH[b]	−2	Iso 95 SmA 80 SmC* 70 SmF/I
		+2	SmF/I 72 SmC* 81 SmA 100 Iso
11	Azo, C18[c] (1:1)	−2	Iso 43 Cr
12	Azo, C18 (1.3)	−2	Iso 42 Cr
13	Azo, C18 (1.9)	−2	Iso 45 Cr

a) Azo, azobenzene arm.
b) OH, Hydroxy residue.
c) C18, octadecanoyl arm.

ratios are averaged values, which was also be confirmed by matrix-assisted laser desorption/ionization time-of-flight mass spectroscopy (MALDI-TOF-MS). All compounds were obtained as solid materials (Table 5.3). The 1/1 mixed arm compound **11** was completely liquefied on irradiation with the UV light and solidified on irradiation with visible light. Meanwhile, 1/3 of the mixed arm compound **12** (see Figure 5.5) was not fluidized on irradiation with the UV light, but the softening with color change was confirmed by observation of the deformation under the stress. The 1/9 mixed arm compound **13** exhibited the color change on UV irradiation, but it remained a hard solid. Consequently, introduction of linear alkyl arms stabilized the solid states of the mixed arm hexamers. For comparison, the linear alkyl arm was changed to bent arm. This mixed arm hexamer was synthesized using azobenzene arms and oleic acid at 1/3 ratio in feed. The mixed arm hexamer with azobenzene and bent alkyl arm (1/2.3) **13** was obtained by purification with chromatography. Thin-layer chromatography and NMR spectroscopy of the compound revealed that it is a mixture of the number of azobenzene arms and their position. This compound was a fluidic liquid independent of photoirradiation. As described in Section 5.2, this is because introduction of the bent arm reduces the melting point below room temperature.[1]

5.3.4 Structure of Sugar Alcohol

A linear sugar alcohol moiety in azobenzene hexamers was converted to a cyclic *scyllo*-inositol to investigate the effect of the material structure at the center position. The azobenzene hexamer made of *scyllo*-inositol **15** has no liquid crystalline phase, and melts at 160–170 °C. On irradiation with UV light, the color of the compound was changed, but it was not liquefied [15b]. The melting point of the cis isomer is evidently higher than ambient temperature. There are several differences between the linear and cyclic alcohol derivatives in respect to the photo phase-transition behavior, even though the molecular weights and

---

1 Unpublished data. The materials 11–14 were synthesized in a similar manner described in the literature [15].

chemical components of the two hexamers are almost same. The difference indicates that total molecular shape is important to realize the photoinduced phase transition.

Many sugar alcohols have stereo isomers. Hitherto, D-mannitol was mainly used for a scaffold of hexamers. D-sorbitol is a cheaper alternative, while allitol is a very expensive meso isomer. To investigate the effect of the stereo structure on the sugar alcohol moieties, in addition to D-mannitol, D-sorbitol and allitol are used as starting materials for the synthesis of azobenzene hexamers [19]. As a result, the isomers **16**, **17** (see Figure 5.6) exhibit not only similar liquid crystallinity but also similar photoinduced phase-transition behavior to the mannitol derivative (Table 5.4) [15]. The thermal phase behaviors are slightly different among stereo isomers. D-Sorbitol and D-mannitol derivatives are monotropic liquid crystals, whereas the allitol derivative is an enantiotropic liquid crystal. All hexamers have a glass transition temperature; thus, the solid states are basically liquid crystalline glass. Interestingly the glass transition temperatures are different by 10 °C. They are 80, 70, and 60 °C for D-mannitol, D-sorbitol, and allitol derivatives, respectively. The melting points of D-mannitol, D-sorbitol, and allitol themselves are 170, 100, and 150 °C, respectively. The difference probably corresponds to the symmetry of molecules; however, the tendency is unrelated to the glass transition temperature of each hexamer. In any case, glass transition is an important parameter for practical application of the materials because it directly relates to thermal stability. In this sense, the stereo structure of the sugar alcohol moieties is an important factor also.

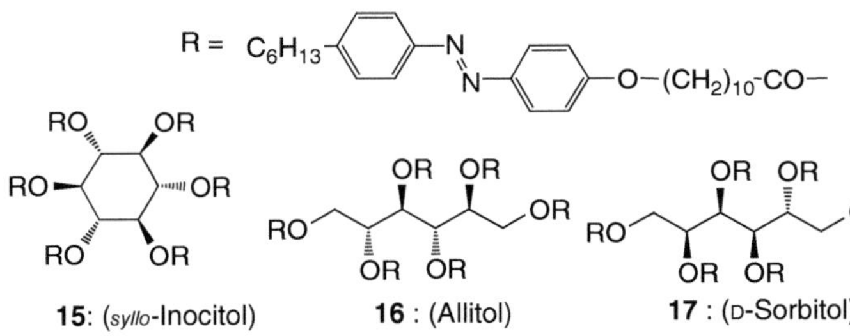

Figure 5.6 Chemical structures of azobenzene hexamers (**15, 16,** and **17**) with different scaffolds where the sugar alcohols used as starting materials are shown in parentheses.

Table 5.4 Thermal phase behaviors of azobenzene hexamers (**15, 16,** and **17**).

	Alcohol	Rate (°C min^{-1})	Transition temperature (°C)
15	scyllo-Inositol	−2	Iso 162 Cr (Cr 171 Iso)
16	D-Sorbitol	−2	Iso 109 SmC* 84 SmF/I 70Tg
17	Allitol	−2	Iso 109 SmC* 83 SmF/I 60Tg
		+2	SmF/I 84 SmC* 111 Iso

5.4 Summary

Photoinduced solid–liquid phase transition is based on controlling the transition temperature of the materials. We exhibited how molecular structures such as alkyl chain length and number of azobenzene arms in sugar alcohol derivatives influence photoreactivity and phase-transition behavior. Recently, reports on solid–liquid phase-transition materials by other groups besides ours have increased. One of the applications of such materials is as photochemically reversible adhesives, which enables a repeatedly reworkable adhesion system. Recently, we investigated the phase transition of polymeric materials, which offers an advantage in the realization of strong reversible adhesives [20]. The knowledge of the molecular structure effect is used effectively for the molecular design of such high-performance polymeric materials.

Acknowledgments

The author thanks Prof. Nobuyuki Tamaoki, Ms. Tanaka Asuka, Mr. Satoshi Kanazawa, Ms. Yoko Okuyama, Ms. Tamaki Fukata, Dr. Masaru Yoshida, Dr. Hideyuki Kihara, Dr. Hideki Nagai, and Dr. Yasuo Norikane. The unpublished work on mixed arms was supported by the JSPS KAKENHI Grant Numbers 26288096.

References

1 (a) Tazuke, S., Kurihara, S., and Ikeda, T. (1987). Amplified image recording in liquid-crystal media by means of photochemically triggered phase-transition. *Chem. Lett.* 16: 911–914. (b) Ikeda, T. and Tsutsumi, O. (1995). Optical switching and image storage by means of azobenzene liquid-crystal films. *Science* 268: 1873–1875.
2 Murata, K., Aoki, M., Suzuki, T. et al. (1994). Thermal and light control of the sol–gel phase-transition in cholesterol-based organic gels – novel helical aggregation modes as detected by circular-dichroism and electron-microscopic observation. *J. Am. Chem. Soc.* 116: 6664–6676.
3 (a) Kim, D.Y., Tripathy, S.K., Li, L., and Kumar, J. (1995). Laser-induced holographic surface-relief gratings on nonlinear-optical polymer-films. *Appl. Phys. Lett.* 66: 1166–1168. (b) Rochon, P., Batalla, E., and Natansohn, A. (1995). Optically induced surface gratings on azoaromatic polymer-films. *Appl. Phys. Lett.* 66: 136–138.
4 Lee, S., Kang, H.S., and Park, J.K. (2012). Directional photofluidization lithography: micro/nanostructural evolution by photofluidic motions of azobenzene materials. *Adv. Mater.* 24: 2069–2103.
5 Karageorgiev, P., Neher, D., Schulz, B. et al. (2005). From anisotropic photo-fluidity towards nanomanipulation in the optical near-field. *Nat. Mater.* 4: 699–703.

6 Uchida, K., Izumi, N., Sukata, S. et al. (2006). Photoinduced reversible formation of microfibrils on a photochromic diarylethene microcrystalline surface. *Angew. Chem. Int. Ed.* 45: 6470–6473.

7 Norikane, Y., Hirai, Y., and Yoshida, M. (2011). Photoinduced isothermal phase transitions of liquid-crystalline macrocyclic azobenzenes. *Chem. Commun.* 47: 1770–1772.

8 Uchida, E., Sakaki, K., Nakamura, Y. et al. (2013). Control of the orientation and photoinduced phase transitions of macrocyclic azobenzene. *Chem. Eur. J.* 19: 17391–17397.

9 Okui, Y. and Han, M.N. (2012). Rational design of light-directed dynamic spheres. *Chem. Commun.* 48: 11763–11765.

10 Ishiba, K., Morikawa, M., Chikara, C. et al. (2015). Photoliquefiable ionic crystals: a phase crossover approach for photon energy storage materials with functional multiplicity. *Angew. Chem. Int. Ed.* 54: 1532–1536.

11 Norikane, Y., Uchida, E., Tanaka, S. et al. (2014). Photoinduced crystal-to-liquid phase transitions of azobenzene derivatives and their application in photolithography processes through a solid–liquid patterning. *Org. Lett.* 16: 5012–5015.

12 Kim, D.Y., Lee, S.A., Kim, H. et al. (2015). An azobenzene-based photochromic liquid crystalline amphiphile for a remote-controllable light shutter. *Chem. Commun.* 51: 11080–11083.

13 Uchida, E., Azumi, R., and Norikane, Y. (2015). Light-induced crawling of crystals on a glass surface. *Nat. Commun.* 6: 7310.

14 Norikane, Y., Tanaka, S., and Uchida, E. (2016). Azobenzene crystals swim on water surface triggered by light. *CrystEngComm* 18: 7225–7228.

15 (a) Akiyama, H. and Yoshida, M. (2012). Photochemically reversible liquefaction and solidification of single compounds based on a sugar alcohol scaffold with multi azo-arms. *Adv. Mater.* 24: 2353–2356. (b) Akiyama, H., Kanazawa, S., Okuyama, Y. et al. (2014). Photochemically reversible liquefaction and solidification of multiazobenzene sugar-alcohol derivatives and application to reworkable adhesives. *ACS Appl. Mater. Interfaces* 6: 7933–7941.

16 (a) Nakanishi, T. (2014). Room temperature liquid formulation by attaching alkyl chains on pi-conjugated molecules. *J. Synth. Org. Chem. Jpn.* 72: 1265–1270. (b) Babu, S.S. and Nakanishi, T. (2013). Nonvolatile functional molecular liquids. *Chem. Commun.* 49: 9373–9382.

17 Akiyama, H. and Kihara, H., (2016). Photochemically reversible adhesives, *39th Annual Meeting The Adhesion Society, Extended Abstract*.

18 Akiyama, H., Yoshida, M., Kihara, H. et al. (2014). Organic photofunctional materials composed of azobenzene derivatives: liquid–solid phase transition in multi azobenzene compounds with partially substituted structures. *J. Photopolym. Sci. Technol.* 27: 301–305.

19 Akiyama, H., Kanazawa, S., Yoshida, M. et al. (2014). Photochemical liquid–solid transitions in multi-dye compounds. *Mol. Cryst. Liq. Cryst.* 604: 64–70.

20 Akiyama, H., Fukata, T., Yamashita, A. et al. (2017). Reworkable adhesives composed of photoresponsive azobenzene polymer for glass substrates. *J. Adhes.* 93: 823–830.

6

Functional Organic Supercooled Liquids

Kyeongwoon Chung[1], Da Seul Yang[2], and Jinsang Kim[2,3,4,5,6]

[1] Korea Institute of Materials Science (KIMS), 3D Printing Materials Center, Changwon, Gyeongsangnamdo 51508, South Korea
[2] University of Michigan, Macromolecular Science and Engineering, Ann Arbor, MI 48109, USA
[3] University of Michigan, Department of Materials Science and Engineering, Ann Arbor, MI 48109, USA
[4] University of Michigan, Department of Chemical Engineering, Ann Arbor, MI 48109, USA
[5] University of Michigan, Department of Biomedical Engineering, Ann Arbor, MI 48109, USA
[6] University of Michigan, Department of Chemistry, Ann Arbor, MI 48109, USA

6.1 Organic Supercooled Liquids

It is natural that a material exists in liquid phase above its melting temperature (T_m). On the other hand, there is a liquid phase which is maintained below the T_m of a material: "supercooled liquid." In general, an organic crystalline solid transforms to its liquid phase at its T_m upon heating, and thermodynamics drive the material to crystallize below its T_m upon cooling. Only under specific conditions such as rapid temperature quenching can organic crystalline molecules exist as supercooled liquid (above glass transition temperature, T_g) or glass (below T_g). Even in this case, the supercooled liquid easily turns into its crystalline solid phase upon subsequent heating since the supercooled liquid is generally in a kinetically trapped metastable state (Figure 6.1).

The research on the organic supercooled liquid has been focused on the theoretical aspect of supercooled liquid and glass formation such as kinetics and thermodynamics of glass transition [1–6], viscosity of supercooled liquid and viscous slowdown [1, 7–10], and molecular dynamic simulation on crystal nucleation [2, 11]. On the other hand, practical application and functionality of organic supercooled liquids are quite rare due to their short-lived metastable nature. However, there is an emerging research applying supercooled liquid as a functional material. In this chapter, a few unique examples of functional organic supercooled liquid are discussed. Distinct stimuli-responsive optical properties from thermally stable organic supercooled liquid and potential optical and biomedical application have been reported based on a thorough and systematic investigation on the correlation between thermal stability of supercooled liquid and molecular design of organic molecules [12]. Furthermore, an interesting example of highly emissive organic supercooled liquid is presented [13].

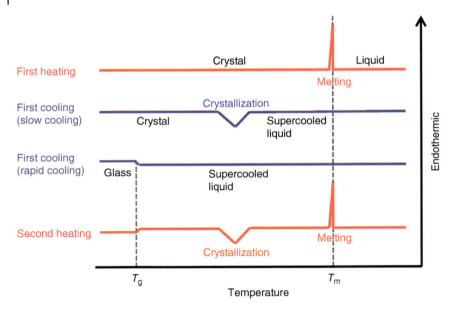

Figure 6.1 Thermal trace of conventional organic crystalline molecule under differential scanning calorimetry (DSC) measurement; T_g and T_m are glass transition and melting temperature, respectively.

6.2 Stimuli-Responsive Organic Supercooled Liquids

6.2.1 Shear-triggered Crystallization

The optical properties such as absorption and emission of conjugated organic molecules can vastly differ in crystalline solid and amorphous supercooled liquid. A few examples have demonstrated unique stimuli-responsive optical property changes upon the phase transfer between crystalline solid and supercooled liquid state, especially fluorescence change under applied force.

In phenomenological view, the materials which show color or fluorescence change upon external mechanical stimuli such as pressure and shear are called mechanochromic materials, and many organic conjugated materials are reported for their mechanochromic behavior. The general mechanism for the mechanochromism of organic conjugated molecules is based on the crystal-to-amorphous transformation (and/or crystal-to-crystal transformation), and the applied force acts as a destructive force to break the energetically favorable crystalline phase into an energetically unstable amorphous state [14–18].

Similar to the conventional mechanochromism, the shear force can trigger a fluorescence emission property change of organic supercooled liquids through shear-triggered crystallization of the supercooled liquids. However, the characteristic of shear-triggered crystallization of organic supercooled liquid is distinguished from the usual mechanochromism of organic materials since its phase transformation is from the metastable (supercooled liquid) to energetically favorable stable state (crystal). The effect of this difference in mechanism for optical property change upon applied stimuli is discussed later in this chapter.

The prerequisite to utilize supercooled liquid in a functional application is its thermal stability. It can be viewed as a paradox because we have already defined organic supercooled liquid as a short-lived metastable phase. Only some crystalline organic molecules are reported to form thermally stable supercooled liquid without spontaneous crystallization upon thermal cycles despite sufficient molecular mobility above T_g as well as energetically favorable crystalline phase below T_m [19–22]. It is a challenging but interesting question how we can correlate the thermal stability of organic supercooled liquid and chemical structure design of organic materials.

6.2.2 Scratch-Induced Crystallization of Trifluoromethylquinoline Derivatives

Koga et al. introduced a series of N,N-R-phenyl-7-amino-2,4-trifluoromethylquinoline (TFMAQ) derivatives which can access supercooled liquid phase upon cooling of the melt [23]. Four derivatives with R = methyl, ethyl, isopropyl, and phenyl group at amino moiety (compounds **1**, **2**, **3**, and **4**, Figure 6.2a) are synthesized and characterized to exhibit polymorphism as well as formation of supercooled liquid without spontaneous crystallization.

The thermal properties of the compounds were verified with differential scanning calorimetry (DSC) measurement. All four compounds maintain the supercooled liquid phase after cooling of the melts down to room temperature (Figure 6.2c–f). The T_m of compound **1** is measured at 98 °C for both polymorphs, **1α** and **1β**. The polymorphs of compounds **2**, **2α** and **2β**, exhibit different T_m of 70 and 62 °C. The compounds **3** and **4** show melting at 94 and 118 °C, respectively. Since the T_g of the compounds **1**, **2**, **3**, and **4** are observed at c. −13, −20, −9, and 23 °C, the compounds **1**, **2**, and **3** are confirmed to exist in supercooled liquid phase at 20 °C after cooling of the melt. The stability of the supercooled liquid of compounds **1**, **2**, and **3** are observed in sandwiched glasses. Compounds **1** and **2** show crystallization in the supercooled liquid domain after four hours. Compound **3** shows clear crystallization after 18 hours. Even though the authors did not discuss the origin of the difference in stability, it is highly plausible that the molecular design and chemical structure difference between the compounds is correlated with the different stability of the supercooled liquids. On the other hand, it is worth noting that crystallization of all compounds is confirmed upon subsequent heating only after cooling down to −30 to −40 °C, due to possible nucleation at very low temperatures (Figure 6.2c–f).

The optical properties of supercooled liquid and crystalline solid phases of the presented compounds differ significantly. In case of compound **1**, all the polymorphs, **1α**, **1β**, and **1γ**, show strong green emission with fluorescence quantum yield (QY) of 0.52, 0.51, 0.46, and λ_{max} of 509, 517, and 520 nm, respectively. In contrast, the supercooled liquid of compound **1** shows relatively weak emission with QY of 0.22 with λ_{max} of 541 nm.

The supercooled liquid of compound **1** undergoes phase transition to crystalline solid (**1β** and **1γ**) upon scratching with fine wire, accompanied

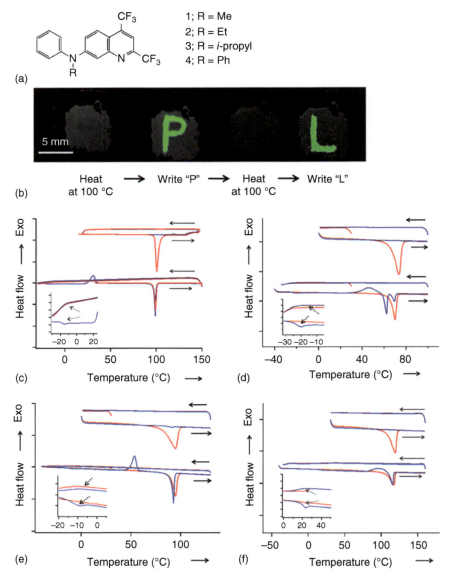

Figure 6.2 (a) Chemical structures of N,N-R-phenyl-7-amino-2,4-trifluoromethylquinoline (TFMAQ) derivatives. (b) Photographs for compound **1** in the cycle of heating at 100 °C, and scratching at 25 °C. (c–e) Thermal traces of crystals **1α** (c) and **2α** (d), and compound **3** (e) and **4** (f) for first (red) and second (blue) cycles with different temperature ranges (upper and bottom). Insets show the expansion in the temperature range observed at the glass transition. Source: Reproduced with permission from Karasawa et al. [23]. Copyright 2014, American Chemical Society.

by an increase of fluorescence efficiency (Figure 6.2b). This scratch-induced crystallization occurred rapidly where the scratching (shearing) was applied and then gradually propagated to the entire connected supercooled liquid domain. The optical property alternation is reversible through writing with scratching and erasing with heating above 100 °C. The results suggest possible applications of the supercooled liquid in a switching device for solid-state emitting materials.

6.2.3 Highly Sensitive Shear-Triggered Crystallization in Thermally Stable Organic Supercooled Liquid of a Diketopyrrolopyrrole Derivative

In the previous section, we discussed the potential application of organic supercooled liquids. We have several important and challenging questions to answer when considering supercooled liquid as a promising functional organic material. How can we make thermally stable supercooled liquids? What is the correlation between molecular design and thermal stability of organic supercooled liquids? What is the role of shearing to trigger crystallization? What is the merit of shear-triggered crystallization compared to conventional mechanochromism of organic conjugated molecules?

To answer these questions, we conducted a systematic investigation on the thermally stable supercooled liquid of a diketopyrrolopyrrole (**DPP**) derivative and its highly sensitive shear-triggered crystallization [12]. A DPP derivative, **DPP8**, is synthesized and obtained as an orange crystal with bright yellow fluorescence with QY of 0.55 (Figure 6.3a,b). Different from conventional organic crystals, the **DPP8** crystal melted at 134 °C and stayed in supercooled liquid state without crystallization upon cooling down to room temperature. The supercooled liquid shows very dim red fluorescence (QY = 0.02) in contrast to the crystalline solid phase (Figure 6.3b), and exhibits excellent thermal stability even after multiple thermal cycles as well as very slow cooling (cooling rate: 5, 1, 0.2 °C min^{-1}) (Figure 6.4a).

This unique thermal stability of supercooled liquid **DPP8** is originated from a very small Gibbs free energy difference (ΔG) between crystal and supercooled liquid (Figure 6.3a). Crystallization begins with nucleation in the supercooled liquid domain, a process which directly correlated with the ΔG between the two phases. When crystal nuclei are generated in a supercooled liquid, there are energy gain for the formation of the energetically favorable crystalline phase and energy barrier for building the interface between crystal and supercooled liquid, as expressed in the following equation.

$$\Delta G^{\text{nucleation}} = \frac{4}{3}\pi r^3 \Delta G_v + 4\pi r^2 \sigma$$

From this equation, we can define critical radius of nucleation, r^*, which means the minimum size of a stable crystal nucleus. If the generated nucleus is smaller than r^*, the nucleus disappears. If the nucleus is bigger than r^*, it has a chance to spontaneously grow under the temperature range where molecular motion is high enough (Figure 6.3c).

$$\text{For } dG/dr = 0, \quad r^* = -\frac{2\sigma}{\Delta G_v} \sim -\frac{2\sigma T_m}{\Delta H_v (T_m - T)}$$

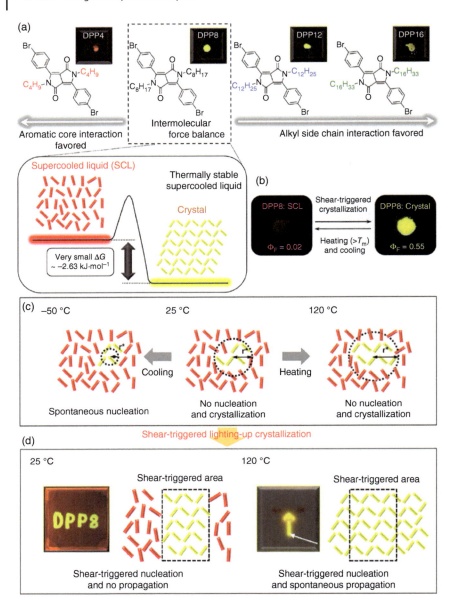

Figure 6.3 Thermally stable supercooled liquid and shear-triggered lighting-up crystallization of **DPP8**. (a) **DPP8** and its derivatives. An intermolecular force balance between two different interactions acting in opposite directions makes **DPP8** exhibit small ΔG between supercooled liquid and crystalline solid phases, resulting in thermally stable supercooled liquid.
(b) Reversible phase transformation with large fluorescence change between the two phases.
(c) Nucleation is restricted in **DPP8** supercooled liquid due to an unattainable yet required large critical radius (r^*) at 25 and 120 °C. In contrast, subsequent heating developed crystallization when molten DPP8 was cooled down to −50 °C. (d) Shear-triggered lighting-up crystallization of **DPP8**. Source: Reproduced with permission from Chung et al. [12]. Copyright 2015, American Chemical Society.

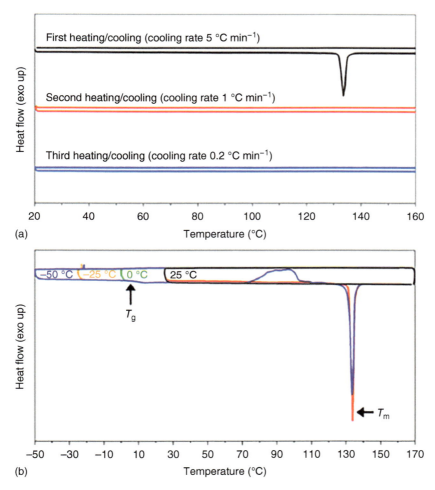

Figure 6.4 (a) DSC curves at different cooling rates (heating rate: 10 °C min^{-1}). (b) DSC curves with different cooling temperatures (heating rate: 10 °C min^{-1}, cooling rate: 5 °C min^{-1}). Source: Reproduced with permission from Chung et al. [12]. Copyright 2015, American Chemical Society.

From the equation for r^*, one can figure out that r^* is inversely proportional to ΔG between crystalline solid and supercooled liquid phases, and, therefore, if ΔG is extremely small, the crystal nucleation is restricted due to unattainable but required large r^* in the supercooled liquid domain (Figure 6.3c).

The fact that nucleation is forbidden in **DPP8** supercooled liquid is verified by two different experiments. First, even though nucleation is forbidden due to the required large r^*, there should be a threshold temperature at which crystal nucleation occurs because the r^* is decreasing upon cooling of the supercooled liquid. Indeed, in DSC characterization, the supercooled liquid **DPP8** shows excellent stability in thermal cycles down to 25, 0, and −25 °C without crystallization along either cooling and heating, but shows crystallization and melting peaks in the subsequent heating after being cooled down to −50 °C (Figure 6.4b). Furthermore, the forbidden nucleation is confirmed with a crystal seeding test. At 120 °C, the **DPP8** supercooled liquid was crystallized by adding **DPP8** crystals which act as crystal nuclei.

The ΔG between supercooled liquid and crystalline solid **DPP8** is estimated as small as −2.6 to −6.4 kJ mol^{-1} by means of multiple methods such as relative solubility measurement [24], Hoffman equation [25], and Hess's law. The obtained ΔG value (2.63–6.24 kJ mol^{-1}) is comparable to the ΔG between polymorphs and is even much smaller than 50.21 kJ mol^{-1} (12 kcal mol^{-1}) of the ΔG between cis- and trans-azobenzene, a well-known photoisomerizable compound [24, 26]. Therefore, it is concluded that the very small ΔG of **DPP8**, which requires an unattainable large r^*, results in lack of nucleation and crystallization and thereby thermally stable supercooled liquid can be formed.

The very small ΔG of **DPP8** is directly correlated with the molecular chemical structure design. From the investigation with DPP derivatives having different alkyl chain lengths, we found that a subtle intermolecular force balance is built in the **DPP8** structure which enables unusually small ΔG even at supercooled temperature more than 100 °C below from its T_m (Figure 6.3a). When alkyl chain length decreases (**DPP4**, Figure 6.3a), the molecule shows a completely different X-ray diffraction (XRD) pattern: In contrast to **DPP8** that exhibits lamellar packed alkyl chains and weakly coupled DPP aromatic cores, **DPP4** shows stacked aromatic cores and quite different fluorescence property due to favorable interactions between aromatic DPP cores. As a result of increased intermolecular interactions, **DPP4** shows spontaneous crystallization upon cooling of the melt. On the other hand, if alkyl chain length increases (**DPP12** and **DPP16**, Figure 6.3a), higher tendency for crystallization than **DPP8** is also observed due to increased Van der Waals interactions between lamellar packed alkyl chains. Indeed, the subtle force balance between aromatic core interactions and aliphatic side chain interactions in **DPP8** exhibits minimum ΔG and consequent minimum tendency to crystallize among the derivatives. A recent work of Nakanishi et al. demonstrates that the crystallization propensity of supercooled liquids based on 9,10-diphenylanthracene derivatives is also governed by the alkyl chain and aromatic core ratio and the bulkiness of the aromatic component of the derivatives [27].

Very interestingly, external mechanical agitation such as shear force can trigger crystallization of the thermally stable supercooled liquid **DPP8** with tremendous optical property changes: a more than 25-fold increase in QY and a color change from dim red to bright yellow (Figure 6.3b). Shearing induces density fluctuation of DPP molecules, which enables rearrangement of molecule for the formation of larger nuclei than the critical radius of nucleation, r^*. Since the molecular mobility can be manipulated with temperature control, two different types of fluorescence switching via shear-triggered crystallization can be achieved. At a low temperature (25 °C), molecular motion is restricted and therefore only the area where shearing is applied can be crystallized with turned-on fluorescence; thereby, a fine fluorescence patterning can be accomplished (Figure 6.3d). In contrast, at a high temperature (120 °C), massive fluorescence amplification

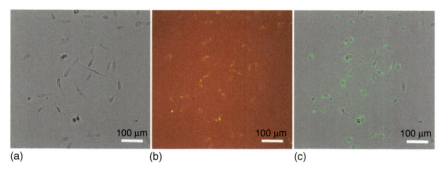

Figure 6.5 Shear-triggered crystallization by living cell attachment. (a) Bright field image and (b) fluorescence image of HS-5 bone marrow fibroblast cells on the supercooled liquid **DPP8** film after two-day incubation. (c) In order to visualize the alignment of cell positions relative to the fluorescent signals from the induced **DPP8** crystals by cell adhesion, the greenish yellow crystalline fluorescent signals were isolated from panel b and labeled as green, and overlaid with panel a. Source: Reproduced with permission from Chung et al. [12]. Copyright 2015, American Chemical Society.

is observed by propagation of crystallization through the entire connected supercooled liquid domain (Figure 6.3d).

As described, shear-triggered crystallization is a unique phase transformation from a metastable supercooled liquid to an energetically more stable crystalline solid. Differing from the conventional mechanochromism of organic molecules, the shear-triggered crystallization of **DPP8** reveals a much higher sensitivity. The threshold shear stress for shear-triggered nucleation in supercooled liquid domain is characterized as low as 0.9 kPa that is about million times sensitive than conventional organic mechanochromism systems (~1 GPa) [28, 29].

A potential biomedical application of the highly sensitive shear-triggered crystallization is suggested since the threshold shear stress for shear-triggered crystallization is in the range of traction force that living cells express to a substrate when they are attached and spread. Interestingly, when HS-5 bone marrow fibroblast cells are seeded on top of supercooled liquid **DPP8** films, the cells induce crystallization and the location of the cell is exactly matched with the fluorescence pattern from the shear-triggered crystallization (Figure 6.5), which demonstrates that the shear-triggered crystallization of supercooled liquid **DPP8** is sensitive enough to detect cell attachment and cell contraction. Therefore, the supercooled liquid **DPP8** system is highly plausible to be utilized as a new type of biomedical characterization platform.

6.3 Highly Emissive Supercooled Liquids

Sada and Kokado et al. reported interesting molecules that can form highly emissive supercooled liquids [13]. The superior emissive property of the developed tetraphenylethene derivatives (Figure 6.6a) is originated from aggregation-induced emission (AIE) phenomenon. In contrast to the concentration-induced quenching, a certain class of organic molecules show

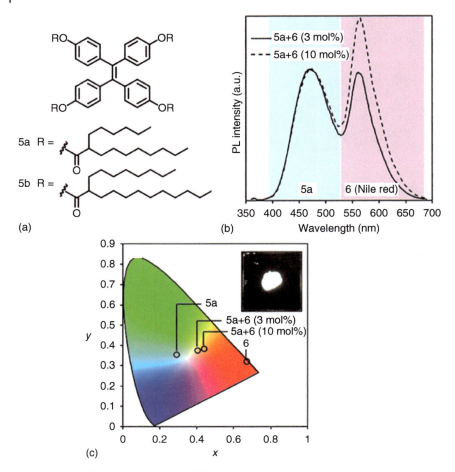

Figure 6.6 (a) Chemical structures of AIE-active tetraphenylethene derivatives. (b) Photoluminescence spectra for samples of **5a** and **6** (Nile red) with 3 and 10 mol% concentrations. (c) Commission Internationale de l'Eclairage (CIE) chromaticity diagram of **5a** (bulk state), **6** (solution state), and the mixture of **5a** and **6** (bulk state). Inset: a photograph of **5a+6** (3 mol%) under 365 nm UV light. Source: Reproduced with permission from Machida et al. [13]. Copyright 2017, The Royal Society of Chemistry.

weak fluorescence in isolated state but strong fluorescence under aggregated conditions due to the restricted molecular motion or specific types of intermolecular packing.

Tetraphenylethene is one of the representative examples of AIE dyes. The tetraphenylethene core is combined with long alkyl chains for liquefaction of the resulting compounds. The synthesized compounds (**5a** and **5b**) show their T_g at −64 and −60.2 °C, T_c at 6.2 and −32.7 °C, and T_m at 29.7 and 3.2 °C, respectively. The compounds do not crystallize under cooling and maintain their supercooled liquid phase even though the compounds crystallize upon subsequent heating due to the lack of thermally stability. In a dilute tetrahydrofuran solution, **5a** shows very weak fluorescence QY of 0.01. However, similar to conventional AIE compounds, the supercooled **5a** and **5b** shows intensive fluorescence with QY of 0.67 and 0.65 (λ_{max} = 475 nm), respectively. Even though the compounds are not crystalline solid or aggregate, the high viscosity (c. 6000 cP) of **5a** and **5b** liquids can restrict the molecular motion and enable AIE process.

Utilizing the virtue of liquid, **5a** is introduced as solvent for different dyes. By dissolving red-emitting organic dye Nile red (**6**) in **5a**, white light emission is realized. An efficient energy transfer (85%) from **5a** to Nile red is observed when 3 mol% of Nile red is dissolved in **5a**, and the mixture of **5a** and Nile red emits white light with Commission Internationale de l'Eclairage (CIE) coordinate of (0.41, 0.38) (Figure 6.6b,c). Possible fine-tuning in color and inkjet printing application are suggested to utilize the developed highly emissive organic supercooled liquid mixture.

6.4 Conclusion

In this chapter, a few unique examples of functional organic supercooled liquids were introduced. Because of its short-lived metastable character, organic supercooled liquids have not generally considered as functional materials for feasible applications. However, superior properties such as excellent film formation, sensitivity to stimuli, and acting as solvent for different materials possibly make supercooled liquid an attractive candidate for unique functional applications. To expand the application potential, it is important to build better understanding on how we can secure a certain amount of stability in an organic supercooled liquid. The systematic investigation on the correlation between molecular design and thermal stability of supercooled liquid DPP8 is a great starting point for the future investigation on universal molecular design principles for functional supercooled liquids. Beyond the suggested applications in this chapter, functional organic supercooled liquids are believed to have a broader impact on various applications including organic electronics, sensors, biomedical applications, and pharmaceutical science.

References

1 Farmer, T. (2015). *Structural Studies of Liquids and Glasses Using Aerodynamic Levitation*, 7–12. Cham, Heidelberg, New York, Dordrecht, and London: Springer International Publishing.
2 Lu, M. and Taylor, L.S. (2016). Vemurafenib: a tetramorphic system displaying concomitant crystallization from the supercooled liquid. *Cryst. Growth Des.* 16: 6033–6042.
3 Mirigian, S. and Schweizer, K.S. (2015). Dynamical theory of segmental relaxation and emergent elasticity in supercooled polymer melts. *Macromolecules* 48: 1901–1913.
4 Mallamace, F., Corsaro, C., Leone, N. et al. (2014). On the ergodicity of supercooled molecular glass-forming liquids at the dynamical arrest: the o-terphenyl case. *Sci. Rep.* 4: 3747.

5 Powell, C.T., Paeng, K., Chen, Z. et al. (2014). Fast crystal growth from organic glasses: comparison of o-terphenyl with its structural analogs. *J. Phys. Chem. B* 118: 8203–8209.
6 Greet, R.J. and Turnbull, D. (1967). Glass transition in o-terphenyl. *J. Chem. Phys.* 46: 1243–1251.
7 Carpentier, L., Decressain, R., and Descamps, M. (2004). Relaxation modes in glass forming *meta*-toluidine. *J. Chem. Phys.* 121: 6470–6477.
8 Rah, K. and Eu, B.C. (2003). Theory of the viscosity of supercooled liquids and the glass transition: fragile liquids. *Phys. Rev. E* 63: 051204.
9 Hansen, C., Stickel, F., Berger, T. et al. (1997). Dynamics of glass-forming liquids. III. Comparing the dielectric α– and β– relaxation of 1-propanol and o-terphenyl. *J. Chem. Phys.* 107: 1086–1093.
10 Cicerone, M.T. and Ediger, M.D. (1996). Enhanced translation of probe molecules in supercooled o-terphenyl: signature of spatially heterogeneous dynamics? *J. Chem. Phys.* 104: 7210–7218.
11 Sosso, G.C., Chen, J., Cox, S.J. et al. (2016). Crystal nucleation in liquids: open questions and future challenges in molecular dynamics simulations. *Chem. Rev.* 116: 7078–7116.
12 Chung, K., Kwon, M.S., Leung, B.M. et al. (2015). Shear-triggered crystallization and light emission of a thermally stable organic supercooled liquid. *ACS Cent. Sci.* 1: 94–102.
13 Machida, T., Taniguchi, R., Oura, T. et al. (2017). Liquefaction-induced emission enhancement of tetraphenylethene derivatives. *Chem. Commun.* 53: 2378–2381.
14 Sagara, Y. and Kato, T. (2009). Mechanically induced luminescence changes in molecular assemblies. *Nat. Chem.* 1: 605–610.
15 Luo, X., Li, J., Li, C. et al. (2011). Reversible switching of the emission of diphenyldibenzofulvenes by thermal and mechanical stimuli. *Adv. Mater.* 23: 3261–3265.
16 Kwon, M.S., Gierschner, J., Yoon, S.-J., and Park, S.Y. (2012). Unique piezochromic fluorescence behavior of dicyanodistyrylbenzene based donor–acceptor–donor triad: mechanically controlled photo-induced electron transfer (eT) in molecular assemblies. *Adv. Mater.* 24: 5487–5492.
17 Yoon, S.-J., Chung, J.W., Gierschner, J. et al. (2010). Multistimuli two-color luminescence switching via different slip-stacking of highly fluorescent molecular sheets. *J. Am. Chem. Soc.* 132: 13675–13683.
18 Guo, Z.-H., Jin, Z.-X., Wang, J.-Y., and Pei, J. (2014). A donor–acceptor–donor conjugated molecule: twist intramolecular charge transfer and piezochromic luminescent properties. *Chem. Commun.* 50: 6088–6090.
19 Okamoto, N. and Oguni, M. (1996). Discovery of crystal nucleation proceeding much below the glass transition temperature in a supercooled liquid. *Solid State Commun.* 99: 53–56.
20 Okamoto, N., Oguni, M., and Sagawa, Y. (1997). Generation and extinction of a crystal nucleus below the glass transition temperature. *J. Phys. Condens. Matter* 9: 9187–9198.

21 Baird, J.A., Van Eerdenbrugh, B., and Taylor, L.S. (2010). A classification system to assess the crystallization tendency of organic molecules from undercooled melts. *J. Pharm. Sci.* 99: 3787–3806.
22 Whitaker, C.M. and McMahon, R.J. (1996). Synthesis and characterization of organic materials with conveniently accessible supercooled liquid and glassy phases: isomeric 1,3,5-tris(naphthyl)benzenes. *J. Phys. Chem.* 100: 1081–1090.
23 Karasawa, S., Hagihara, R., Abe, Y. et al. (2014). Crystal structures, thermal properties, and emission behaviors of N,N-R-phenyl-7-amino-2,4-trifluoromethylquinoline derivatives: supercooled liquid-to-crystal transformation induced by mechanical stimuli. *Cryst. Growth Des.* 14: 2468–2478.
24 López-Mejías, V., Kampf, J.W., and Matzger, A.J. (2009). Polymer-induced heteronucleation of tolfenamic acid: structural investigation of a pentamorph. *J. Am. Chem. Soc.* 131: 4554–4555.
25 Hoffman, J.D. (1958). Thermodynamic driving force in nucleation and growth processes. *J. Chem. Phys.* 29: 1192–1193.
26 Cembran, A., Bernardi, F., Garavelli, M. et al. (2004). On the mechanism of the cis–trans isomerization in the lowest electronic states of azobenzene: S_0, S_1, and T_1. *J. Am. Chem. Soc.* 126: 3234–3243.
27 Lu, F., Jang, K., Osica, I. et al. (2018). Supercooling of functional alkyl-π molecular liquids. *Chem. Sci.* 9: 6774–6778.
28 Dong, Y., Xu, B., Zhang, J. et al. (2012). Piezochromic luminescence based on the molecular aggregation of 9,10-Bis((E)-2-(pyrid-2-yl)vinyl)anthracene. *Angew. Chem. Int. Ed.* 51: 10782–10785.
29 Woodall, C.H., Brayshaw, S.K., Schiffers, S. et al. (2014). High-pressure crystallographic and spectroscopic studies on two molecular dithienylethene switches. *CrystEngComm* 16: 2119–2128.

7

Organic Liquids in Energy Systems

Pengfei Duan[1], Nobuhiro Yanai[2], and Nobuo Kimizuka[2]

[1] CAS Center for Excellence in Nanoscience, National Center for Nanoscience and Technology (NCNST), CAS Key Laboratory of Nanosystem and Hierarchical Fabrication, No. 11 ZhongGuanCun BeiYiTiao, Beijing 100190, China
[2] Kyushu University, Department of Chemistry and Biochemistry, Graduate School of Engineering, Center for Molecular Systems (CMS), 744 Moto-oka, Nishi-ku, Fukuoka 819-0395, Japan

7.1 Introduction

Solving the global energy issue is one of the most important research topics in these decades [1]. The development of renewable energy technologies is the focus point toward the goal of a sustainable society. Hydropower and wind turbines have been already implemented on a large scale, but limitations on their geographical location imply that these technologies alone cannot produce the required energy output [2]. Therein, efficient utilization of solar energy has attracted wide attention in science and technology.

In this context, functional organic liquids possess several unique advantages [3–5]. These liquid materials exhibit nonvolatility and tunable and well-defined optical/electronic properties, with the ability to dissolve organic or inorganic dopants. In addition, they can be directly placed and filled in any shape and geometry, which allows their use in the form of thin liquid films with the inherent performances maintained even under mechanical strain. In terms of molecular design, the following points are highlighted. (i) Functional organic chromophores can be introduced in these liquids by surrounding them with bulky and soft side chains. These substituents effectively isolate the chromophores, which enhance their thermal as well as photostability [4]. This durability is in remarkable contrast to the common organic dye molecules that in many cases decompose gradually under photoirradiation. (ii) Their properties in the bulk liquid state are predictable from those of isolated species in solution because the intrinsic properties of individual chromophores are well preserved in the core-isolated environment. When the liquid chromophores show intermolecular interactions, it is also possible to control their optical/electronic properties by tuning the inter-chromophore interactions. (iii) Various organic and inorganic functional compounds can be dissolved without aggregation, when the cohesive energy of the solutes is properly controlled by considering the interaction with host

Functional Organic Liquids, First Edition. Edited by Takashi Nakanishi.
© 2019 Wiley-VCH Verlag GmbH & Co. KGaA. Published 2019 by Wiley-VCH Verlag GmbH & Co. KGaA.

organic liquids. Thus, the properties and functionality of these solvent-free systems are highly tunable [4–6].

Based on these remarkable features, functional organic liquids have tremendous potential for energy-related applications. In this chapter, we introduce a new perspective of functional organic liquids toward the application to photon energy storage and conversion, by highlighting two intriguing functions: the molecular solar thermal fuels and photon upconversion (UC) [7, 8]. In both topics, it is of importance to maximize the unique potential of π-liquids. The findings and understandings obtained in these new areas would fertilize the emerging field of functional organic liquids for their applications in the photon energy conversion technologies.

7.2 Photoresponsive π-Liquids for Molecular Solar Thermal Fuels

The capture, conversion, and storage of solar energy have been considered as promising strategies for solving the global energy demands. Substantial efforts have been devoted to the development of endergonic chemical transformations such as photocatalytic splitting of water to H_2 and O_2, and conversion of CO_2 and H_2O to CH_3OH and O_2, respectively [1, 9–11]. These open-system solar fuel cycles receive widespread attention in the field of chemistry and materials science. On the other hand, solar thermal molecular fuel has recently attracted increasing attention as an alternative means for solar energy storage (Figure 7.1) [12–15]. Photoisomerization reaction of photoresponsive molecules provides metastable high-energy compounds, which possess higher energy stored as chemical bonding. By applying external stimuli such as heat, voltage, or photoirradiation, these metastable forms undergo isomerization to the stable ground state and the stored energy can be released on demand as heat. The sequence of photoisomerization followed by exothermic isomerization can be repeated many times with no emission of chemical by-products, thus realizing a closed-cycle storage of solar energy as high-energy chemical bonds.

The potential of storing solar energy by reversible photoisomerizations was recognized in 1909 when the photodimerization of anthracene was proposed for this purpose [16]. So far, the closed-system chemical storage of solar energy has been investigated for the photoinduced structural changes between C=C and C—C bonds and E/Z isomerizations of N=N or C=C bonds. In the early stages, norbornadiene derivatives (NBDs) attracted attention because of the relatively large energy stored in the corresponding quadricyclanes [17]; however, norbornadienes suffer from the limited absorption in the UV region and lower chemical stability [12, 18]. Meanwhile, azobenzenes [12, 19] and, more recently, organometallic photochromic compounds such as ruthenium–fulvalene complexes have attracted attention because of high cyclability without degradation [20, 21]. Azobenzene chromophores have been popularly employed as molecular switches because of their high quantum yields to undergo reversible trans–cis isomerization upon sequential absorption of two different wavelengths of light

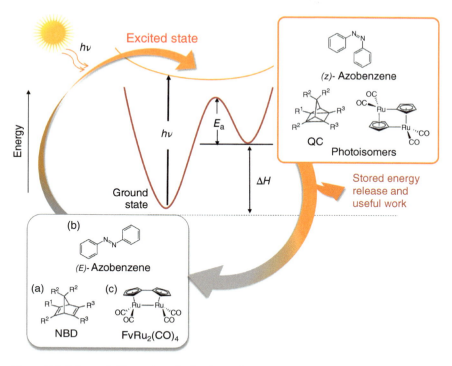

Figure 7.1 Schematic illustration of solar thermal energy storage in typical molecular fuels. (a) Norbornadiene derivative (NBD), (b) (*E*)-azobenzene, (c) fulvalene diruthenium complex. Source: Adapted with permission from Kimizuka et al. [7]. Copyright 2016, The American Chemical Society.

[22–24]. The photoisomerization of *trans*-azobenzenes to a higher energy metastable cis isomer allows the storage of photon energy in the form of molecular strain energy (c. 50 kJ mol^{-1}) [12, 15, 19]. This stored energy can be released as heat in the course of the cis–trans thermal isomerization, and these features rendered azobenzenes as potential candidates for the closed-cycle solar fuels.

Meanwhile, the use of azobenzenes is confronted with the following major problems. First, the exothermicity of 50 kJ mol^{-1} for *cis*-azobenzene is comparable to or smaller than the minimum gravimetric energy density desired for thermal storage materials (c. 100 J g^{-1}) [25]. Second, the trans–cis photoisomerization of azobenzene chromophores is a process accompanied by the increase in molecular volume, which is significantly suppressed within crystals [26]. Consequently, the photochromism of azobenzene chromophores has been mostly investigated in dilute solutions [12, 19]. Although trans–cis photoisomerization of azobenzene chromophores in the condensed state has been investigated for thin crystalline needles [27], liquid crystals [28], and liquid crystalline polymers [29, 30], the use of thin crystalline needles is apparently not suited to the application to solar thermal molecular fuels. Likewise, the use of polymer matrices inevitably decreases the volumetric energy density. Thus, there exist lethal problems for azobenzene chromophores to be used as solar fuels. Although azobenzene-functionalized carbon nanotubes have recently been reported to thermally store solar energy [15, 31], the use of nanotubes is undesirable because their presence not only limits the light absorbed by azobenzene chromophores but the need for time-consuming modification processes also retracts from the appeal.

With these points in mind, we came up with a reasonable solution to using the solvent-free liquid system with azobenzene chromophores that could satisfy the important requirements for solar molecular thermal fuels [32, 33]. The potential of liquid azobenzenes for molecular solar thermal storage was first described for the nonionic compound **1** (Figure 7.2) [32]. A branched 2-ethylhexyl group was introduced to the azobenzene core to fluidify the chromophore, similar to 2-ethylhexyl-*p*-methoxycinnate, a popular ingredient in sunscreens over several decades [34], and to the chromophoric π-liquids [3–5]. Here, the small molecular weight of the 2-ethylhexyl group is advantageous for maintaining the high gravimetric and volumetric energy density.

Irradiation of neat liquid **1** with UV light caused a color change from bright orange (*trans*-**1**) to dark red (Figure 7.2a), indicating that the structural conversion process is visually monitorable. Figure 7.1a shows UV–vis absorption spectral changes of neat liquid **1** in the course of UV and visible light illumination. The observed π–π* absorption band for *trans*-**1** (λ_{max} = 344 nm, neat condition)

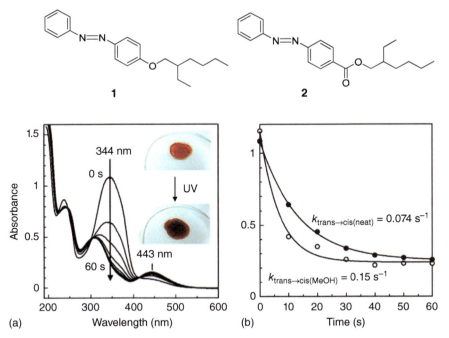

Figure 7.2 Chemical structure of azobenzene liquids. (a) UV–vis spectral changes in **1** upon irradiation at 365 nm under neat conditions (no solvent). Inset: Pictures of liquid **1** in the trans form (top) and cis form (bottom). (b) Dependence of the peak absorbance on the irradiation time. Irradiation at 365 nm: (●) neat liquid **1**; (○) **1** in methanol (7.5 × 10^{-5} M). The rate constants of each reaction were determined by theoretical fitting according to the equation $A_t = A_\infty + (A_0 - A_\infty)\exp(-kt)$, where A_t, A_∞, and A_0 represent the absorbance at irradiation time t, in the photostationary state, and before irradiation, respectively. Source: Adapted with permission from Masutani et al. [32]. Copyright 2014, The Royal Society of Chemistry.

is almost identical to that observed for dilute methanol solutions, indicating the absence of appreciable chromophore–chromophore interactions among the azobenzene chromophores. Upon UV light illumination (365 nm), the π–π* absorption peak at around 344 nm showed a decrease within 60 seconds and an n–π* absorption peak appeared at around 443 nm, which are typical of trans-to-cis photoisomerization of azobenzene chromophores. Figure 7.2b compares time courses of the absorbance change at 344 nm observed for the neat liquid and **1** in methanol solution. The trans-to-cis (and cis-to-trans) photoisomerization of **1** in neat liquid (filled circle) proceeded facilely as observed for that of the monomeric species in methanol (open circle). It demonstrates that the photoisomerization of liquid azobenzene chromophores is not suppressed despite the neat liquid state. This is in remarkable contrast with azobenzene liquid **2**, in which the 2-ethylhexyl group is introduced via an ester bonding. Although photoisomerization of **2** occurs in methanol solution, that in the neat liquid was suppressed (K. Masutani, unpublished data in this laboratory). Compound **2** gives a melting point at 6.8 °C (ΔH, 16.3 kJ mol^{-1}), which indicates the larger intermolecular cohesive energy originating from the ester group. This would be ascribed to the ester group directly attached to the azobenzene chromophore, which contributed to enhancing the dipole–dipole interactions in the liquid state. Besides the enhanced intermolecular interactions, the relaxation pathway of excited singlet state in the neat liquid state could be affected by the carbonyl group directly attached to the azobenzene chromophore. These results clearly indicate the importance of molecular design to tailor the photofunctionality of π-liquids in the condensed state.

In Figure 7.3, *trans*-**1** showed larger shear viscosity compared to that observed for *cis*-**1** in the lower shear velocity regime ($\gamma \leq 0.2$). This indicates the presence

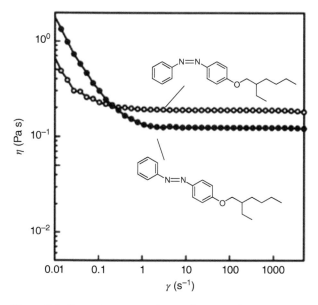

Figure 7.3 Shear viscosity vs. shear velocity on double-logarithmic scale for *trans*-**1** (●) and *cis*-**1** (○) at 25 °C. The molar content of cis isomer was 79%. Source: Adapted with permission from Masutani et al. [32]. Copyright 2014, The Royal Society of Chemistry.

of larger cohesive interactions between the trans-isomers as compared to the bent cis-isomers. This is potentially correlated with the larger static intermolecular interactions revealed by *trans*-azobenzene derivatives in various self-assemblies formed under thermodynamic equilibrium. It is noteworthy that in the smaller shear velocity regime, these *trans*- and *cis*-azobenzene liquids behave as non-Newtonian fluids and exhibit pseudoplastic flow, i.e. the decrease in viscosity observed with increasing the rate of shear. The observed property as shear thinning liquids reflects the decrease in the intermolecular interactions with increasing the shear velocity, which feature has not been reported for the previously reported alkylated liquid π-compounds [3–5].

Meanwhile, at the higher shear velocity, the shear viscosity for these azobenzene liquids becomes independent of the shear velocity, which is a behavior typical of the Newtonian fluid. Thus, the rheological behavior of the azobenzene liquid **1** is a combination of non-Newtonian and Newtonian behavior, and the observed changes suggest a shear-induced transition from the stick-to-slip boundary condition, indicative of the presence of fluid clusters [32]. The shear viscosities η of *trans*-**1** and *cis*-**1** were 0.12 Pa s, 0.19 Pa s, respectively, above the shear velocities of $3.0\,S^{-1}$ (*trans*-**1**) and $0.57\,S^{-1}$ (*cis*-**1**). These η values are comparable to that of olive oil (0.13 Pa s at 10 °C). The viscosity of liquid arises from the friction between neighboring molecules, and, consequently, it depends on the size and shape of the molecules and their intermolecular cohesive interactions. It is to be noted that, at higher shear velocity regime, the *cis*-**1** displayed higher shear viscosity (0.19 Pa s) compared to that shown by *trans*-**1** (0.12 Pa s). The larger shear viscosity η observed for *cis*-**1** at higher shear velocity regime suggests the presence of molecular friction force specific to the cis-isomer. The *cis*-azobenzene has a dipole moment of 3 D, while the nearly planar trans-isomer has a dipole moment near zero [24]. The dipole–dipole interaction shows potential dependence on the inverse third power of the intermolecular distance r, and this strength (c. 2–8 kJ mol^{-1}) often surpasses that of the van dear Waals interactions whose potential shows dependence on the inverse sixth power of the intermolecular distance (<5 kJ mol^{-1}) [35]. It is reasonable that the dipole–dipole interactions play an important role in the rotational friction experienced by the cis isomer [32]. It is one of the essential features of photoisomerizable liquids that enables to explore the intermolecular interactions of stereoisomers without the interference by solvent molecules.

The photon-energy storage performance, thermal properties, and heat storage capacity of liquid **1** were investigated using differential scanning calorimetry (DSC). *trans*-**1** showed a glass transition temperature at around −63 °C and no thermal decomposition peak was observed in the measured temperature range from −63 to 220 °C, demonstrating the high thermal stability of the azobenzene liquid **1**. Figure 7.4a shows DSC thermograms of *cis*-**1** obtained at varied photoisomerization degrees. In these DSC isotherms, exothermic peaks were observed at around 120 °C with varied ΔH values depending on the molar ratio of cis isomers. At the cis-molecular content of 68%, isomerization enthalpy (ΔH) was determined as 35 kJ mol^{-1} (energy density, 32 W h kg^{-1}). The ΔH corresponding to the 100% cis isomer was estimated as 52 kJ mol^{-1} (47 W h kg^{-1}) by extrapolation of the plot (Figure 7.4b), which value is consistent

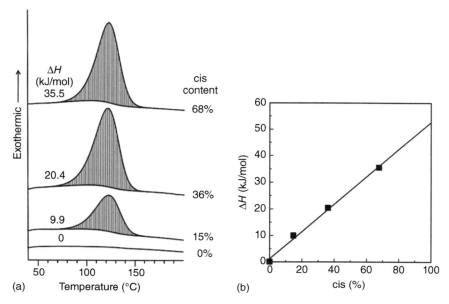

Figure 7.4 (a) DSC thermograms of **1** at varied molar ratio of cis isomer. Heating rate, 50 K min^{-1}. (b) Dependence of cis → trans isomerization enthalpy ΔH against the molar content of cis isomer. Source: Adapted with permission from Masutani et al. [32]. Copyright 2014, The Royal Society of Chemistry.

with the reported isomerization enthalpy of unsubstituted *cis*-azobenzene [12]. It indicates that the basic thermal characteristics are retained for liquefied azobenzene chromophores in the solvent-free state.

This liquid azobenzene system has a remarkable advantage over the existing solid-state heat storage materials in terms of the nonrequirement of solvent dilution, which is a promising feature for application to the flow molecular solar thermal energy storage (MOST) systems. Although the charging energy and bulk energy density reported for the state-of-the-art molecular storages such as the ruthenium–fulvalene complexes ($\Delta H = 83$ kJ mol^{-1}, 52 W h kg^{-1}) [21] and the azobenzene-functionalized single-walled carbon nanotubes [31] are numerically higher than those observed for **1**, these previous solid materials need to be dissolved or dispersed in solvents for use. Consequently, their overall energy densities are significantly reduced by dilution. In contrast, photoisomerization of liquid azobenzene **1** facilely occurs in the neat state and a decrease in the net energy density by dilution is successfully avoided. These features render the liquid **1** a unique and potential solar thermal storage material.

7.3 Azobenzene-Containing Ionic Liquids and the Phase Crossover Approach

Room-temperature ionic liquids (ILs) have been attracting much interest as environmentally benign solvents for a variety of applications because of their unique properties including their very low vapor pressure and high ionic conductivity. Their applications include the media for organic chemical reactions [36], separations [37], and electrochemical and energy applications [38, 39]. We have reported the first examples of ordered molecular self-assembly [40–42], polysaccharides, and supramolecular ionogels [40, 43] and launched an interfacial materials chemistry in ILs [44–46], which offered new perspectives and possibilities in the research field of ILs. As expected, it was a natural course of the ILs' research field to develop photofunctional ILs. Figure 7.5 shows the examples of azobenzene-containing ILs, in which ionic moieties are introduced directly or indirectly to azobenzene chromophores [47–49]. These azobenzene-based IL molecules display trans-to-cis photoisomerization, as exemplified by **5** (Figure 7.5), which showed reversible photoisomerization between the trans- and cis-form where the both isomers were in the viscous liquid state [49]. However, the photoisomerization of azobenzene ILs **3–5** has not been investigated with a view to developing solar thermal IL fuels.

In the previous section, the azobenzene-containing liquid **1** allowed storage and release of photon energy without dilution, where the energy density of 52 kJ mol^{-1} is restricted by the intrinsic nature of azobenzene chromophore. To push the limit, we introduced the concept of self-assembly and developed the photoliquefiable ionic crystals (ICs) which showed increased energy-storage capacity by combining the latent heat [33]. A series of cationic azobenzene derivatives **6**(*n*,*m*)-X was developed, which comprises a set of structural modules: alkyl tails (*n*), spacer methylene length (*m*), an oligo(ethylene oxide)-based ammonium group, and a counterion (X) (Figure 7.5). The physical state of compound **6**(*n*,*m*) can be regulated by tuning the chemical structure of each module. The compounds with short alkyl chains, **6**(1,2)-Cl and **6**(1,4)-Cl, gave an IL phase at ambient temperature, indicating that the bulky oligoether-based ammonium group significantly lowered the crystallinity by weakening van der Waals interactions, aromatic stacking interactions, and the Madelung energy. In contrast, the longer alkyl chain compounds, **6**(4,6)-Br, **6**(6,4)-Cl, **6**(6,4)-Br, and **6**(8,2)-Br, were obtained in crystalline form. The substitution of the halide counterion for bis(trifluoromethanesulfonyl)imide (Tf$_2$N) significantly lowered the melting point and **6**(6,4)-Tf$_2$N gave an IL at ambient temperature. This feature is ascribed to weakened electrostatic interactions as a result of the enhanced delocalization of anionic charge on the Tf$_2$N ion.

The neat ILs **6**(1,2)-Cl, **6**(1,4)-Cl, and **6**(6,4)-Tf$_2$N showed reversible photoisomerization characteristics, as revealed by color changes and UV/vis spectra. Very interestingly, all the IC compounds **6**(6,4)-Br, **6**(6,4)-Cl, **6**(4,6)-Br, and **6**(8,2)-Br showed photoinduced IC–IL phase transition. Figure 7.6 shows the effect of photoillumination on X-ray diffraction (XRD) patterns of **6**(6,4)-Br (a) and the corresponding polarizing optical microscopy (POM) images under crossed polarizers (b) as a typical example. Before UV irradiation, the yellow IC of **6**(6,4)-Br showed birefringence in POM images under the crossed polarizer (Figure 7.6b-i), which is consistent with the crystalline lamellar structure as observed in XRD measurements. After illumination of UV light by a high-pressure mercury lamp, the yellow crystal sample melted to give a red liquid phase within a few minutes (Figure 7.6a-ii, inset). The irradiated part of the sample became dark in the POM

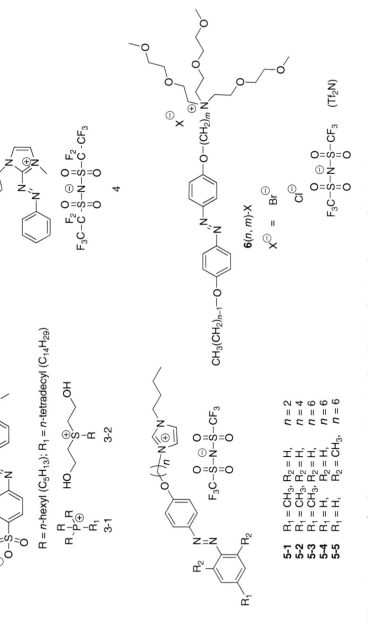

Figure 7.5 Chemical structures of azobenzene-containing ionic liquids **3–5** and ionic azobenzene derivative **6**.

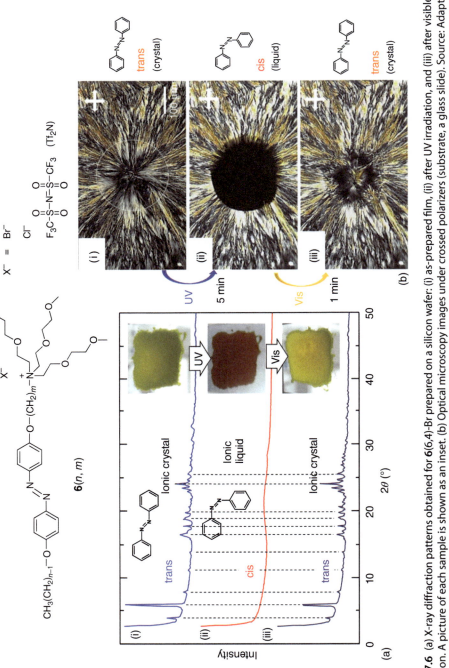

Figure 7.6 (a) X-ray diffraction patterns obtained for **6**(6,4)-Br prepared on a silicon wafer: (i) as-prepared film, (ii) after UV irradiation, and (iii) after visible irradiation. A picture of each sample is shown as an inset. (b) Optical microscopy images under crossed polarizers (substrate, a glass slide). Source: Adapted with permission from Ishiba et al. [33]. Copyright 2015, John Wiley & Sons.

images under the crossed polarizer (Figure 7.6b-ii) and the photoinduced liquefaction to the isotropic IL phase was further confirmed by the disappearance of XRD patterns (Figure 7.6a-ii). Meanwhile, upon illumination of the visible light to the photoliquefied ILs, crystallization occurred reversibly within one minute, and reappearance of the birefringence in POM image and XRD patterns characteristic to the *trans*-azobenzene compound (Figure 7.6a-iii,b-iii). The reversible photoliquefaction and photocrystallization were similarly observed for the other ICs at room temperature, and these are the first examples of reversible photoliquefaction and crystallization of ICs.

The photon energy storage capacity was investigated for the ILs *cis*-**6**(6,4)-Br and *cis*-**6**(6,4)-Tf$_2$N prepared by photoisomerization, respectively. These samples were photoisomerized in advance and stored in the cis form. Figure 7.7a,b shows DSC thermograms obtained for ILs of *cis*-**6**(6,4)-Tf$_2$N and *cis*-**6**(6,4)-Br,

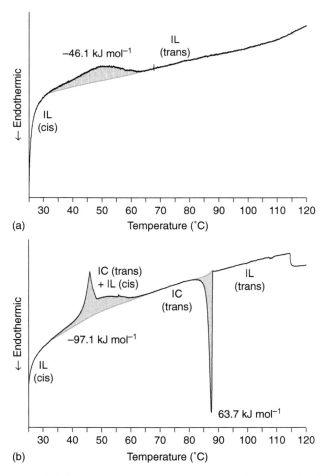

Figure 7.7 DSC thermograms of *cis*-azobenzene derivatives: (a) *cis*-**6**(6, 4)-Tf$_2$N (IL) and (b) *cis*-**6**(6, 4)-Br. Each sample contained the cis-form in 89 mol% as determined by ^{1}H NMR spectroscopy in CDCl$_3$. Heating rate, 0.2 °C min^{-1}. Source: Adapted with permission from Ishiba et al. [33]. Copyright 2015, John Wiley & Sons.

respectively. Upon heating cis-**6**(6,4)-Tf$_2$N, a broad exothermic peak was observed around at 49 °C with a ΔH value of 46.1 kJ mol^{-1}, which corresponds to the ΔH value of 51.8 kJ mol^{-1} for the pure cis isomer (Figure 7.7a).

This peak is associated with the thermally induced cis–trans isomerization of the IL, and the observed change in enthalpy (ΔH) is consistent with the reported energy ($\Delta H_{cis-trans}$) stored in a cis-azobenzene chromophore [12]. In contrast, when the photoliquefied cis-**6**(6,4)-Br was heated, two exothermic peaks were found (Figure 7.7b). In addition to the broad exothermic peak associated with the thermal cis–trans isomerization (temperature range: 34–65 °C), a sharp exothermic peak is observed around at 46 °C. This peak is ascribed to the crystallization of trans-**6**(6,4)-Br formed by thermally induced isomerization of cis-**6**(6,4)-Br IL. The total ΔH determined for the sharp and broad exothermic peaks observed in Figure 7.7b amounts to 97.1 kJ mol^{-1}, which is almost double the energy stored in a cis-azobenzene chromophore (Figure 7.8). It goes over the necessary standards for the specific energy required for thermophysical storages (the specific energy requirement of 100 J g^{-1} corresponds to 75.7 kJ mol^{-1} for **6**(6,4)-Br) [25]. Meanwhile, further heating of the sample gave a sharp endothermic peak at 87 °C (ΔH, 63.7 kJ mol^{-1}), which is due to the melting of IC (trans-**6**(6,4)-Br) phase.

The photoisomerization of ICs shows multiple functionalities, and changes in ionic conductivity were observed in addition to the reversible liquefaction phenomena and changes in optical properties. The trans-**6**(6,4)-Br in the IC phase showed a considerably lower ionic conductivity of 4.2 × 10^{-10} S cm^{-1}, thus reflecting suppressed migration of ionic species in the crystalline state. Meanwhile, upon photoliquefaction of trans-**6**(6,4)-Br, a salient increase in ionic conductivity

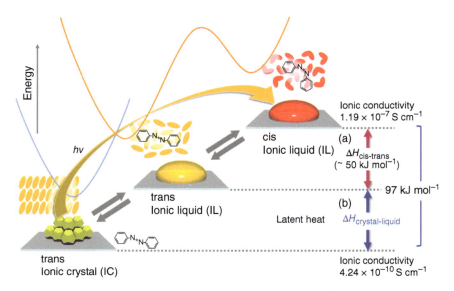

Figure 7.8 Schematic illustration of a photoinduced ionic crystal (IC) to ionic liquid (IL) phase transition observed for **6**(6,4) and enthalpy changes associated with the changes in (a) cis-IL to trans-IL and (b) trans-IL to trans-IC. Ionic conductivities for cis-IL and trans-IC are also shown. Source: Adapted with permission from Kimizuka et al. [7]. Copyright 2016, The American Chemical Society.

was observed to be 1.2×10^{-7} S cm^{-1}, which value is 280 times higher than that observed for the IC state. The observed remarkable change in ionic conductivity is reasonably explained by the enhanced mobility of ionic species in the cis-IL. The change in ionic conductivity showed good reversibility, reflecting the totally reversible IC–IL phase transition. Thus, the reversible photoliquefaction of ICs leads to concomitant control on the ionic conductivity at ambient temperature.

Although photoinduced solid–liquid phase-transition phenomena have been recently reported for nonionic azobenzene derivatives of macrocyclic azobenzenophane [50–52] and carbohydrate conjugates [53], oligoether-appended azobenzene compounds [54] were also shown to display similar phenomena. However, in these studies, photon energy storage properties were not the subject of research. The integration of self-assembly and photoinduced IC–IL transition dissolves the limitations in the current solar thermal storage technology, providing a new perspective of phase crossover chemistry which is defined as the proactive and simultaneous control of multiple physical properties or functions of molecular self-assembled systems based on the triggered changes in their physical states [33].

7.4 Photon Upconversion and Condensed Molecular Systems

Photon UC through annihilation between two long-lifetime excited triplets, namely, the sensitized triplet–triplet annihilation-based photon upconversion (TTA-UC), has recently come into the spotlight because of its wide applications that range from renewable energy production to bioimaging and phototherapy [55–62]. Due to its occurrence with low-intensity, noncoherent incident light, TTA-UC has been widely studied. Generally, the TTA-UC process requires a bimolecular system consisting of triplet donor (D) and acceptor (A) (Figure 7.9). This mechanism starts with the generation of donor (sensitizer) triplets by intersystem crossing (ISC) from the photogenerated singlet state, followed by D-to-A triplet–triplet energy transfer (TTET) to form optically dark, metastable acceptor triplets. The subsequent diffusion and collision of two excited acceptor triplets generate a higher energy excited singlet state through triplet–triplet annihilation (TTA), from which the unconverted fluorescence is emitted. Because of the long lifetime of the triplet species, the excitation power density can be reduced to as low as a few milliwatts per square centimeter, that is on the order of sunlight intensity at a specific excitation wavelength.

To date, efficient TTA-UC has been achieved for deaerated organic solutions with molecularly dissolved donors and acceptors, as it allows fast diffusion and collision of the excited molecules. Furthermore, the removal of dissolved molecular oxygen from the solution is imperative to observe TTA-UC, since triplet excited states are very sensitive to oxygen and facilely undergo quenching. However, the use of such volatile organic solvents and the requirement for oxygen-free atmosphere hamper the practical applications of TTA-UC, and thus it is highly desired to develop solvent-free TTA-UC systems which desirably work under

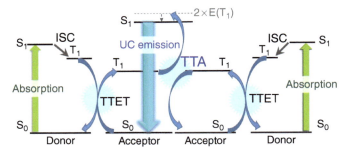

Figure 7.9 Schematic energy-level diagram of TTA-based upconversion for a model donor/acceptor pair. Solid arrows indicate transitions. The process involves the population of the singlet excited state of a donor (S_1) upon absorption of incident light, which is followed by intersystem crossing (ISC) to the triplet excited state (T_1). When the triplet donor encounters the ground-state acceptor, triplet–triplet energy transfer (TTET) from the donor to an acceptor yields an acceptor triplet state. When two acceptors in their triplet state undergo triplet–triplet annihilation (TTA), one of the acceptors is excited to its excited singlet state while the other acceptor is relaxed to its ground state. The photon emitted from the acceptor singlet state (UC emission, turquoise blue arrow) has a higher energy than that of the initially absorbed photons (green arrow). Source: Adapted with permission from Kimizuka et al. [7]. Copyright 2016, The American Chemical Society.

air. Although amorphous polymer films have been employed as solid matrixes [63], they inevitably restrict the diffusion of excited triplet molecules, thereby necessitating the undesired use of high-incident light intensity to ensure effective concentration of excited triplets. To circumvent these problems, condensed chromophore systems are attractive since they are expected to allow high mobility of triplet excitation energy in the assembled structures. Such a UC mechanism can be called triplet energy migration-based upconversion (TEM-UC) [7, 8]. As one of the TEM-UC systems, solvent-free chromophoric liquid systems have been developed, which consists of liquid acceptors doped with appropriate sensitizers. In the following sections, we introduce two nonvolatile liquid systems exhibiting photon UC under ambient conditions.

7.5 TTA-UC Based on the Amorphous π-Liquid Systems

Nakanishi and coworkers reported that the covalent modification of 9,10-diphenylanthracene (DPA) with multiple branched alkyl chains provides room-temperature fluorescent liquids [4]. As DPA has been a benchmark acceptor in TTA-UC, we took advantage of the liquid chromophores and reported the first example of condensed liquid TTA-UC system using a fluorescent liquid acceptor **7** and a Pt(II) porphyrin donor **8** modified with branched alkyl chains (Figure 7.10) [6].

The donor **8** having branched alkyl chains was synthesized because the commercially available donor **10** was not molecularly dissolved in the acceptor liquid **7** as observed by POM. The lipophilic donor **8** was homogeneously miscible with **7** in the examined concentration range up to donor/acceptor = 1 mol%

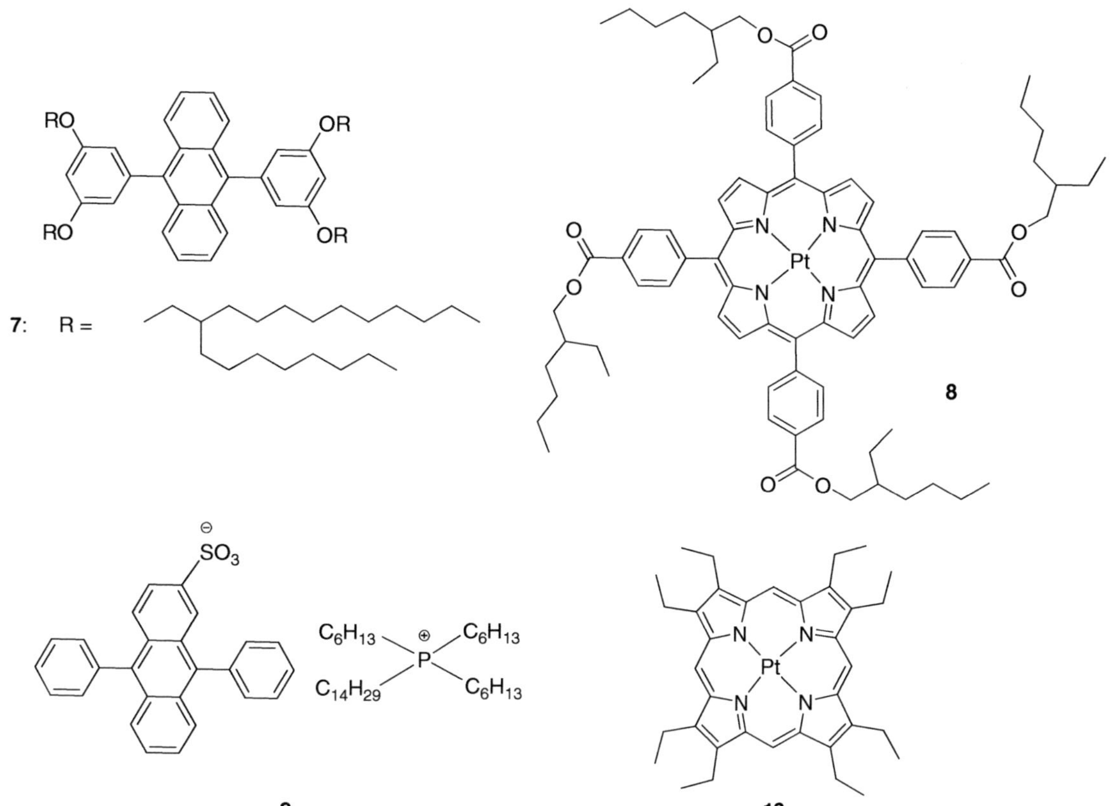

Figure 7.10 Chemical structures for liquid acceptors (emitters) **7**, **9** and donors (triplet sensitizers) **8**, **10**.

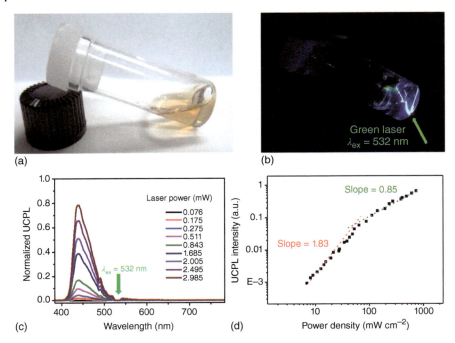

Figure 7.11 Photographs of the doped liquid (8/7 = 0.01 mol%) upon exposure to (a) white light and (b) a 532-nm green laser (the incident laser direction is indicated by a green arrow). Bright blue luminescence is due to UC emission, and green spots are the scattered incident laser. (c) Photoluminescence spectra of the doped liquid with different incident intensities of the 532-nm laser in air. (d) Dependence of UC emission intensity at 433 nm on the incident light intensity. The dashed lines are the fitting results with slopes of 1.83 (red) and 0.85 (green) in the low- and high-power regimes, respectively. UCPL, upconversion photoluminescence. Source: Adapted with permission from Duan et al. [6]. Copyright 2013, The American Chemical Society.

(Figure 7.11a,b). The acceptor liquid showed a high fluorescence quantum yield of 68%, while it was quenched upon increasing the molar ratio of the donor notably above 0.1 mol%, indicating the occurrence of acceptor-to-donor singlet–singlet back energy transfer. Accordingly, the low donor/acceptor ratio of 0.01 mol% was employed in the UC experiments. Very interestingly, blue UC emission was clearly observed from the donor-doped liquid upon excitation by a 532-nm green laser even in air (Figure 7.11b,c). It was a great surprise that the UC emission was observable without deaeration since excited triplets are very sensitive to air. It indicates that in the viscous acceptor liquid system, the permeation of molecular oxygen and collision with acceptors are restrained by the surrounding branched alkyl chains. It highlights the important feature of these liquid acceptor systems. In Figure 7.11c, it is notable that the donor phosphorescence at 660 nm was completely quenched regardless of the excitation intensity, indicating the quantitative TTET from the donor to the surrounding acceptor liquid. The excitation intensity dependence of UC emission intensity showed the quadratic-to-linear transition (Figure 7.11d). In addition, the UC emission of the donor-doped liquid showed a decay in the millisecond time scale. These observations give clear evidence for the occurrence of TTA-UC. The UC properties of the deaerated liquid specimens were almost identical to those obtained in air.

A useful figure of merit of TTA-UC is given by a threshold excitation intensity I_{th}, at which the spontaneous decay rate of the excited acceptor triplets equals the TTA rate ($\Phi_{TTA} = 0.5$) [60]. Above the incident power of I_{th}, TTA provides the main deactivation channel for the acceptor triplet and consequently the UC efficiency Φ_{UC} is maximized. In Figure 7.11d, it is to be noted that the threshold intensity I_{th} was observed at a relatively low excitation intensity of around 50 mW cm^{-2}, which is comparable to those reported for the solvent-free polymer systems. Remarkably, the donor-doped liquid showed a high UC quantum yield of 14% (theoretical maximum = 50%) at around 300 mW cm^{-2}, which is close to the highest records reported for solvent-free UC systems. There was no obvious loss in the quantum efficiency within 20 days of air exposure, and a good quantum yield as high as 7.5% was still observed even after a longer period of 80 days. These results are explicable by suppressed diffusion of oxygen molecules into dense and highly viscous liquid chromophores. We presume that further enhancement in the durability of DPA chromophore toward the photo-oxidation would improve the long-term stability of the liquid TTA-UC system.

To prove the contribution of triplet energy migration in the liquid TTA-UC system, low-temperature measurement was conducted. In general, TTA-UC systems based on molecular diffusion in amorphous polymer matrices is inhibited below the glass transition temperature of the matrixes [63], which suppresses the translational movement of dyes and accordingly the molecular collision required for the classical TTA-UC. When the TTA-UC system shown in Figure 7.11 was conducted below the glass transition temperature of the liquid acceptor 7 (T_g, −59 °C), UC emission was observed with an intensity around 20% of that observed at ambient temperatures. This result clearly demonstrates that the triplet energy migration occurs among the frozen acceptor molecules (molecular glass), without the help of molecular diffusion. This is a distinct feature from the conventional solution/polymer dispersion systems [63].

It is well understood that triplet energy transfer and energy migration occur via the electron exchange mechanism (Dexter energy transfer) [64]. Although the involvement of triplet energy migration is a unique characteristic of the photon UC in condensed chromophores, the triplet diffusion constant D_T in the liquid acceptor 7 was determined from the equation:

$$D_T = (8\pi a_0 \alpha \Phi_{ET} I_{th})^{-1} \tau_T^{-2} \qquad (7.1)$$

where α is the absorption coefficient at the excitation wavelength, Φ_{ET} is the donor-to-acceptor TTET efficiency, D_T is the diffusion constant of acceptor triplet (in three-dimensional diffusion system), a_0 is the annihilation distance between acceptor triplets (usually ~10 Å; 9.1 Å for DPA [65]), and τ_T is the lifetime of the acceptor triplet. Using the I_{th} value of 50 mW cm^{-2} (Figure 7.11d), D_T in the liquid 7 was determined to be on the order of 10^{-7} cm^2 s^{-1}. This fell short of expectations and is smaller than the molecular diffusion constant of DPA observed in a low-viscosity solvent (1.2×10^{-5} cm^2 s^{-1}) [65]. This is primarily ascribed to the large interchromophore distance of ~2.1 nm in liquid **7** as determined by small-angle X-ray scattering [6]. Although the molten alkyl chains fortunately showed sealing ability toward molecular oxygen, the large separation of chromophores imposed by them and lack of molecular order are the drawbacks for attaining high mobility of the excited triplet energy.

7.6 Photon Upconversion Based on Bicontinuous Ionic Liquid Systems

The relatively low triplet diffusion constant observed for the amorphous π-acceptor liquid directed our attention to introducing molecular order and to minimize the interchromophore distances. Although the enhancement of molecular order while maintaining the liquid phase is a conflicting issue, the recent notion that ILs display nanometer-scale bicontinuous network structures [66–69] prompted us to design acceptor-containing ILs. It was expected that such a bicontinuous chromophore alignment in chromophoric IL systems facilitate the diffusion of triplet excitons (Figure 7.12).

A new chromophoric IL **9** was developed by introducing 9,10-diphenylanthracene-2-sulfonate anion (DPAS) as a counterion to the tetra-alkylated phosphonium ion (Figure 7.10) [70]. Note that no alkyl chains were introduced in the DPAS chromophore. It showed redshifts in absorption and fluorescence spectra in the neat IL state as compared to those obtained for methanol solutions, indicating the presence of interchromophore interactions. This feature is in contrast with the nonionic acceptor liquid **7**. Interestingly, the fluorescence quantum yield of **9** in the neat IL state (70%) is almost the same as that observed in methanol (67%). This indicates that self-absorption in the neat IL state is not serious, reflecting the small overlap between the absorption and emission bands. Thus, the chromophoric IL satisfies both the interchromophore interactions and the high quantum yield, which are the desirable characteristics for TEM-UC in the condensed state.

Under the excitation of the donor **10** dissolved in IL **9** (λ_{em} at 532 nm), an upconverted blue emission was clearly detected around at 450 nm with a long triplet lifetime on the millisecond time scale ($\tau \sim 2.5$ ms). The phosphorescence of the donor **10** at 650 nm was not detected regardless of the excitation power, indicating the quantitative donor-to-acceptor TTET process. While the measurements of these IL systems were carried out under deaerated conditions to properly evaluate the effect of chromophore arrangement on TTA-UC, the UC emission was detected even after being exposed to air for many days, similar to that observed for the nonionic upconverting liquid **7**. The triplet diffusion constant D_T in IL **9** estimated by Eq. (7.1) was significantly larger than the molecular diffusion constant estimated from the Stokes–Einstein equation (1.76×10^{-12} cm^2 s^{-1}) by considering the viscosity of IL **9** (2760 Pa s) and the molecular size of DPAS (0.45 nm). It indicates that the triplet excited states generated in IL **9** diffuse predominantly by energy migration and not by molecular diffusion (Figure 7.12c). Although some contributions of local

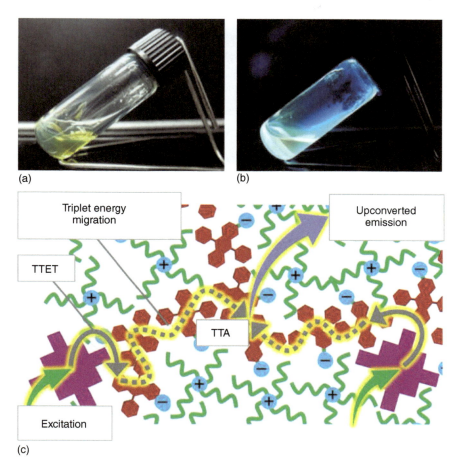

Figure 7.12 Photographs of IL **9** under (a) white light and (b) UV light (λ_{ex} = 365 nm), (c) a schematic representation of TEM-UC in the bicontinuous chromophore networks in IL **9**. Source: Adapted with permission from Hisamitsu et al. [70]. Copyright 2015, John Wiley & Sons.

molecular reorientation should be involved in the liquid state, these results are reasonably explained by the formation of contiguous ionic chromophore arrays that provide efficient TEM pathways in the IL.

The relationship between the nanoscale structural order in chromophoric ILs and their TEM-UC properties has been scrutinized for IL **11**, which consists of a long-chained, nonspherical phosphonium ion and a DPAS anion (Figure 7.13) [71]. While **11** was obtained as a liquid, it underwent solidification to IC after standing for several days at room temperature. The rate of crystallization can be increased by keeping it several hours (>3 hours) at 115 °C (Figure 7.13b). Long alkyl chains naturally assume various conformations and the high viscosity of IL stabilized the metastable supercooled IL phase at room temperature, as confirmed by DSC measurement and POM. The birefringence observed for the IC sample (Figure 7.13b, left) completely disappeared by heating above the melting temperature (141 °C). Upon subsequent cooling to room temperature, the

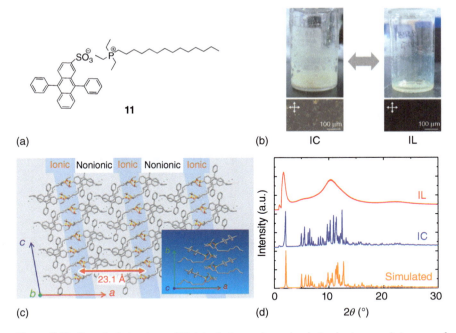

Figure 7.13 Chemical structure of **11** (a), photographs and polarized microscopic images of **11** in the IC phase (left) and supercooled IL phase (right) (b), crystal structure of **11** viewed along the b-axis, showing a lamellar structure consisting of ionic and nonionic layers. Inset: The structure viewed along the c-axis to see a long-tail alkyl chain sandwiched between two anthracene rings. P, orange; S, yellow; O, red; C, gray. Hydrogen atoms are omitted for clarity (c), PXRD patterns of **11** in the supercooled IL phase (top) and the IC phase (middle) at room temperature. A simulated pattern from the single-crystal structure of **11** is also shown (bottom) (d). Source: Adapted with permission from Hisamitsu et al. [71]. Copyright 2018, Royal Society of Chemistry.

birefringence did not recover from the optically isotropic supercooled IL phase (Figure 7.13b, right). Further cooling below the glass transition temperature (∼15 °C) to 0 °C gave a glassy state without any birefringence.

Single-crystal structural analysis **11** showed a space group assigned to the $P2_1/c$ monoclinic system (Figure 7.13c) [71]. The IC phase shows a lamellar structure consisting of ionic and nonionic layers, and the distance between two ionic layers of **11** is 23.1 Å. In the ionic layers, Coulomb interactions operate between the cationic phosphonium and anionic sulfonate moieties, while long-tail alkyl chains and DPA moieties are packed in the nonionic layers through van der Waals and CH–π interactions. The center-to-center intermolecular distance between anthracene units along the b axis is 9.2 Å. Powder X-ray diffraction (PXRD) measurement for the IC phase showed a good match with the pattern simulated from the crystal structure (Figure 7.13d, middle and bottom). The peak at 2.08° observed for the IC phase is ascribed to the (100) peak that corresponds to the distance between the ionic layers. The observation of low-angle peak at 1.63° in the IL phase (Figure 7.13d, top) indicates the presence of an ordered ionic layer arrangement. The interlayer distance was larger for the IL phase (29.1 Å) than for the IC phase (23.1 Å), probably due to the volume expansion caused by the disordering of long alkyl chains in cations. The broad peaks around 5° and 10° of the IL phase also reflect the peaks of the IC phase, and thus the nano-segregated layer structure is maintained even in the IL phase. These nano-segregated structure in the IL phase would be directed by ionic chain formation and limited miscibility between the ionic and nonionic layers. Diameters of structured domains in the IC and IL phases as estimated by Scherrer's equation using full-width at half maximum (FWHM) of (100) peaks were estimated as 418.9 (IC) and 9.3 nm (IL), respectively.

The Φ_{FL} value of **11** in the methanol solution was 70%, while those in the IC and IL phases showed slightly lower values of 69% and 65%, respectively, showing the minor effect of aggregation-caused quenching as similarly observed for **9**. Under excitation at λ_{ex} = 532 nm, both the donor (**10**)-doped IC and IL samples of **11** showed clear TTA-UC emissions at 450–470 nm. The triplet lifetime (τ_T) of **11** in the IL and IC phases was estimated as 1.9 and 5.0 ms, respectively. The TTA-UC measurement for the IL phase gave an I_{th} value of 308 mW cm^{-2}, a diffusion constant D_T of 2.33 × 10^{-9} cm^2 s^{-1} and a triplet diffusion length L_D of 21.1 nm (**10**, 0.01 mol%). Notably, this L_D value is larger than the domain size obtained from the XRD peak width (9.3 nm), indicating that the long-lived triplet excitons can migrate beyond the size of nanodomains in the IL phase [71]. Meanwhile, in contrast to the IC phase, the nanostructured IL phase showed superior performance to disperse the donor **10** in high concentration (~0.1 mol%), which sample gave an I_{th} value as low as 31 mW cm^{-2}. The finding of two-dimensional (2D) nanostructural order in chromophoric IL and the occurrence of triplet energy transportation well over the IL nanodomain size provide a new perspective to the chemistry of ILs.

7.7 Conclusion and Outlook

In this chapter, we discussed the potential of organic liquids in energy-related functions, with particular focus on the molecular solar thermal fuel and photon UC. From the viewpoint of maximizing gravimetric energy density desired for thermal storage materials, it is an essential development that the photoisomerization reaction of azobenzene chromophore takes place in the absence of solvents [32]. Moreover, the heat storage capacity can be significantly enhanced by utilizing phase transition between liquid and crystal states, illustrating the advantage of liquid-based phase crossover chemistry [33]. The optimization of chemical structure and phase-transition properties would further increase the heat storage ability. Toward real-world applications, it is also important to develop catalytic reaction systems for converting metastable photoisomers to the most stable states (e.g. cis-to-trans isomerization in the case of azobenzene derivatives) without any heating treatments, so that the stored energy can be utilized on demand.

The unique characteristics of organic liquids have been also appreciated in the field of photon UC. It is to be noted that molecularly isolated or weakly interacting chromophores in neutral liquids or ILs show high fluorescent quantum yield.

These condensed liquids are capable of molecularly dissolving donor molecules depending on the chemical structure and concentration. Under optimum conditions, the triplet excitation energy was effectively transferred from the donors to the acceptor liquids, which was followed by energy migration among acceptors and efficient upconverted emission at relatively weak excitation light intensity. It is also notable that the suppressed concentration and diffusion of oxygen molecules in these highly viscous liquids enabled stable UC emission even in the ambient condition.

In terms of the photophysical functions based on the short-range interactions as exemplified by TTA, it is crucial to design structures in condensed liquids. This was clearly demonstrated by the acceptor IL system which showed better performance as compared to the neutral acceptor liquid. Apparently, the formation of bicontinuous molecular networks in ILs would play an important role in the further development of triplet-related functions. For photochemical processes based on long-range interactions such as Förster-type resonance energy transfer and singlet energy migration [72, 73], the potential of neutral π-liquids would be more eminent. It will, however, become more and more important to understand the complex and hierarchical structure of organic liquids and their relationship with physical properties and new functions. Obviously, the future development of functional condensed liquids requires both experimental and theoretical approaches and their synergistic collaborations. It would be also essential to take a comprehensive, panoramic view of the condensed matter and molecular self-assemblies as a whole, and ask ourselves what makes the difference between each molecular system.

References

1 Lewis, N.S. and Nocera, D.G. (2006). Powering the planet: chemical challenges in solar energy utilization. *Proc. Natl. Acad. Sci. U.S.A.* 103: 15729–15735.
2 Lindley, D. (2010). The energy storage problem. *Nature* 463: 18–20.
3 Ribierre, J.C., Aoyama, T., Kobayashi, T. et al. (2007). Influence of the liquid carbazole concentration on charge trapping in C-60 sensitized photorefractive polymers. *J. Appl. Phys.* 102: 033106.
4 Babu, S.S., Hollamby, M.J., Aimi, J. et al. (2013). Nonvolatile liquid anthracenes for facile full-colour luminescence tuning at single blue-light excitation. *Nat. Commun.* 4: 1969.
5 Babu, S.S. and Nakanishi, T. (2013). Nonvolatile functional molecular liquids. *Chem. Commun.* 49: 9373–9382.
6 Duan, P.F., Yanai, N., and Kimizuka, N. (2013). Photon upconverting liquids: matrix-free molecular upconversion systems functioning in air. *J. Am. Chem. Soc.* 135: 19056–19059.
7 Kimizuka, N., Yanai, N., and Morikawa, M. (2016). Photon upconversion and molecular solar energy storage by maximizing the potential of molecular self-assembly. *Langmuir* 32: 12304–12322.

8 Yanai, N. and Kimizuka, N. (2016). Recent emergence of photon upconversion based on triplet energy migration in molecular assemblies. *Chem. Commun.* 52: 5354–5370.
9 Pagliaro, M., Konstandopoulos, A.G., Ciriminna, R., and Palmisano, G. (2010). Solar hydrogen: fuel of the near future. *Energy Environ. Sci.* 3: 279–287.
10 Blankenship, R.E., Tiede, D.M., Barber, J. et al. (2011). Comparing photosynthetic and photovoltaic efficiencies and recognizing the potential for improvement. *Science* 332: 805–809.
11 Osterloh, F.E. (2008). Inorganic materials as catalysts for photochemical splitting of water. *Chem. Mater.* 20: 35–54.
12 Kucharski, T.J., Tian, Y.C., Akbulatov, S., and Boulatov, R. (2011). Chemical solutions for the closed-cycle storage of solar energy. *Energy Environ. Sci.* 4: 4449–4472.
13 Borjesson, K., Lennartson, A., and Moth-Poulsen, K. (2013). Efficiency limit of molecular solar thermal energy collecting devices. *ACS Sustainable Chem. Eng.* 1: 585–590.
14 Borjesson, K., Dzebo, D., Albinsson, B., and Moth-Poulsen, K. (2013). Photon upconversion facilitated molecular solar energy storage. *J. Mater. Chem. A* 1: 8521–8524.
15 Kolpak, A.M. and Grossman, J.C. (2011). Azobenzene-functionalized carbon nanotubes as high-energy density solar thermal fuels. *Nano Lett.* 11: 3156–3162.
16 Jones, G. and Ramachandran, B.R. (1976). Catalytic activity in reversion of an energy storing valence photoisomerization. *J. Org. Chem.* 41: 798–801.
17 Yoshida, Z.I. (1985). New molecular-energy storages-systems. *J. Photochem.* 29: 27–40.
18 Gray, V., Lennartson, A., Ratanalert, P. et al. (2014). Diaryl-substituted norbornadienes with red-shifted absorption for molecular solar thermal energy storage. *Chem. Commun.* 50: 5330–5332.
19 Olmsted, J., Lawrence, J., and Yee, G.G. (1983). Photochemical storage potential of azobenzenes. *Sol. Energy* 30: 271–274.
20 Boese, R., Cammack, J.K., Matzger, A.J. et al. (1997). Photochemistry of (fulvalene)tetracarbonyldiruthenium and its derivatives: efficient light energy storage devices. *J. Am. Chem. Soc.* 119: 6757–6773.
21 Kanai, Y., Srinivasan, V., Meier, S.K. et al. (2010). Mechanism of thermal reversal of the (fulvalene)tetracarbonyldiruthenium photoisomerization: toward molecular solar–thermal energy storage. *Angew. Chem. Int. Ed.* 49: 8926–8929.
22 Natansohn, A. and Rochon, P. (2002). Photoinduced motions in azo-containing polymers. *Chem. Rev.* 102: 4139–4175.
23 Yager, K.G. and Barrett, C.J. (2006). Novel photo-switching using azobenzene functional materials. *J. Photochem. Photobiol., A* 182: 250–261.
24 Beharry, A.A. and Woolley, G.A. (2011). Azobenzene photoswitches for biomolecules. *Chem. Soc. Rev.* 40: 4422–4437.
25 Gur, I., Sawyer, K., and Prasher, R. (2012). Searching for a better thermal battery. *Science* 335: 1454–1455.

26 Irie, M. (2008). Photochromism and molecular mechanical devices. *Bull. Chem. Soc. Jpn.* 81: 917–926.

27 Koshima, H., Ojima, N., and Uchimoto, H. (2009). Mechanical motion of azobenzene crystals upon photoirradiation. *J. Am. Chem. Soc.* 131: 6890–6891.

28 Soberats, B., Uchida, E., Yoshio, M. et al. (2014). Macroscopic photocontrol of ion-transporting pathways of a nanostructured imidazolium-based photoresponsive liquid crystal. *J. Am. Chem. Soc.* 136: 9552–9555.

29 Ikeda, T. and Tsutsumi, O. (1995). Optical switching and image storage by means of azobenzene liquid-crystal films. *Science* 268: 1873–1875.

30 Hosono, N., Kajitani, T., Fukushima, T. et al. (2010). Large-area three-dimensional molecular ordering of a polymer brush by one-step processing. *Science* 330: 808–811.

31 Kucharski, T.J., Ferralis, N., Kolpak, A.M. et al. (2014). Templated assembly of photoswitches significantly increases the energy-storage capacity of solar thermal fuels. *Nat. Chem.* 6: 441–447.

32 Masutani, K., Morikawa, M., and Kimizuka, N. (2014). A liquid azobenzene derivative as a solvent-free solar thermal fuel. *Chem. Commun.* 50: 15803–15806.

33 Ishiba, K., Morikawa, M., Chikara, C. et al. (2015). Photoliquefiable ionic crystals: a phase crossover approach for photon energy storage materials with functional multiplicity. *Angew. Chem. Int. Ed.* 54: 1532–1536.

34 Tan, E.M.M., Hilbers, M., and Buma, W.J. (2014). Excited-state dynamics of isolated and microsolvated cinnamate-based UV-B sunscreens. *J. Phys. Chem. Lett.* 5: 2464–2468.

35 Xu, L., Miao, X., Ying, X., and Deng, W. (2012). Two-dimensional self-assembled molecular structures formed by the competition of van der Waals forces and dipole–dipole interactions. *J. Phys. Chem. C* 116: 1061–1069.

36 Hallett, J.P. and Welton, T. (2011). Room-temperature ionic liquids: solvents for synthesis and catalysis. 2. *Chem. Rev.* 111: 3508–3576.

37 Joshi, M.D. and Anderson, J.L. (2012). Recent advances of ionic liquids in separation science and mass spectrometry. *RSC Adv.* 2: 5470–5484.

38 Armand, M., Endres, F., MacFarlane, D.R. et al. (2009). Ionic-liquid materials for the electrochemical challenges of the future. *Nat. Mater.* 8: 621–629.

39 MacFarlane, D.R., Tachikawa, N., Forsyth, M. et al. (2014). Energy applications of ionic liquids. *Energy Environ. Sci.* 7: 232–250.

40 Kimizuka, N. and Nakashima, T. (2001). Spontaneous self-assembly of glycolipid bilayer membranes in sugar-philic ionic liquids and formation of ionogels. *Langmuir* 17: 6759–6761.

41 Nakashima, T. and Kimizuka, N. (2002). Vesicles in salt: formation of bilayer membranes from dialkyldimethylammonium bromides in ether-containing ionic liquids. *Chem. Lett.* 1018–1019.

42 Nakashima, T. and Kimizuka, N. (2012). Controlled self-assembly of amphiphiles in ionic liquids and the formation of ionogels by molecular tuning of cohesive energies. *Polym. J.* 44: 665–671.

43 Ikeda, A., Sonoda, K., Ayabe, M. et al. (2001). Gelation of ionic liquids with a low molecular-weight gelator showing T-gel above 100 degrees C. *Chem. Lett.* 1154–1155.
44 Nakashima, T. and Kimizuka, N. (2003). Interfacial synthesis of hollow TiO_2 microspheres in ionic liquids. *J. Am. Chem. Soc.* 125: 6386–6387.
45 Nakashima, T. and Kimizuka, N. (2011). Water/ionic liquid interfaces as fluid scaffolds for the two-dimensional self-assembly of charged nanospheres. *Langmuir* 27: 1281–1285.
46 Morikawa, M.-A., Takano, A., Tao, S., and Kimizuka, N. (2012). Biopolymer-encapsulated protein microcapsules spontaneously formed at the ionic liquid–water interface. *Biomacromolecules* 13: 4075–4080.
47 Branco, L.C. and Pina, F. (2009). Intrinsically photochromic ionic liquids. *Chem. Commun.* 6204–6206.
48 Kawai, A., Kawamori, D., Monji, T. et al. (2010). Photochromic reaction of a novel room temperature ionic liquid: 2-phenylazo-1-hexyl-3-methylimidazolium bis(pentafluoroethylsulfonyl)amide. *Chem. Lett.* 39: 230–231.
49 Zhang, S.G., Liu, S.M., Zhang, Q.H., and Deng, Y.Q. (2011). Solvent-dependent photoresponsive conductivity of azobenzene-appended ionic liquids. *Chem. Commun.* 47: 6641–6643.
50 Norikane, Y., Hirai, Y., and Yoshida, M. (2011). Photoinduced isothermal phase transitions of liquid-crystalline macrocyclic azobenzenes. *Chem. Commun.* 47: 1770–1772.
51 Uchida, E., Sakaki, K., Nakamura, Y. et al. (2013). Control of the orientation and photoinduced phase transitions of macrocyclic azobenzene. *Chem. Eur. J.* 19: 17391–17397.
52 Hoshino, M., Uchida, E., Norikane, Y. et al. (2014). Crystal melting by light: X-ray crystal structure analysis of an azo crystal showing photoinduced crystal-melt transition. *J. Am. Chem. Soc.* 136: 9158–9164.
53 Akiyama, H. and Yoshida, M. (2012). Photochemically reversible liquefaction and solidification of single compounds based on a sugar alcohol scaffold with multi azo-arms. *Adv. Mater.* 24: 2353–2356.
54 Okui, Y. and Han, M.N. (2012). Rational design of light-directed dynamic spheres. *Chem. Commun.* 48: 11763–11765.
55 Baluschev, S., Miteva, T., Yakutkin, V. et al. (2006). Up-conversion fluorescence: noncoherent excitation by sunlight. *Phys. Rev. Lett.* 97: 143903.
56 Cheng, Y.Y., Khoury, T., Clady, R. et al. (2010). On the efficiency limit of triplet–triplet annihilation for photochemical upconversion. *Phys. Chem. Chem. Phys.* 12: 66–71.
57 Gray, V., Dzebo, D., Abrahamsson, M. et al. (2014). Triplet–triplet annihilation photon-upconversion: towards solar energy applications. *Phys. Chem. Chem. Phys.* 16: 10345–10352.
58 Kim, J.H. and Kim, J.H. (2012). Encapsulated triplet–triplet annihilation-based upconversion in the aqueous phase for sub-band-gap semiconductor photo-catalysis. *J. Am. Chem. Soc.* 134: 17478–17481.

59 Liu, Q., Yin, B.R., Yang, T.S. et al. (2013). A general strategy for biocompatible, high-effective upconversion nanocapsules based on triplet–triplet annihilation. *J. Am. Chem. Soc.* 135: 5029–5037.

60 Monguzzi, A., Tubino, R., Hoseinkhani, S. et al. (2012). Low power, non-coherent sensitized photon up-conversion: modelling and perspectives. *Phys. Chem. Chem. Phys.* 14: 4322–4332.

61 Singh-Rachford, T.N. and Castellano, F.N. (2010). Photon upconversion based on sensitized triplet–triplet annihilation. *Coord. Chem. Rev.* 254: 2560–2573.

62 Vadrucci, R., Weder, C., and Simon, Y.C. (2015). Organogels for low-power light upconversion. *Mater. Horiz.* 2: 120–124.

63 Singh-Rachford, T.N., Lott, J., Weder, C., and Castellano, F.N. (2009). Influence of temperature on low-power upconversion in rubbery polymer blends. *J. Am. Chem. Soc.* 131: 12007–12014.

64 Turro, N.J., Ramamurthy, V., and Scaiano, J.C. (2010). *Modern Molecular Photochemistry of Organic Molecules*. Sausalito, CA: University Science Books.

65 Monguzzi, A., Tubino, R., and Meinardi, F. (2008). Upconversion-induced delayed fluorescence in multicomponent organic systems: role of Dexter energy transfer. *Phys. Rev. B* 77: 155122.

66 Lopes, J.N.A.C. and Pádua, A.A.H. (2006). Nanostructural organization in ionic liquids. *J. Phys. Chem. B* 110: 3330–3335.

67 Pádua, A.A.H., Gomes, M.F.C., and Lopes, J.N.A.C. (2007). Molecular solutes in ionic liquids: a structural, perspective. *Acc. Chem. Res.* 40: 1087–1096.

68 Weingärtner, H. (2008). Understanding ionic liquids at the molecular level: facts, problems, and controversies. *Angew. Chem. Int. Ed.* 47: 654–670.

69 Pott, T. and Méléard, P. (2009). New insight into the nanostructure of ionic liquids: a small angle X-ray scattering (SAXS) study on liquid tri-alkyl-methyl-ammonium bis(trifluoromethanesulfonyl)amides and their mixtures. *Phys. Chem. Chem. Phys.* 11: 5469–5475.

70 Hisamitsu, S., Yanai, N., and Kimizuka, N. (2015). Photon-upconverting ionic liquids: effective triplet energy migration in contiguous ionic chromophore arrays. *Angew. Chem. Int. Ed.* 54: 11550–11554.

71 Hisamitsu, S., Yanai, N., Kouno, H. et al. (2018). Two-dimensional structural ordering in a chromophoric ionic liquid for triplet energy migration-based photon upconversion. *Phys. Chem. Chem. Phys.* 20: 3233–3240.

72 Kimizuka, N. and Kunitake, T. (1989). Specific assemblies of the naphthalene unit in monolayers and the consequent control of energy transfer. *J. Am. Chem. Soc.* 111: 3758–3759.

73 Nakashima, T. and Kimizuka, N. (2002). Light-harvesting supramolecular hydrogels assembled from short-legged cationic L-glutamate derivatives and anionic fluorophores. *Adv. Mater.* 14: 1113–1116.

8

Organic Light Emitting Diodes with Liquid Emitters

Jean-Charles Ribierre[1,2], Jun Mizuno[2,3], Reiji Hattori[4], and Chihaya Adachi[1,2]

[1] *Kyushu University, Center for Organic Photonics and Electronics Research (OPERA), 744 Motooka, Nishi, Fukuoka 819-0395, Japan*
[2] *Kyushu University, Japan Science and Technology Agency (JST), ERATO, Adachi Molecular Exciton Engineering Project, 744 Motooka, Nishi, Fukuoka 819-0395, Japan*
[3] *Waseda University, Institute of Nanoscience and Nanotechnology, 513 Wasedatsurumakicho, Shinjuku, Tokyo 162-0041, Japan*
[4] *Kyushu University, Art, Science and Technology Center for Cooperative Research (KASTEC), 6-1 Kasugakoen, Kasuga, Fukuoka 816-8580, Japan*

8.1 Introduction

During the past decades, organic light-emitting diodes (OLEDs) have attracted considerable research interest owing to their potential for the next-generation flat-panel displays and low-cost solid-state lighting. Unlike other conventional light sources, OLEDs have indeed some outstanding advantages including surface emitting, ease for large-area manufacturing, suitability for flexible and transparent display applications, low energy consumption, and the potential to be low-cost. In fact, the first observation of organic electroluminescence was reported in 1969 from anthracene single crystal, but the high operating voltage required to turn on these devices prevented at that time the possibility to think about the development of any practical applications [1]. In 1987, the first breakthrough in the OLED field was the realization of an efficient low-voltage operating OLED based on an organic bilayer heterostructure by Kodak [2]. One year later, the first trilayer OLED structure was proposed in which a hole transport layer (HTL), an emissive layer (EML), and an electron transport layer (ETL) were sandwiched between two electrodes in order to facilitate charge injection and transfer into the emissive region of the device [3]. Another important finding was the first demonstration of conjugated polymer light-emitting diodes (LEDs) fabricated via a solution-processing method in 1990 by Burroughes et al. [4] The external quantum efficiency (EQE) of this first generation of OLEDs was however still low. This parameter can be expressed as the product of the solid-state photoluminescence quantum yield (PLQY), the charge balance factor defined as the number of excitons generated within the device per injected charge carrier, the exciton branching ratio, and the light-outcoupling efficiency. Excitons produced via electrical pumping may exist as singlets or as triplets,

Functional Organic Liquids, First Edition. Edited by Takashi Nakanishi.
© 2019 Wiley-VCH Verlag GmbH & Co. KGaA. Published 2019 by Wiley-VCH Verlag GmbH & Co. KGaA.

depending on the spins of the injected charge carriers. Based on spin statistics, only 25% of the excitations generated under electrical injection of charge carriers are singlets and the 75% remaining excitons are triplets. In the case of the first generation of OLEDs based on conventional fluorescent materials, this spin statistic implies that 75% of the injected charges are lost, resulting in an upper limit value for the EQE of about 5%. One of the major discoveries in the OLED research field was then the demonstration of phosphorescent OLEDs in 1998, which increased the upper limit of the internal quantum efficiency from 25% to 100% using phosphorescent heavy metal complexes [5]. Heavy atoms such as iridium or platinum can increase considerably spin–orbit interactions within a molecule which leads to a situation where the relaxation of the triplets to the ground state becomes partially allowed. Following these pioneering works, a large variety of phosphorescent dyes was synthesized and used in high-efficiency OLEDs emitting in the blue, green, and red region of the visible electromagnetic spectrum [6]. Although external quantum efficiencies higher than 35% have been reported in phosphorescent OLEDs [7], the shortage and high cost of heavy atoms such as iridium are strong disadvantages for their use in commercialized products. In order to overcome the spin statistics of fluorescent materials without the use of heavy metals, the most successful recent approach has been based on the use of thermally activated delayed fluorescent (TADF) materials as emitters [8]. The high efficiency of this third generation of OLEDs is due to an effective upconversion from triplets to singlets using thermal energy, which allows the triplet excitons to contribute to the electroluminescence. In the past few years, a large variety of TADF molecules have been reported and have been successfully utilized in OLEDs, with performances comparable to those obtained in phosphorescent devices [9–11].

Although OLEDs based on organic solid thin films have strong merits for flat-panel display and lighting applications, there are several issues that still need to be sorted out, especially in terms of the device lifetime, the backplane stability and uniformity, as well as the production yield and cost. One promising approach to resolve these problems is based on the development of OLEDs with a solvent-free liquid organic light-emitting layer. Such liquid devices exhibit some unique features compared to conventional solid-state OLEDs. The use of liquid organic semiconductors is expected to provide mechanical flexibility and robustness to the light-emitting displays since liquid materials prevent detachment and peeling of the active layer from the electrodes even under sharp and extreme folding. Another significant advantage of this class of organic semiconductors is the possibility of recovering the emission from the active layer by replacing the decomposed liquid emitter with a fresh material without breaking the device. Although liquid OLEDs combined with microfluidic circuitry show great promise for the development of degradation-free organic light-emitting displays, it should be mentioned that the development of this technology is still at a primitive stage and further investigations are needed to improve our understanding of their fundamental characteristics and enhance their electroluminescence performances.

In this chapter, we provide an overview of the current state of the art of the liquid OLED technology. After a brief review of the fundamentals about conventional solid-state OLEDs, the engineering and physics of liquid OLEDs as well as the liquid materials used in these devices are described. We also report on recent research works that are being developed to improve the performance of liquid OLEDs in an application-oriented view. In particular, we show the possibility of combining liquid OLEDs with microfluidic technology to fabricate large-area flexible microfluidic OLEDs, multicolor light-emitting electroluminescent devices, as well as white liquid OLEDs.

8.2 Organic Light-emitting Diodes with a Solvent-Free Liquid Organic Light-emitting Layer

8.2.1 Basics of Conventional Solid-state OLEDs

A conventional solid-state OLED structure consists of single or multiple organic layers sandwiched between an anode and a cathode electrode. At least one of these two electrodes should be transparent for an effective light outcoupling. For this purpose, in most cases, indium tin oxide (ITO) glass substrates are used as transparent anode. The total thickness of the organic layer(s) in OLEDs is typically around 100–150 nm. When a bias voltage is applied across the device, holes and electrons are injected into the organic layer(s) from the anode and cathode, respectively. The charge carriers are then transported in the organic materials generally via a hopping process and recombine to form excitons. Some of these excitons then decay radiatively to emit photons. The color of the emission depends on the energy difference between the highest occupied molecular orbital (HOMO) and the lowest unoccupied molecular orbital (LUMO) levels of the light-emitting molecules. The emission spectrum can thus be controlled by the extent of the conjugation length of the organic emitter. For an efficient hole and electron injection into the organic layer(s), the work function of the electrode materials must be carefully selected to lower the injection energy barrier as much as possible. For this purpose, while ITO is generally chosen for the anode, low work function metals such as Al and Ca are typically preferred for the cathode. An essential parameter for achieving high electroluminescence efficiency is the charge balance factor defined as the number of excitons formed within the device per injected charge carrier. This factor may approach unity when a multilayer structure is used, typically containing an EML between HTL and ETL with the appropriate HOMO and LUMO levels to ensure exciton formation and radiative recombination in the EML.

Electroluminescent organic materials are often divided into two types: small molecules and oligomers/polymers. While small molecules are generally deposited into thin films by thermal evaporation, polymers/oligomers are processed from solutions using spin-coating or inkjet printing techniques. It should be noticed that small-molecule devices can have more sophisticated multilayer structures than polymer LEDs, due to the fact that orthogonal solvents should be used to prepare a multilayer stack by a wet process. In fact, in both cases, the main requirements for efficient OLED materials are high PLQY in film, good charge mobilities (for both types of carriers), excellent film-forming properties, good thermal and oxidative stability and, if possible, good color purity. As is shown here, some of these requirements can be fulfilled using solvent-free liquid organic semiconductors.

8.2.2 First Demonstration of a Fluidic OLED Based on a Liquid Carbazole Host

The first demonstration of a solvent-free liquid OLED used a 9-(2-ethylhexyl) carbazole (EHCz), the so-called liquid carbazole, as liquid host material [12]. The chemical structure of this molecule is shown in Figure 8.1a. This material has a glass transition temperature well below 0 °C and is liquid at room temperature. In addition, this liquid organic molecule is transparent in the visible region of the spectrum, emits blue fluorescence, and shows hole mobility of about 4×10^{-6} cm^2 V^{-1} s^{-1} under an electric field of 2.5×10^5 V cm^{-1} [13]. In order to achieve electroluminescence, the liquid carbazole host was doped with about 1 wt% of rubrene. This light-emitting guest molecule was selected due to its PLQY near unity in highly diluted systems and its ability to transport electrons. A schematic representation of the device structure is displayed in Figure 8.1a. In such architecture, poly(3,4-ethylenedioxythiophene):poly(styrenesulphonate) (PEDOT:PSS) and Cs$_2$CO$_3$ are used as hole and electron injection layers, respectively. The fabrication method is summarized in Figure 8.1b. A glass substrate with a row-by-row patterned ITO electrode was prepared using a conventional photolithography technique. After cleaning, a conductive layer of PEDOT:PSS was spin coated on top of ITO and then annealed at 120 °C for 30 minutes. Ag spacers with a thickness of 80 nm were then thermally evaporated through a shadow mask. In parallel, Cs$_2$CO$_3$ layer was deposited by spin coating on another precleaned ITO glass substrate and then annealed at 150 °C for one hour in a nitrogen-filled glove box. Following the fabrication of the anode and cathode substrates, a droplet of the liquid active organic material was placed on top of PEDOT:PSS film, covered with the cathode substrate, and clipped tightly. The thickness of the active layer was estimated to be typically in the range between 200 and 300 nm.

The current density–voltage–luminance (J–L–V) and the EQE–J characteristics of a representative liquid OLED are shown in Figure 8.1c. The current density is found to increase linearly with voltage, indicating that there is no space charge layer formed in the liquid layer. Since the liquid carbazole is a hole transport material [13], the onset of the electroluminescence corresponds to the onset of electron injection in the blend. The highest luminance and EQE measured in this device were 0.35 cd m^{-2} and 0.03%, respectively. From the comparison of the photoluminescence and electroluminescence spectra (data not shown here [12]), it was concluded that holes and electrons directly recombine on rubrene molecules to form excitons. This scenario is well supported by the HOMO and LUMO levels of EHCz and rubrene, as depicted in the inset of Figure 8.1c. While the performance of this first liquid OLED was quite poor, especially if we compare it with the current state-of-the-art efficiency of solid-state OLEDs, this pioneering study demonstrated that liquid OLEDs can be an alternative for the development of light-emitting applications.

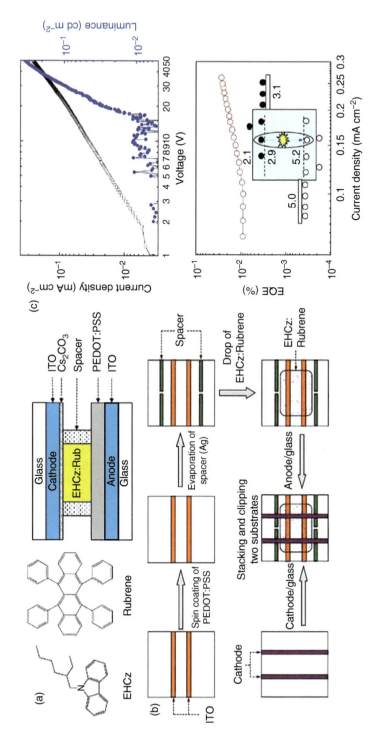

Figure 8.1 (a) Chemical structure of liquid carbazole and rubrene. A schematic representation of the liquid OLED structure is also displayed. (b) Device fabrication procedure. (c) Current density–voltage–luminance (J–L–V) and EQE–J characteristics. The proposed mechanism of device operation is shown in inset. Source: Reproduced with permission from Xu and Adachi [12] Copyright 2009, AIP Publishing.

8.2.3 Introduction of an Electrolyte to Improve the Liquid OLED Performance

The performance of liquid OLEDs based on a liquid carbazole host was considerably enhanced by incorporating an electrolyte into the liquid-emitting layer and a titanium dioxide (TiO_2) hole-blocking layer (HBL) [14]. Liquid OLEDs with the following architectures: ITO/PEDOT:PSS/Liquid EML/TiO_2/ITO were fabricated where the liquid EML is a blend containing EHCz as liquid host, 16.7 wt% of 6-{5-[3-methyl-4-(methyl-octyl-amino)-phenyl]-thiophen-2-yl}-naphthalene-2-carboxylic acid hexyl ester (BAPTNCE) as emitter, and 0.1 wt% of tetrabutylammonium hexafluorophosphate (TBAHFP) as electrolyte. Chemical and device structures with the energy diagram are represented in Figure 8.2a,b. The absorption and PL spectra show that an efficient Förster-type energy transfer of singlet excitons can occur from the liquid host to the guest molecules. In addition, the EML was found to show green emission with a PLQY of 55%. The J–L–V curves in Figure 8.2c demonstrate that the PEDOT:PSS layer improves the hole injection in the liquid OLEDs. However, device 2 (ITO/PEDOT:PSS/Liquid EML/ITO) was found to not show any electroluminescence due to a large energy barrier for electron injection. As shown in Figure 8.2c,d, doping the EML with an electrolyte resulted in a reduction of the driving voltage and led to intense electroluminescence. Very importantly, these results demonstrated that the incorporation of an electrolyte in the liquid light-emitting layer provides ambipolar charge injection and transport, which is essential to achieve efficient electroluminescence. Regarding the mechanism, electrolyte anions and cations can move to the anode and cathode sides, respectively, when an applied electric field is applied across the liquid material. This leads to the formation of electric dipole layers at both electrode interfaces, resulting in a reduction of the injection barrier and a significant reduction of the turn-on voltage.

As explained, the use of multilayer structures has been used successfully to improve the charge-transport properties of solid-state OLEDs. Because EHCz is a hole transport material, the use of an HBL between the liquid EML and the cathode should be highly beneficial to the liquid OLED performances. A TiO_2 layer cannot be dissolved by the liquid material, show unipolar electron transport and provides a large energy barrier between the HOMO of 5.8 eV of EHCz and its valence band of 7.5 eV. Figure 8.2e shows the influence of such a TiO_2 layer on the liquid OLED properties. While the J–V curve was not significantly modified by the insertion of the HBL, the maximum luminance and EQE values were substantially enhanced, due to an improvement in the charge-carrier balance. Although the electron energy barrier is still high between EHCz and TiO_2, it can be seen that the use of HBL resulted in an increase in the EQE value. Finally, in Figure 8.2f, the maximum EQE value from the liquid OLEDs is plotted against the thickness of the TiO_2 HBL. For thickness larger than 10 nm, EQE was found to saturate to the value of 0.31%. The enhancement of the liquid OLED performance was found, however, to be limited due to an additional exciton quenching at the TiO_2 interface [14].

The incorporation of an electrolyte was found to dramatically improve the performance of liquid OLEDs. It is worth noting, however, that an improvement

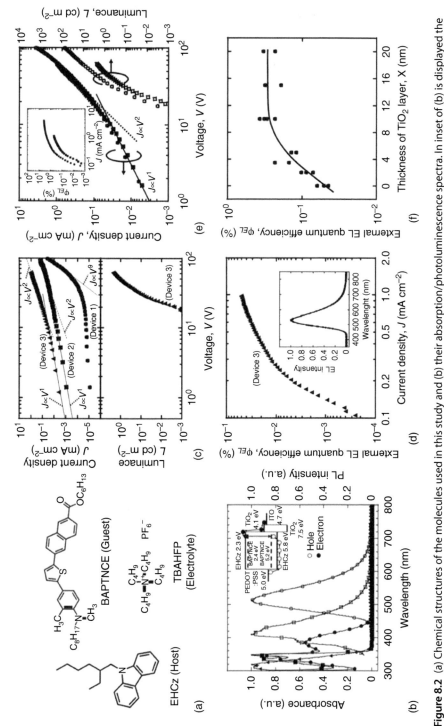

Figure 8.2 (a) Chemical structures of the molecules used in this study and (b) their absorption/photoluminescence spectra. In inset of (b) is displayed the device architecture and its energy diagram. Influence of the uses of PEDOT:PSS and an electrolyte on the (a) J–L–V and (b) EQE–J characteristics. Device 1: ITO/EML without electrolyte/ITO; device 2: ITO/PEDOT:PSS/EML without electrolyte/ITO; and device 3: ITO/PEDOT:PSS/EML with electrolyte/ITO. The electroluminescence spectrum of 3 is shown in inset of (d). (e) Effects of the insertion of a TiO$_2$ hole-blocking layer (HBL) on the J–L–V and EQE–J characteristics. Squares and circles correspond to data obtained with and without 10-nm-thick HBL. (f) Effect of the HBL thickness on the EQE value. Source: Reproduced with permission from Hirata et al. 2011 [14]. Copyright 2011, Wiley-VCH.

in the charge-carrier injection properties was also obtained by the molecular engineering of a new liquid carbazole material. Kubota et al. synthesized a liquid carbazole substituted with a poly(ethylene oxide) (PEO) chain [15]. Due to the orientation of the large dipole of PEO units, formation of electric double layers took place at the electrode interfaces, resulting in an efficient carrier injection into a liquid organic semiconductor without addition of electrolyte. The fabricated device using this liquid host material exhibited a maximum EQE value of 0.75%, which is 25 times higher than that obtained in the first liquid OLED study [12].

8.2.4 Liquid OLED Material Issues

While the performances of liquid OLEDs based on EHCz host were improved by introducing an electrolyte into the liquid EML and by inserting an HBL between the liquid EML and the cathode [14], these devices still show some problems including low maximum luminance, low EQE, high driving voltage, and short device lifetime. In particular, the short device lifetime of these devices could be due to the instability of radical cations in the liquid light-emitting blends. In that context, it was necessary to develop novel liquid π-conjugated host materials and liquid emitters with improved semiconducting properties and excellent electrochemical stability. For instance, Hirata et al. synthesized two liquid carbazole-based host compounds: 9-{2-[2-(2-methoxyethoxy)ethoxy]ethyl}-9H-carbazole (TEGCz) and 9,90-{2-[2-(2-methoxyethoxy)ethoxy]ethyl}-3,30-bis(9H-carbazole) [(TEGCz)2] as well as one liquid light-emitting guest: 3-phenylpropionic acid 2-ethylhexyl ester (PLQ) [16]. Their results provide clear evidence that the device lifetime of liquid OLEDs is strongly improved when electrochemically stable molecules are used as liquid host. The introduction of stable radical cations in the liquid host was obtained by the dimerization of the liquid carbazole such as (TEGCz)2. In addition, it was shown that the use of a guest light-emitting compound with a HOMO level energy higher than that of the electrochemically stable host can lead to a substantial improvement in the device lifetime without decreasing the electroluminescence efficiency and luminance.

To improve our understanding of the degradation mechanism in liquid organic semiconductors, Fukushima et al. carried out nuclear magnetic resonance (NMR) spectroscopy measurements in TEGCz and (TEGCz)2 [17]. ^{1}H and DOSY NMR experiments clarified the molecular reactions during the operation of the devices and enabled to identify the characteristics of the degraded materials. The results demonstrated that the degradation of liquid carbazole–based OLEDs is essentially due to a dimerization reaction. Other possible chemical reactions involving the ethylene glycol chains as well as other C—C and C—N bonds in the carbazole moiety were not observed indeed. In addition, this study suggested that the introduction of substituents groups at the para position of the carbazole unit could be an effective way to improve the device lifetime.

It should be finally mentioned that a series of solvent-free liquid fluorene derivatives with siloxane side chains was recently reported [18]. These compounds show high PLQY, good electrochemical stability, good ambipolar charge mobilities, and excellent lasing properties. However, OLEDs based on these fluorene liquids did not show good electroluminescent performances, mainly due to the LUMO of these materials being around 2.8 eV, resulting in a poor electron injection from the cathode even with the use of an electrolyte. It is evident, however, that the future of liquid OLEDs necessitates the development of new host and guest liquid organic materials with suitable physical and electrochemical features. Since solvent-free liquid organic semiconductors functionalized with siloxane side chains can compete in terms of photophysical and charge-transport properties with organic glassy semiconductors, this approach based on the high-fluidity-promoting character of siloxane chains, which makes possible solid-to-liquid transformation of quite large molecular systems using only a low siloxane content, seems to be very promising for the development of a new generation of high-efficiency liquid organic electroluminescent devices.

8.3 Microfluidic OLEDs

8.3.1 Refreshable Liquid Electroluminescent Devices

The replacement of locally decomposed liquid light-emitting materials with fresh materials by capillarity, enabling to maintain the original emission characteristics of liquid OLEDs, was first demonstrated by Hirata et al. [14] However, due to the high viscosity of the liquid active material at room temperature, the replacement of the viscous liquid in the devices took a significant long time. While the demonstration of electroluminescence based on a moving liquid emitter already proved the potential of this approach for light-emitting applications, the development of novel device structures allowing a fast emission recovery by replacement of the liquid emitter and uniform emission over large area was still clearly required. To reach this objective, Shim et al. proposed and fabricated a new device structure with a mesh-structured cathode and a liquid reservoir at the back of the cathode [19]. The schematic representation of the device architecture and the fabrication method are provided in Figure 8.3a. In such a structure, the liquid material can be circulated through the holes of the mesh cathode. Another important feature of this device is the uniform small gap between the electrodes at a submicron level over a large area, which should enable efficient charge injection into the liquid active material at low applied bias voltages. As explained in Figure 8.3a, the structure was prepared by a multistep microelectromechanical system processing. Patterned ITO was used as the anode and Al as cathode materials. After completion of the empty device, the liquid active material was dropped into the structure and filled the gap between the electrodes via capillarity effect. The liquid material used in this work was 2-[2-(2-methoxyethoxy)ethoxy]ethyl 4-(pyren-1-yl)butanoate (PLT) doped with 0.25% (mass fraction) of the electrolyte tributylmethylphosphonium bis (trifluoromethanesulfonyl)imide (TMP-TFSI). The J–L–V characteristics measured in a representative liquid OLED are shown in Figure 8.3b. Electroluminescence was observed above 5 V, indicating that both types of charge carriers can be injected above this voltage. The maximum EQE value of this device was around 0.2%. The

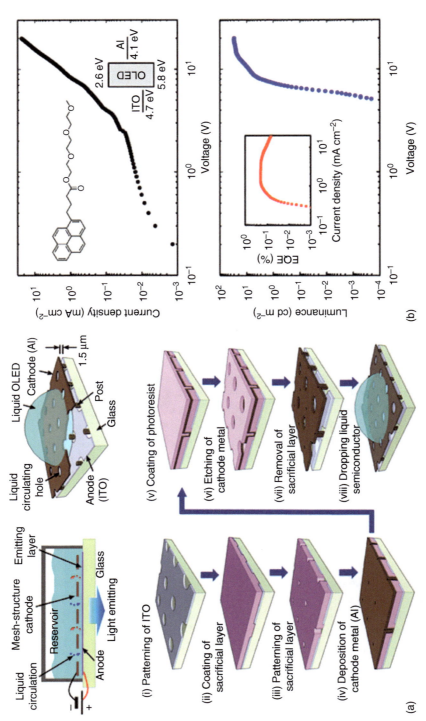

Figure 8.3 (a) Architecture and fabrication method of a liquid OLED with a mesh-structured cathode and a reservoir at the back. (b) *J–V–L* and EQE–*J* characteristics of a representative liquid OLED with the mesh-structured cathode and a reservoir at the back. The structure of the liquid emitter and the energy diagram of the device are displayed in the inset. Source: Reproduced with permission from Shim et al. [19]. Copyright 2012, AIP Publishing.

luminance of the device was then found to decrease under a constant driving current density because of a degradation of the liquid active material. However, it was shown that the emission could be quickly recovered to the initial state using a convectional circulation of the liquid emitter through the c cathode.

8.3.2 Fabrication of Microfluidic Organic Light-Emitting Devices

Microfluidic technology has been successfully used in the past decade for a broad range of applications including chemical and biological analysis and lab-on-a-chip devices. It can be anticipated that the next generation of microfluidic devices will require the incorporation of electrodes in the microchannels. In that context, Kasahara et al. proposed and fabricated the first prototype microfluidic OLED combining liquid OLED with microfluidic technologies [20]. For this purpose, a microfluidic chip based on the negative photoresist SU-8 and containing ITO anode and cathode electrodes was fabricated using photolithography, vacuum ultraviolet treatment and low-temperature bonding techniques. The design of this prototype device and its fabrication process are schematically represented in Figure 8.4A, B. The liquid light-emitting material was PLQ doped with 0.25 wt% of the electrolyte TMP-TFSI. Fresh doped PLQ was then continuously injected from the inlets into the active device areas through the microchannels using syringe pumps. As shown in Figure 8.4C, blue electroluminescence was produced by the radiative recombination of excitons in the liquid microchannels using the appropriate applied bias voltages. After passing through the microchannels, the liquid material was collected at the outlets. The continuous injection of the liquid emitters was found to not affect the efficiency of the device. Overall, this first demonstration of a microfluidic OLED indicates that this technology has considerable potential for future organic light-emitting display applications.

8.3.3 Large-Area Flexible Microfluidic OLEDs

In contrast to conventional solid-state OLEDs, liquid light-emitting organic layers used in microfluidic OLEDs are formed by filling the gaps between anode and cathode electrodes with the liquid semiconductor. This kind of device structure should provide excellent mechanical flexibility since the liquid material can change its shape and adapt perfectly well to the flexible device. To be one step ahead in the development of microfluidic OLED technology, the demonstration of highly flexible large-area liquid-based devices was required. In addition, it was also essential to find an improved method for patterning liquid emitters on a single chip. For this purpose, Tsuwaki et al. reported on a simple method based on four different processes to fabricate large-area flexible microfluidic OLED [21]. A schematic representation of this device is displayed in Figure 8.5A. A screen-printing technique with wet etching was first used for the deposition of ITO onto flexible polyethylene terephthalate (PET) substrates. In the second step, a new photolithography method, the so-called the belt-transfer exposure technique, was used for the patterning of large-area microchannels. Then, for the third step, heterogeneous low-temperature bonding was used through the

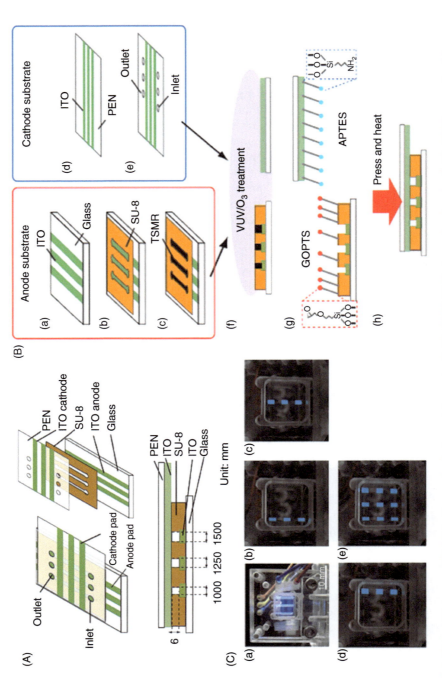

Figure 8.4 (A) Architecture and (B) fabrication method of a prototype microfluidic OLED. (C) Photographs of a microfluidic OLED generating blue electroluminescence from the microchannels. Source: Reproduced with permission from Kasahara et al. [20]. Copyright 2013, Elsevier.

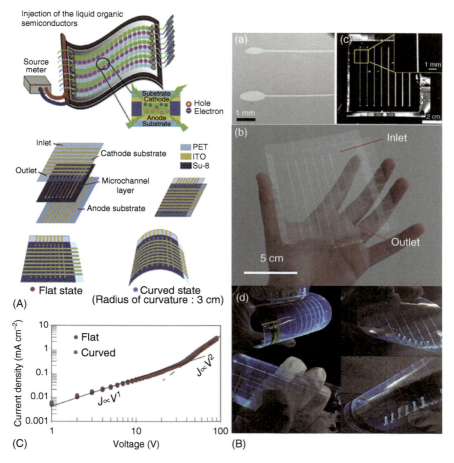

Figure 8.5 (A) Concept and design of a large-area flexible microfluidic OLED. (B) (a) SEM image of the SU-8 microchannel layer, (b) photograph image of the microchannel, (c) scanning acoustic microscope image of the bonded interface, and (d) photographs of the fabricated large-area flexible microfluidic OLED under UV irradiation. (C) J–V characteristics of the flexible OLED in flat and curved states under stopped flow conditions. Source: Reproduced with permission from Tsuwaki et al. [21]. Copyright 2014, Elsevier.

utilization of amine- and epoxy-terminated self-assembled monolayers. Finally, the liquid emitter (PLQ doped with 0.25 wt% of the electrolyte TMP-TFSI) was injected into the fabricated flexible microchannels. As shown by the scanning electron microscope (SEM) image in Figure 8.5B-a, the microchannel layer was successfully prepared onto the flexible PET substrate without any exposure failure. All the microchannels were formed as expected and without destruction as confirmed by the photograph and the scanning acoustic microscope image in Figures 8.5B-bc. Finally, the photographs showing the microfluidic OLED under ultraviolet (UV) irradiation after the liquid active material was injected (Figure 8.5B-d) provide evidence that the liquid was injected from the inlets to the outlets without leakage, even when the device was curved.

The electroluminescence properties of the large-area flexible microfluidic OLED were then characterized in both flat and curved states. The results including the J–V curves in Figure 8.5C provided evidence that charge-carrier injection and recombination occurred in both states. It is also worth noting that this device offered stable electroluminescence even in the high-voltage region up to 100 V, indicating that the device microchannel structure is sufficiently strong to not be affected by the applied high voltages. Maximum luminance value of this flexible device was around 5.5 cd m^{-1} at 100 V. Such a low luminance value and the high driving voltage were attributed to the large thickness of the liquid active layer (around 4.5 μm). While the demonstration of a large-area flexible microfluidic OLED is a relevant achievement, smaller gap of liquid-emitting layer and deposition of extra organic materials on anode and cathode are certainly necessary to boost the overall performance of this kind of devices.

8.3.4 Multicolor Microfluidic OLEDs

Demonstration of multicolor microfluidic OLEDs on the same chip is evidently essential for providing evidence of the potential of this technology for display applications. In fact, there were some challenging issues related to multicolor electroluminescence in liquid OLEDs. First, there are only few fluidic materials available for use as the emitting layer. Second, liquid OLEDs were designed to have only a single liquid-emitting layer sandwiched between two electrode-coated glass substrates. To obtain such a multicolor liquid OLED, Kasahara et al. proposed to use PLQ liquid as blue emitter in the neat layer and also as host doped with green, yellow, and red fluorescent guest molecules (2 wt% of 5,12-diphenyltetracene (DPT) for green, 2 wt% of 5,6,11,12-tetraphenyltetracene (rubrene) for yellow, and 0.4 wt% of tetraphenyldibenzo periflanthene (DBP) for red emission) [22]. Chemical structures of these compounds are shown in Figure 8.6A. The light-emitting liquids were also doped with 0.25 wt% of the electrolyte TMP-TFSI. It is important to mention that the overlap between the emission spectrum of PLQ and the absorption spectra of these dyes indicate that an efficient Förster-type energy transfer can take place from the liquid host to the emitters. In parallel, the fabricated devices had a 3 × 3 matrix of light-emitting pixels in SU-8 microchannels sandwiched between ITO glass substrate and amine-terminated self-assembled monolayer-coated ITO deposited onto PET substrate. A schematic representation of this device architecture is provided in Figure 8.6A. Note that the anode and cathode substrates were prepared separately using photolithography and wet etching processes. Heterogeneous bonding techniques were then used to prepare the sandwiched device structure before filling it with the liquid light-emitting materials. It is also worth noting that the work functions of the untreated ITO and self-assembled monolayer-treated ITO were measured using a photoelectron spectroscopy technique. The results indicated that the modified ITO cathode had a lower work function due to the presence of amine groups in the self-assembled monolayer.

The photographs shown in Figure 8.6B confirm that an efficient energy transfer occurs from the liquid host to the different emitters. They also provide evidence that the prepared emitting materials remain in their liquid phase after the

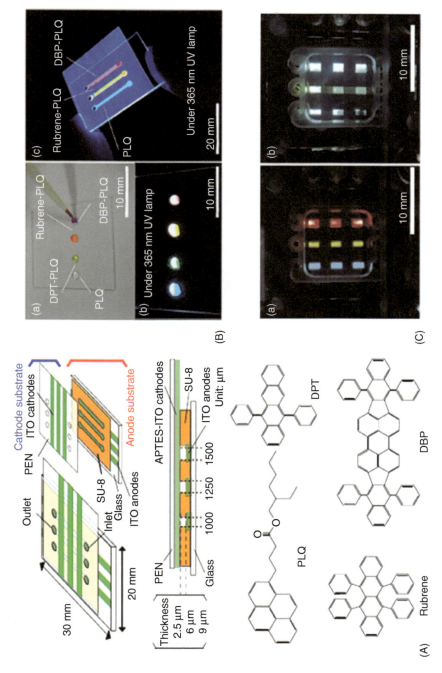

Figure 8.6 (A) Schematic representation of the multicolor liquid OLED structure containing SU-8 microchannels. The chemical structures of the materials used in this study are also displayed. (B) Photographs of the blue-, green-, yellow-, and red-emitting liquid materials under ambient and UV illuminations. It can be seen that different liquid emitting layers can be patterned individually on a single microfluidic chip. (C) Photographs showing the multicolor electroluminescence from a microfluidic OLED device based on PLQ. (D) $J-V$ and (E) $L-V$ characteristics of the microfluidic OLED devices based on PLQ. (F) Photographs showing (a) the fast luminance degradation and (b) the fast recovery characteristics of a PLQ-based microfluidic OLED. Source: Reproduced with permission from Kasahara et al. [22]. Copyright 2015, Elsevier.

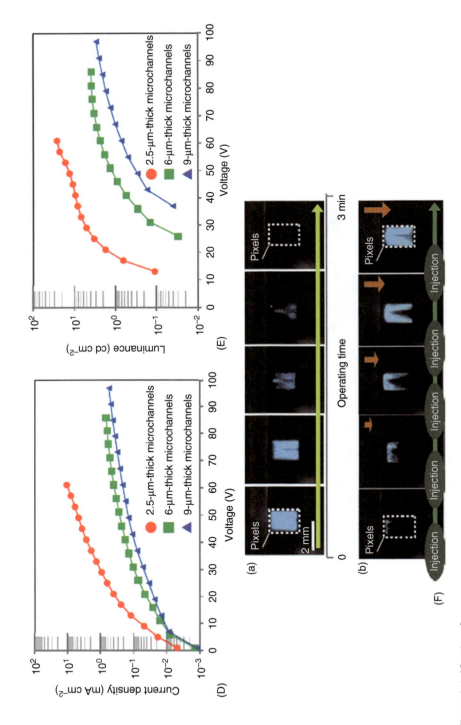

Figure 8.6 (*Continued*)

doping process. In addition, it can be seen that the liquid emitting layers were well patterned in the microfluidic OLED without any leakage. Figure 8.6C shows a photograph of a 6-μm-thick microfluidic OLED with green, yellow, and red emitting pixels operating under a DC applied voltage. This result indicates that holes and electrons were injected into the liquid layers from the ITO and modified ITO electrodes, respectively. This clearly demonstrates that the patterned liquid emitting layers could be obtained on-demand by simply injecting small amounts of liquid materials in the microchannels. It was also shown that this type of devices can be controlled using a passive matrix addressing. On the other hand, nonuniform electroluminescence was observed from the liquid OLED based on the DBP-doped PLQY green emitting blend, presumably due to a decomposition of the guest molecules under applied voltages. This supports again the need to optimize carefully the concentration of dyes in the emitting liquid layer and the need to develop new liquid materials with improved performance and stability. Regarding the mechanism of electroluminescence in these liquid devices, it should be highlighted that the electroluminescence spectra and the energy diagram of the devices suggested that charge carriers were trapped by the guest emitters and hence that electroluminescence came from the direct recombination on guest emitters.

The J–V and L–V characteristics of 2.5-, 6-, and 9-μm-thick microfluidic OLEDs based on a PLQ neat layer are displayed in Figures 8.6D,E. These curves clearly demonstrate that the current density and luminance were both improved when the thickness of the liquid emitting layer in the microchannels was decreased. Noticeably, the turn-on voltage, which was defined at the luminance of $0.01\,\text{cd}\,\text{m}^{-2}$, decreased when reducing thickness, indicating that the bulk resistance of the liquid layer decreased with the film thickness. However, the luminance remained very low compared with conventional solid-state OLEDs based on around 100-nm-thick organic multilayer structures. In that context, future studies on liquid OLEDs should focus on the fabrication of submicron-thick microchannels to achieve highly luminescent devices.

As shown in Figure 8.6F, when a constant DC voltage of 70 V was applied across the liquid OLED based on PLQ neat layer, the luminance of the device clearly decreased quickly with increasing operating time. Nearly no electroluminescence could be observed after only three minutes, which was attributed to a degradation of the PLQ layer during the charge-carrier injection. On the other hand, a fresh liquid PLQ layer could be injected manually from a syringe into the targeted microchannel under no-voltage condition. Using this method, a full recovery of the electroluminescence emission can be seen in Figure 8.6F. This important achievement suggested that such refreshable electroluminescence feature in microfluidic device could be applied in the near future for the realization of degradation-free liquid display technology.

8.3.5 Microfluidic White OLEDs

In order to generate white emission, a mixture of two complementary colors (yellow and blue, for instance) or three primary colors (blue, green, and red) has been used to achieve a broad emission spectrum covering all the visible light wavelengths. In parallel, three main OLED structures have been used to produce white electroluminescence: stacked multilayers, blended guest–host layers, and striped layers. Note that high-resolution patterns of less than hundreds of microns in width are also required for the subpixels of light-emitting displays. Such device architectures seem to be compatible with microfluidic OLEDs if sub-100-μm-wide microchannels can be formed between two electrodes.

In that context, Kobayashi et al. demonstrated a microfluidic white OLED based on integrated patterns of greenish-blue and yellow solvent-free liquid emitters [23]. The chemical structures of the molecules used in the devices are shown in Figure 8.7a. PLQ was again used either as greenish-blue liquid emitter or as liquid host doped by 2 wt% of 2,8-di-tert-butyl-5,11-bis(4-tert-butylphenyl)-6,12-diphenyltetracene (TBRb). Note that the liquid electroluminescent materials were also doped with 0.25 wt% of the electrolyte TMP-TFSI. The structure of the microfluidic white OLEDs and the fabrication method using photolithography and heterogeneous bonding technologies are also displayed in Figure 8.7a. In these devices, ITO-coated substrate and ITO-coated polyethylene naphthalate (PEN) plastic substrate were used as anode and cathode, respectively. Twelve SU-8 microchannels were prepared and sandwiched between a 3-glycidodyloxypropyltriethoxysilane (GOPTS)-treated ITO anode and aminopropyltriethoxysilane (APTES)-treated ITO cathode. The epoxy and amine terminations were employed for the heterogeneous bonding of the two substrates. The microchannels were designed to have a width of 60 μm and they were separated by a distance of 40 μm. Microchannel thickness was 6 μm. Neat PLQY and TBRb-doped PLQ were alternatively injected into the microchannels to achieve the integrated light-emitting patterns.

Based on the emission and absorption spectra of PLQ and TBRb, it can be stated that Förster-type energy transfer takes place from the liquid host to the guest molecules. Figure 8.7b shows the electroluminescence spectra of neat and doped PLQ as well as pictures of the operating liquid OLEDs, confirming the blue-greenish and the yellow emission from these liquid materials. In addition, the obtained electroluminescence spectrum of the PLQ:TBRb blend and considering as well the position of the HOMO and LUMO levels of the molecules suggest that the excitons of TBRb were not only generated by energy transfer from the host but also by direct recombination of electrons and holes trapped by the guest molecules. Noticeably, the sum of the electroluminescence spectra from PLQ and doped PLQ layers leads to white emission, as displayed in the inset of Figure 8.7b. It is also worth mentioning that the $J-V$ curves of these devices (data not shown here) confirmed that TBRb molecules act as traps. They also proved that both PLQ and doped PLQ layers in the microchannels can be controlled by the same driving voltage. Finally, it is worth noting that the PLQ and doped PLQ OLEDs exhibited a luminance value of 2.6 and 5.5 cd m^{-2} at 100 V.

A picture of the fabricated microfluidic white OLED is displayed in Figure 8.7c. The liquid emitting layers in the microchannels were obtained by injecting, alternatively, the greenish-blue and yellow emitting liquids from the inlets. This photograph of the device taken under UV illumination confirms that there was no leakage between the microchannels and fully validates the proposed fabrication methodology for realizing white highly integrated microfluidic OLEDs.

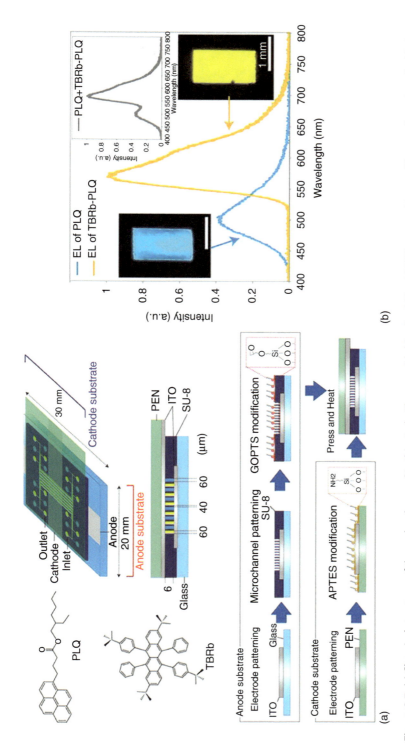

Figure 8.7 (a) Chemical structures of the molecules used in this study. The greenish-blue liquid host PLQY was used in neat film or as host doped with the yellow fluorescent guest emitter TBRb. Schematic representations of the microfluidic white OLED and its fabrication process. (b) Electroluminescence spectra of PLQ and PLQ doped with 2 wt% of TBRb in microfluidic OLED. In the inset is shown the superposition of the sum of these electroluminescence spectra leading to white emission. (c) Photograph of the microfluidic white OLED under UV irradiation. (d) Electroluminescence spectrum and photograph of the microfluidic white OLED under an applied voltage of 100 V. Source: Reproduced with permission from Kobayashi et al. [23]. Copyright 2015, Springer Nature.

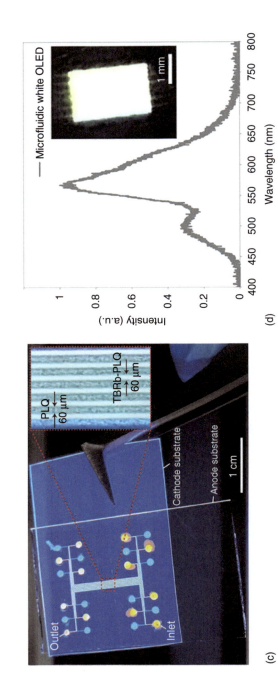

Figure 8.7 (*Continued*)

Figure 8.7d shows the electroluminescence spectrum and a photograph of the microfluidic OLED under an applied bias voltage of 100 V. These data clearly demonstrate that white electroluminescence could be generated from the microchannels. In addition, the white emission spectrum was composed of the greenish-blue emission band from PLQ and the yellow emission band from the doped PLQ, covering the visible region of the spectrum from 420 to 750 nm. The commission internationale de l'eclairage (CIE) coordinates obtained with this device at 100 V were (0.40, 0.42), nearly corresponding to a warm-white emission. Overall, the realization of this prototype microfluidic white OLED demonstrated the potential of liquid OLED technology for lighting applications.

8.4 Conclusions

Although in comparison with conventional solid-state OLEDs, the performances of solvent-free liquid OLEDs are still at an early research stage, the studies already reported in the literature demonstrate the promising potential of this new technology for display and lighting applications. The first liquid OLEDs were realized by simply sandwiching the liquid emitting layers between two glass substrates coated with electrodes. More recently, several works successfully combined liquid OLEDs with microfluidic technologies to demonstrate a new kind of organic light-emitting displays. In particular, it was shown that this new microfluidic OLED technology based on the use of liquid organic semiconductors is promising for on-demand multicolor electroluminescent devices, flexible large-area light-emitting displays, and white OLEDs. Several studies about liquid OLEDs have also clearly provided evidence of the great opportunity offered by liquid materials to realize refreshable degradation-free light-emitting devices. The liquid state of these electroluminescent materials enables indeed to replace deteriorated materials in the devices by reinjection of fresh liquids into the light-emitting pixels. The unique properties of liquid semiconductors in terms of mechanical flexibility, on-demand multicolor emission and refreshable luminance features can be seen as strong advantages of this technology over conventional solid-state OLEDs. In the current context, we can expect that microfluidic OLED technology may be applicable for future practical liquid based light-emitting display and lighting applications. However, in order to reach such ambitious objectives, two main problems must be fully overcome. First, this will necessitate the development of new liquid materials with improved charge-carrier mobilities and/or light-emitting properties. In contrast to conventional solid-state OLED materials, there are only a few liquids that have been successfully used in devices so far. The development of novel functional liquid materials should thus lead to improved electroluminescence efficiency and stability. Another important issue that needs to be sorted out is related to the thickness of the liquid active layer in the current liquid OLEDs. This thickness needs to be reduced below 1 μm in order to lower the turn-on voltage and improve the overall electroluminescent performance to a level close to that achieved with solid-state OLEDs. It is worth noting that solvent-free liquid organic materials have also been used successfully in other types of organic optoelectronic devices, including bistable memories [24] and optically pumped organic semiconductor lasers [25–27]. The development of a new generation of liquid semiconductors with improved photophysical and charge-transport properties is thus not only relevant for the realization of a true liquid OLED technology but should be also of strong interest for a variety of other optoelectronic and photonic applications.

References

1 Dresner, J. (1969). Double injection electroluminescence in anthracene. *RCA Rev.* 30: 322–334.
2 Tang, C. and VanSlyke, S. (1987). Organic electroluminescent diodes. *Appl. Phys. Lett.* 51: 913–916.
3 Adachi, C., Tokito, S., Tsutsui, T., and Saito, S. (1988). Electroluminescence in organic films with 3-layer structure. *Jpn. J. Appl. Phys.* 27: 269–271.
4 Burroughes, J.H., Bradley, D.D.C., Brown, A.R. et al. (1990). Light-emitting diodes based on conjugated polymers. *Nature* 347: 539–541.
5 Baldo, M.A., O'Brien, D.F., You, Y. et al. (1998). Highly efficient phosphorescent emission from organic electroluminescent devices. *Nature* 395: 151–154.
6 Sasabe, H. and Kido, J. (2013). Recent progress in phosphorescent organic light-emitting devices. *Eur. J. Org. Chem.* 2013: 7653–7663.
7 Kim, K.H., Lee, S., Moon, C.K. et al. (2014). Phosphorescent dye-based supramolecules for high efficiency organic light-emitting diodes. *Nat. Commun.* 5: 4769.
8 Uoyama, H., Goushi, K., Shizu, K. et al. (2012). Highly efficient organic light-emitting diodes from delayed fluorescence. *Nature* 492: 234–238.
9 Kaji, H., Suzuki, H., Fukushima, T. et al. (2015). Purely organic electroluminescent material realizing 100% conversion from electricity to light. *Nat. Commun.* 6: 8476.
10 Hirata, S., Sakai, Y., Masi, K. et al. (2015). Highly efficient blue electroluminescence based on thermally-activated delayed fluorescence. *Nat. Mater.* 14: 330–336.
11 Zhang, Q., Li, B., Huang, S. et al. (2014). Efficient blue organic light-emitting diodes employing thermally-activated delayed fluorescence. *Nat. Photon.* 8: 326–332.
12 Xu, D. and Adachi, C. (2009). Organic light-emitting diode with liquid emitting layer. *Appl. Phys. Lett.* 95: 053304.
13 Ribierre, J.C., Aoyama, T., Muto, T. et al. (2008). Charge transport properties in liquid carbazole. *Org. Electron.* 9: 396–400.
14 Hirata, S., Kubota, K., Jung, H.H. et al. (2011). Improvement of electroluminescence performance of organic light-emitting diodes with a liquid-emitting layer by introduction of electrolyte and a hole blocking layer. *Adv. Mater.* 23: 889–893.

15 Kubota, K., Hirata, S., Shibano, Y. et al. (2012). Liquid carbazole substituted with a poly(ethylene oxide) group and its application for liquid organic light-emitting diodes. *Chem. Lett.* 41: 934–936.
16 Hirata, S., Heo, H.H., Shibano, Y. et al. (2012). Improved device lifetime of organic light-emitting diodes with an electrochemically stable π-conjugated liquid host in the liquid emitting layer. *Jpn. J. Appl. Phys.* 51: 041604.
17 Fukushima, T., Yamamoto, J., Fukuchi, M. et al. (2015). Material degradation of liquid organic semiconductors analyzed by nuclear magnetic resonance spectroscopy. *AIP Adv.* 5: 087124.
18 Ribierre, J.C., Zhao, L., Inoue, M. et al. (2016). Low threshold amplified spontaneous emission and ambipolar charge transport in non-volatile liquid fluorene derivatives. *Chem. Commun.* 52: 3103.
19 Shim, C.H., Hirata, S., Oshima, J. et al. (2012). Uniform and refreshable liquid electroluminescent device with a back side reservoir. *Appl. Phys. Lett.* 101: 113302.
20 Kasahara, T., Matsunami, S., Edura, T. et al. (2013). Fabrication and performance evaluation of microfluidic organic light emitting diode. *Sens. Actuators, A* 195: 219–223.
21 Tsuwaki, M., Kasahara, T., Edura, T. et al. (2014). Fabrication and characterization of large-area flexible microfluidic organic light-emitting diode with liquid organic semiconductor. *Sens. Actuators, A* 216: 231–236.
22 Kasahara, T., Matsunami, S., Edura, T. et al. (2015). Multi-color microfluidic organic light-emitting diodes based on on-demand emitting layers of pyrene-based liquid organic semiconductors with fluorescent guest dopants. *Sens. Actuators, B* 207: 481–489.
23 Kobayashi, N., Kasahara, T., Edura, T. et al. (2015). Microfluidic white organic light-emitting diode based on integrated patterns of greenish-blue and yellow solvent-free liquid emitters. *Sci. Rep.* 5: 14822.
24 Ribierre, J.C., Aoyama, T., Muto, T., and Andre, P. (2011). Hybrid organic-inorganic liquid bistable memory devices. *Org. Electron.* 12: 1800–1805.
25 Choi, E.Y., Mager, L., Cham, T.T. et al. (2013). Solvent-free fluidic organic dye lasers. *Opt. Express* 21: 11368–11375.
26 Kim, J.H., Inoue, M., Zhao, L. et al. (2015). Tunable and flexible solvent-free liquid organic distributed feedback lasers. *Appl. Phys. Lett.* 106: 053302.
27 Sandanayaka, A.S.D., Zhao, L., Pitrat, D. et al. (2016). Improvement of the quasi-continuous-wave lasing properties in organic semiconductor lasers using oxygen as triplet quenchers. *Appl. Phys. Lett.* 108: 22301.

9

Liquids Based on Nanocarbons and Inorganic Nanoparticles

Avijit Ghosh and Takashi Nakanishi

National Institute for Materials Science (NIMS), International Center for Materials Nanoarchitectonics (WPI-MANA), 1-1 Namiki, Tsukuba 305-0044, Japan

9.1 Liquid Nanocarbons

9.1.1 Introduction

Many exciting scientific and technological outcomes of nanocarbon materials have advanced simply because of their unique properties of high electrical and thermal conductivity, high mechanical strength, as well as lightness in weight [1–3]. In addition, the high chemical stability, unusual interfacial thermal conductance, and optical performance expanded their range of applications from electronics to biomedical engineering. Graphene, fullerenes, and carbon nanotubes (CNTs) are the most popular members of the nanocarbon family. The rigid molecular skeleton with large π-surface of these nanocarbons are intended to form aggregates through π–π interactions as well as strong van der Waals (vdW) forces. Morphology-assisted bulk properties of these nanometer-size carbon materials can be fine-tuned by controlling their intermolecular interactions [4]. In most cases, the polymer matrices or ionic liquids (ILs) as dispersion medium and/or chemical functionalization by synthetic process were adapted as a common method for controlling their intermolecular interactions [5]. In this context, finding these nanocarbons to be ultimately soft systems such as solvent-free "fluids" would be another way of controlling such intermolecular interactions directly at the molecular level. In contrast to the conventionally functionalized nanocarbons, the solvent-free fluidic nanocarbons would be advantageous for a potential range of applications including lubricant, thermoelectric conducting composites, and others. In this section, we briefly highlight the topic of liquid nanocarbons consisting of fullerenes (C_{60}/C_{70}), CNTs, and graphene/graphene oxides (GOs).

9.1.2 General Synthetic Strategies

Two different strategies of surface modification via chemical reactions have been employed to reduce the intermolecular interactions among adjacent nanocarbon units. Covalent attachment of long alkyl chains to the nanocarbon π-unit was

Functional Organic Liquids, First Edition. Edited by Takashi Nakanishi.
© 2019 Wiley-VCH Verlag GmbH & Co. KGaA. Published 2019 by Wiley-VCH Verlag GmbH & Co. KGaA.

mainly adapted for designing liquid fullerenes [6]. Fullerene π-surface wrapped with alkyl chains essentially disrupt the close contact of adjacent π-surfaces, which leads to the facile production of solvent-free liquid fullerenes at ambient temperature. However, the covalent/noncovalent grafting of flexible oligomer chains (corona–canopy-type species) to the surface of nanocarbons is a general practice for producing liquid-like CNTs and GOs [7, 8]. Although this concept of grafting with polymeric corona–canopy to the nanoparticle (NP) core was originally observed for producing liquid-like nanoscale ionic materials (NIMs) pioneered by Giannelis et al. (*vide infra*: liquid nanoparticles section). The complementary interactions between liquid-like oligomeric corona–canopy species and ionic bonds are the important factors for balancing their liquid-like behavior.

9.1.3 Liquid Fullerenes

The development of liquid fullerenes, in particular alkylated fullerenes, has extensively been focused on by our group since the first report of the experimentally confirmed liquid physical state of 2,4,6-tris(alkyloxy)-phenyl-substituted fulleropyrrolidine derivatives [9]. Later on, further modification of molecular architectures with both linear (**1–6**) and branched (**7–9**) alkyl chains (Figure 9.1) was successfully employed for improving facile synthesis and tuning their liquid properties through a wide range of viscosities [6, 9]. However, only an optimum chain length was found to be essential for perfect balance between intermolecular forces (π–π and vdW), which leads to a decrease in melting points and successfully formed relatively low-viscous liquid fullerenes even at room temperature (RT) [6a]. The substitution pattern of alkyl chains was also an influencing factor for effective wrapping of C_{60} moiety, which controls their liquid properties. Relatively, a high carrier mobility of ∼0.03 cm^2 V^{-1} s^{-1}, observed in **5**, indirectly indicating a certain π–π overlap among C_{60} units still does exist, which resulted in such photoconductive properties. Compared to linear chains, the use of branched alkyl chain (**7–9**) was proved to be more efficient for producing lower-viscosity liquid fullerenes as observed in **7** ($\eta^* = $ ∼1500 Pa s) and **8** ($\eta^* = $ ∼260 Pa s) [6, 9, 10]. Relatively larger π-systems such as C_{70} attached with 2,4,6-tris(alkyloxy)phenyl unit (**10–12**, Figure 9.1) also follow the behavior of liquid properties similar to that observed with C_{60}. Because of electron acceptor properties and additional fluidic nature, liquid **5** was further used as dispersion medium for cadmium selenide (CdSe) nanocrystals (NCs) which leads to a pasty photoconductive composite material (Figure 9.2) [11]. Compared to a film composed of pristine CdSe NCs, a 200% photocurrent enhancement under white-light illumination was observed of the donor–acceptor composite film containing 25 wt% CdSe NCs in **5**.

A hydrophobic amphiphile, i.e. liquid alkylated C_{60} derivatives **13** and **14** (Figure 9.1), attached with branched and hyperbranched alkyl chains were used to construct well-ordered assemblies in the presence of either solvophobic or solvophilic components of the parent liquid derivatives [12]. For example, the addition of *n*-decane to **13** leads to the formation of polydispersed spherical core–shell micelles (average diameter of 2.5 ± 0.3 nm) with C_{60}-rich

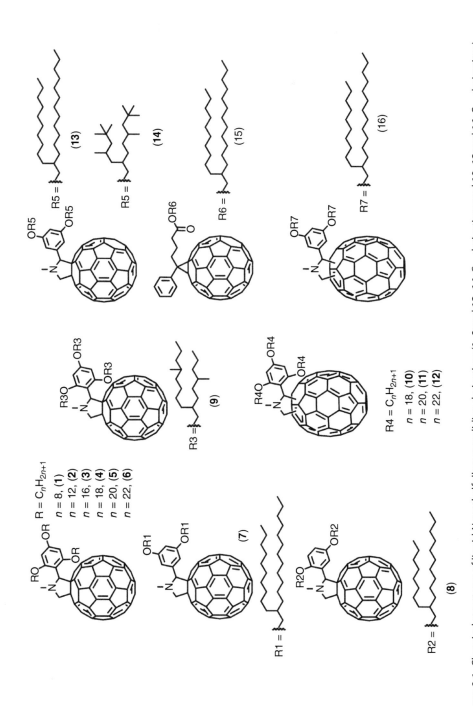

Figure 9.1 Chemical structures of liquid N-methylfulleropyrrolidine derivatives (**1–9** and **13–14**: C_{60}–derivatives and **10–12** and **16**: C_{70}–derivatives) substituted with various alkyloxyphenyl groups and liquid PC_{61} BM (**15**).

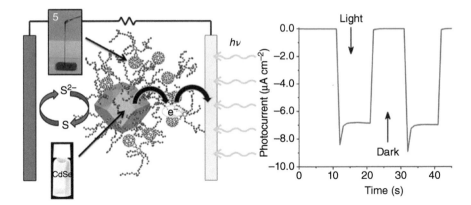

Figure 9.2 (a) Schematic representation of the liquid composite film of CdSe NC (25 wt%) and **5** on an FTO electrode and (b) the corresponding photoelectrochemical performance under white-light illumination (1000 W m^{-2}). Source: Reproduced with permission from Kramer et al. [11]. Copyright 2012, The Royal Society of Chemistry.

core content. However, addition of n-decane to **14** leads to the formation of hexagonally packed gel fibers composed of cylindrical micelles (average diameter 3.2 nm) containing insulated C_{60} nanowires. The extent of favorable ordered self-assembly depends on the solvophobic nature of alkane solvents (n-hexane, n-octane, n-decane, etc.) and is reflected by the difference in cohesive energy between the C_{60} in **13** or **14** and the respective solvent. As with the anticipation, additives with higher affinities for the C_{60} part than alkyl chains are also expected to direct assemble structures. For instance, the addition of pristine C_{60} to **13** (e.g. C_{60}:**13**; 1:2, or 1:10 in ratio) turned the assembly into lamellar-type nanosheets. Moreover, this approach was shown to be applicable to several other alkylated π-conjugated molecular systems such as $PC_{61}BM$ and fullerene-C_{70} derivatives (**15** and **16**, respectively, Figure 9.1).

Giannelis and coworkers reported an ionic fluidic-C_{60} comprising hydroxylated fullerenes (fullerols) attached with an amine-terminated polyethylene/polypropylene oxide oligomer (Jeffamine®) which resulted in a dark colored fluid material [13]. The presence of ionic bonding disrupts significantly the crystallization behavior of the amine molecules. The fluid was stable up to 300 °C with a low glass transition temperature (T_g) of −55 °C. Interestingly, thermal, flow, and viscoelastic properties of this fluid were significantly different from those of a physical mixture of the partially protonated form (sodium form) of fullerol and amine in the same ratio as in the ionic fluid.

9.1.4 Liquid-Like Carbon Nanotubes

The robustness of CNTs imposed by a highly entangled three-dimensional network prevents these materials from easy processability. Many attempts have been made to enhance their processability, but a significant phase separation during processing remains challenging [5a]. Surface functionalization of CNTs with liquid-like physical properties has attracted much attention in recent times. Giannelis and coworkers prepared a highly viscous, tarlike liquid by surface oxidation of CNT with HNO_3 and subsequent grafting of poly(ethylene glycol) (PEG)-substituted tertiary amine [$(C_{18}H_{37})N(CH_2CH_2O)_nH$-$(CH_2CH_2O)_mH$, ($n+m = 50$)] via an acid–base reaction [14a]. The resulted waxy material undergoes a reversible, macroscopic solid-to-liquid transition at 35 °C. Controlled viscoelastic nanofluids were also produced by varying oxidation time during surface functionalization [14b]. The particles and suspending medium in the molten CNT derivatives were combined into a single, homogeneous phase which is quite different from conventional colloidal suspensions. In fact, these liquid CNTs were distributed homogeneously even in a nanocomposite of polyurethane thermoplastic elastomer in the presence of pristine CNTs [14c]. Further modification in the core–shell-type nanostructure leads to the production of liquid CNTs comprising ultralong (100–500 nm long) nanotubes [15]. In this case, the PEG-substituted tertiary amine was replaced with a covalently grafted polysiloxane quaternary ammonium salt [$(CH_3O)_3Si(CH_2)_3N^+(CH_3)_2(C_{18}H_{37})Cl^-$; DC5700], which was further functionalized with sulfonate salt via an ion-exchange process. Similar core–shell-type liquid-like CNTs with different shell thickness were produced simply by varying the organic modifier molecules [7]. Later on, attachment of a canopy consisting of epoxy-terminated silicon to the CNT through covalent bonds was also able to produce liquid-like CNT [16]. In contrast to PEG-terminated hybrid [14a], this silicon-functionalized fluidic CNT contains extremely higher CN content of 85% (in w/w) and even exist as fluid state at RT. An another nonionic solvent-free fluidic hybrid material of CNTs/silica was produced by dispersing colloidal silica in a 3-(trimethoxysilyl)-1-propanethiol and the resulted fluid was fabricated by carboxylic multiwall carbon nanotubes (MWCNTs) and pluronic copolymer (PEO-b-PPO-b-PEO) [17]. The waxy solid can flow at 45 °C.

A variety of corona–canopy-type (representative example: Figure 9.3) liquid-like CNTs were produced and successfully incorporated into epoxy matrix in solvent-free condition to investigate the mechanical performances of the nanocomposites [18]. An improved impact toughness of the nanocomposite was observed; and in some cases, it was even 194% higher than the pure epoxy matrix [18d]. In another system, liquid-like MWCNT reinforcements were also employed into the epoxy matrix for testing the processing and mechanical and interfacial properties of the carbon-filament-wound composite [18f]. The liquid-like MWCNT reinforcements can be uniformly dispersed in the epoxy matrix and participate in the cross-linking reaction with epoxy groups, improving the interfacial bonding. In fact, the influence of fluidity and viscosity of a series of amine-functionalized liquid-like MWCNTs was investigated to test their capability toward CO_2 capture [19]. For producing such CO_2-capturing fluidic MWCNTs, the functionalized trimethoxy silane as corona was grafted with polyether amine canopy to the surface of CNTs. The solvent-free fluidic MWCNTs showed an enhanced CO_2 capture capacity compared to the individual constituent composite components.

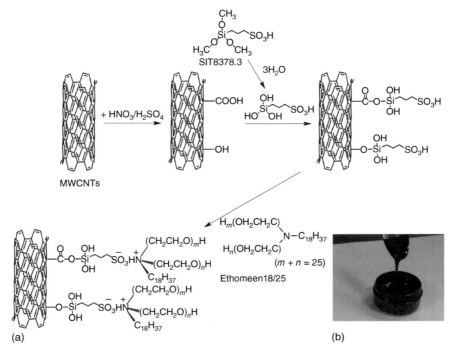

Figure 9.3 (a) Reaction scheme for preparing liquid-like MWCNT derivatives with the attachment of corona and canopy to the surface of MWCNT. In this particular system, sulfonic acid–terminated organosilanes and tertiary amine were used as corona and canopy, respectively. (b) Photo image of liquid physical state of the nanofluid at RT. Source: Adapted with permission from Wu et al. [18c]. Copyright 2013, John Wiley & Sons.

9.1.5 Fluidic Graphene/Graphene Oxide

As with fullerenes or CNTs in multidirectional applications, graphenes also face problems associated with processing, miscibility, and dispersion in solvents or polymer matrices. Surface functionalization of graphene via covalent/noncovalent modification is one way to overcome such problems while maintaining its intriguing properties. The approach of producing liquid-like graphene in recent times has shown its potential, with many advantages [8, 20–27]. Like fluidic CNTs, for producing liquid-like graphene, the graphene oxide surface was also functionalized with a charged corona which is a further tethered counterion canopy via ionic bonding (Figure 9.4) [8]. The resulting hybrid graphene behaved as homogeneous fluid at RT and exhibited very low T_g. The observed higher loss modulus (G'') over storage modulus (G') obtained from rheological measurements confirmed their liquid-like properties. The temperature dependence of G'' and G' indicated that the fluidity of these hybrid materials can be regulated by temperature. Another type of electronic conductive, low-viscosity fluid with graphene@Fe_3O_4 hybrid was reported by attaching organosilanes as corona and polyether amines as canopy onto the surface of the graphene@Fe_3O_4 core [24]. The presence of a paramagnetic counterpart

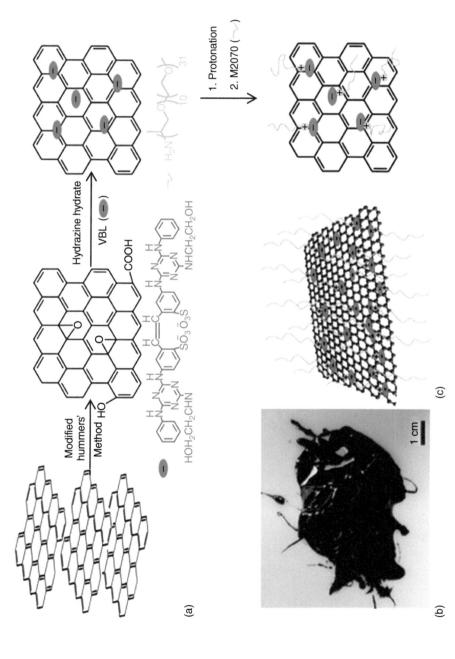

Figure 9.4 (a) A schematic representation for producing liquid-like graphene derivative. (b) A photo image of the physical appearance of this liquid-like graphene. (c) Schematic representation of the overall graphene surface containing its constituent components. Source: Reproduced with permission from Tang et al. [8]. Copyright 2012, The Royal Society of Chemistry.

made this liquid-like hybrid as superparamagnetic material with a specific magnetization of 0.39 emu g^{-1}. Like fluidic CNTs, the liquid-like graphene also exhibited homogeneous dispersion and enhanced impact toughness by incorporating into an epoxy nanocomposite [25]. A recent report described that the introduction of NC into the graphene oxide could lead to a facile and efficient production of liquid graphene at RT via covalent attachment of organosilane corona and polyether amine canopy [26].

It should be noted that the surface oxidation of carbon black (CB) NPs and subsequent grafting of a charged polysiloxane quaternary ammonium salt which was further functionalized with a poly(ethylene) glycol sulfonate salt could also generate a liquid material at RT [27].

9.2 Liquids Based on Inorganic Nanoparticles

9.2.1 Background

Hybrid nanocomposite materials have attracted considerable attention in past decades. This trend showed that a variety of functional NPs with unexpected electronic, optical, magnetic, and biological properties could be achieved by surface engineering of NPs. For example, NPs as flexible sensors [28], colloidal NPs [29], NPs dispersed in a polymer matrix [30], soft NP composite with supramolecular and polymer gels [31], etc. have emerged in the class of hybrid nanocomposites. Poor miscibility, dispersion, and processing are the main challenges for their complete understanding. However, the unique transport properties and wetting behavior of NP suspensions/nanofluids [32] have encouraged scientists in further functionalization toward solvent-free liquid-like behavior. For instance, the attempts of liquid polyoxometalates (POMs) with high conductivity [33a] and fluidic layered organosilicates [33b] were successful in preserving their nanostructures in the liquid state. Later, by grafting an organic modifier to the surface of inorganic NP cores, for example, SiO_2 [34a–c], γ-Fe_2O_3 [34a], TiO_2 [34d], and ZnO [34e], Giannelis and coworkers expanded the gallery of liquid-like NPs which offer solvent-free nanofluids as conducting, lubricating, magnetic, electrorheological, solvent/reaction media. These organic–inorganic hybrids, typically called NIMs, are composed of a charged oligomeric corona attached to the surface of NPs and an oppositely charged canopy to balance the charge [34–39]. Some cases reported a layer of covalently tethered corona–canopy polyme directly anchored to the NP surface for producing liquid-like nanofluids. Results showed that the organic–inorganic hybrid character was suitable for tailoring their physical and optoelectronic properties by controlling the shape, size, and chemical characteristics of the corona and canopy. Furthermore, another generalized and facile route using thiol-containing ionic liquid (IL) as ligand opens a new pathway to self-assembled hybrid liquids containing platinum, gold, palladium, and rhodium NPs at RT [35a]. The solvent-free fluidic character of these liquid-like NPs with zero vapor pressure, high conductivity, and high thermoelectrochemical stability make them potential conductive lubricants, heat-transfer fluids, liquid electrolytes of high temperature electrochemical cells, and many more [33–39]. Here, we briefly introduce the class of liquid-like NPs containing various inorganic cores and their implications in nanoscience.

9.2.2 Liquid-Like Silica Nanoparticles

Silica is one of the widely used NP cores incorporated in the designing of liquid-like NIMs grafted via a variety of soft organic shells (representative example in Figure 9.5) [33b, 34a–c, 36–38]. Self-suspended NP liquids comprising silica NPs functionalized with a dense brush of PEG telomer and oligomer chains were accounted for investigation on the effect of particle core and dynamic flow response in a timeframe of the suspension viscosity [36a]. The viscosity of this fluid can be controlled by manipulating grafting density as well as molecular weight of PEG corona. For example, relatively long telomers and low grafting densities ($\sigma \approx 1$–2 chains nm^{-2}) resulted in a simple liquid, while a shorter chain and more densely grafted ($\sigma \approx 2$ chains nm^{-2}) telomers produced a soft wax material. Interestingly, unlike other suspended NPs, these fluidic silica NPs behaved like a Newtonian fluid. In another type of liquid polymer nanocomposites, Jeffamine M-2070 was used as both grafting shell as well as host liquid in which the grafted silica NPs were suspended [36c]. The suspended nanocomposite exhibited liquid-like behavior only with a polymer content above 30 wt%. The grafting of same host liquid prevents the aggregation of the silica NPs, and small-angle neutron scattering (SANS) revealed that the particles were well dispersed with little clustering. Recently, a white-light-emitting liquefiable

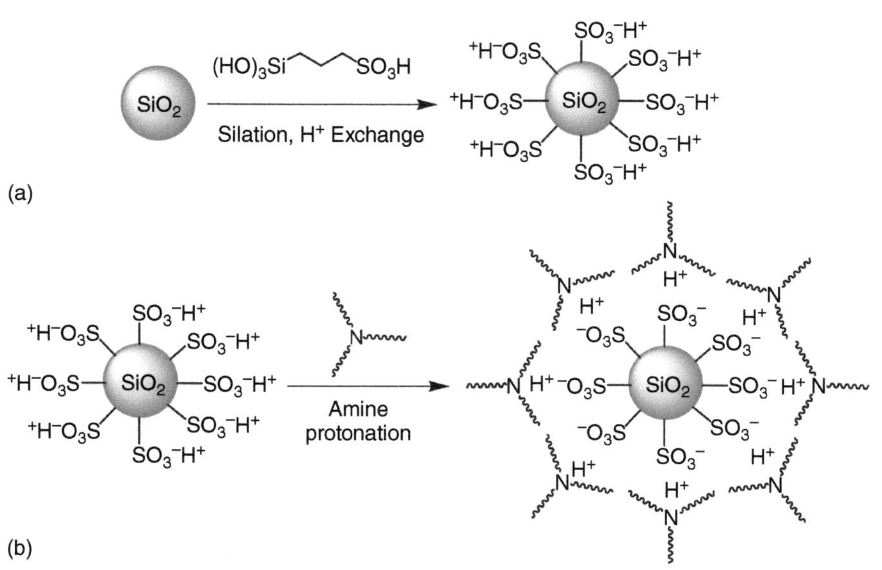

Figure 9.5 Schematic presentation showing the synthetic design of silica nanofluid using sulfonate organosilane as corona and polymeric tertiary amine as canopy. (a) Surface functionalization by condensation with corona. (b) Grafting with canopy through acid–base reaction. Source: Reproduced with permission from Rodriguez et al. [34c]. Copyright 2008, John Wiley & Sons.

silicon nanocrystal (Si-NCs) was prepared by HF etching of Si powder [37]. Covalent attachment of octyl monolayers to the Si surface turned these NCs to a solvent-free liquid physical state. The liquid nature of these NCs provided them high thermally stable photoluminescence (PL) as well as an exceptional stability over a wide pH range (pH 1–13).

Manipulation of bonding type and functional groups turned several silica-based liquid-like NP organic hybrid materials suitable for capturing CO_2 [38]. Task-specific functional groups such as amines were incorporated for capturing CO_2 through chemical interactions between them. Again, the polymeric canopy attached through ionic bonding exhibited high CO_2 capture efficiency over covalently linked canopy. The steric and entropic effects on reorganizing the densely packed canopy influenced significantly the CO_2 capture properties. Thus, by controlling the structure of the polymeric canopy of these hybrid materials, the CO_2 capture properties can be fine-tuned easily. Later, an enhanced CO_2 capture capability was observed using polyhedral oligomeric silsesquioxane (POSS)-based liquid-like nanofluids [38e]. The enhanced thermal stability and porous nature of such hybrid nanofluids contributed to their overall improved CO_2 capture capacity.

A few other type of NP cores such as $CaCO_3$ [39a], Fe_3O_4 [39b], $MnSn(OH)_6$ crystallite [39c], ZrO [39d], and MoS_2 [39e, f] were also successfully used for producing liquid-like nanofluids through an ionic grafting of polymeric organic shell. Incorporation of various cores together with fluidity, tunable mechanical properties, excellent thermal stability, enhanced ionic conductivity, lubricating properties, and improved processability allowed these nanofluids as promising materials for a wide range of applications.

9.2.3 Functional Colloidal Fluids

Further modification in the compositions and synthetic strategy would enable high functionality to these solvent-free hybrid nanofluids. For example, solvent-free metal NPs (Ag and Au) and quantum dots (QDs), produced by successive phase transfer from nonpolar solvents to an IL media, exhibited excellent functionalities of long-term dispersion stability, high electrical conductivity, and rich optical properties [40]. A hydrophobic stabilizer (palmitic acid, tetraoctylammonium bromide, and oleic acid [OA] for Ag, Au, and QD, respectively) was used in nonpolar solvents (such as toluene or chloroform) for the direct synthesis of corresponding NPs (Ag NPs, Au NPs, and CdSe@ZnS). Direct phase transfer of these dispersed NPs to a solvent-free thiol-functionalized imidazolium-type IL (TFI-IL) media produced fluidic NPs (Figure 9.6). The use of short chain length IL helped in stabilization of highly concentrated >60 wt%) NPs, and enhanced electrical conductivity of about $40\,\Omega\,\square^{-1}$ was observed for Ag NP fluids. The significant redshifted PL spectra and non-Newtonian behavior on increasing loading amount of QDs (maximum 48%) in fluidic CdSe@ZnS further concluded that the IL ligand effectively reduced the interparticle distance between NPs. Furthermore, the fluidic Au NPs were turned to magnetic metal fluids in composition with anhydrous $FeCl_3$ powder [41]. Observed higher G'' over G' in the temperature range 25–70 °C and low T_g of −63 °C even with a loading amount of 72 wt% NPs confirmed the liquid-like behavior of these magnetic

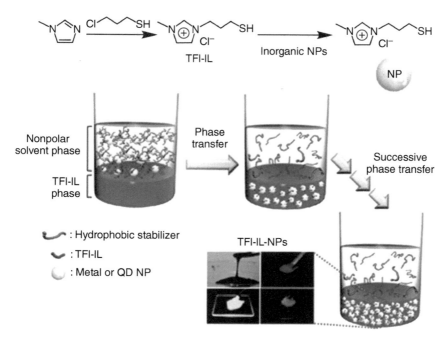

Figure 9.6 Schematic presentation showing the preparation of TFI-IL-stabilized fluidic NPs obtained through phase transfer from nonpolar solvents to the IL media. Inset shows the photo images of nanofluids containing various NP cores. Source: Adapted with permission from Kim et al. [40]. Copyright 2012, The Royal Society of Chemistry.

nanofluids. The presence of high-spin $FeCl_4^-$ ions turned these composites to a paramagnetic, revealed by superconducting quantum interference device (SQUID) measurement. In addition, these magnetic fluids also exhibited a relatively high ionic conductivity of 1.74×10^{-5} S cm^{-1} at 27 °C. The electrical sheet resistance of these fluids was decreased significantly with the increasing loading amount of Au NPs up to 65 wt%. The combining effect of electrical conductivity and paramagnetic response of these magnetic fluids were suitable for reversible ON/OFF electrical switching properties which can be controlled by an external magnetic field in developing magnetic fluid actuators. Another similar type of solvent-free nanocomposite colloidal fluids containing OA-stabilized Fe_3O_4 NPs and CdSe@ZnS QDs was produced using ligand exchange and electrostatic interaction based on layer-by-layer assemblies [42]. Functionalization through phase transfer to the solvent-free TFI-IL media produced these nanocomposites as fluidic. Importantly, these colloidal fluids integrated a variety of desired and tailored functionalities like strong superparamagnetic, fluorescent, rheological, and ionic conduction properties at RT. For instance, ionic conductivity of these non-Newtonian-type fluids was significantly increased from 3.27×10^{-11} to 1.23×10^{-3} S cm^{-1} with the increase in temperature from −153 to 77 °C.

9.2.4 Fluidic Functional Quantum Dots

The earlier discussion revealed that the integrated properties of liquid-like nanofluids are highly dependent on the core functionality. A variety of QD core were efficiently used for producing liquid-like nanofluids with attractive functionalities [43–47]. A synthetic strategy similar to that of producing liquid-like NIMs was also adapted for fluidic QDs. A soft organic polymeric-ionic modifier was grafted to the surface of QDs for preparing single-component solvent-free liquid QDs. Solvent-free ZnO nanofluids of high quantum yield photoluminescence were achieved by attachment of positively charged organosilane to the surface hydroxyl groups of ZnO. Replacement of counterions with PEG-tailed sulfonate anions resulted in a waxy material which undergoes solid–liquid phase transition at 30 °C [34e]. Modification in the ionic counterpart with an ammonium-based IL can also maintain the liquid-like properties of ZnO NCs, which exhibited tunable photoluminescence [43]. Generally, IL media is one of the key components for composing hybrid QD-IL systems where IL can be used as both solvent and capping agent [48]. A series of semiconductor fluids consisting of lead salt (PbS, PbSe, and PbTe) QDs capped with polymeric IL ligands were found to be more stable in solvent-free liquid state than the same QDs dispersed in a solvent [44]. Importantly, the optical properties of PbS QDs were preserved in solvent-free fluidic state, and in some cases even enhanced after capping with IL ligands. Unique solvent-free fluxible QDs of high fluorescence were achieved by ionic grafting of cationic surfactant $C_9H_{19}C_6H_4(OCH_2CH_2)_{10}O(CH_2)_2N^+(CH_3)_3Cl^-$ (NPEQ) on the surface of anionic CdSe/CdS/ZnS QDs through extraction method (Figure 9.7) [45]. However, the multistep synthetic approach as well as fluorescence quenching due to ligand exchange limited the applications of such fluidic QDs. Recently, a simple hydrogen-bonded self-assembly strategy was used to achieved highly fluorescent QD nanofluids containing mercaptoacetic acid–capped CdTe QDs grafted by PEG-substituted tertiary amine [46]. The resulted waxy material exhibited thermoresponsive luminescent properties and undergoes a solid–liquid phase transition at RT or above.

The inclusion interactions between α-cyclodextrin (α-CD) and PEG chains were accounted for producing a series of supramolecular QDs with tunable liquid-like properties [47]. The attachment of flexible PEG chains on the surface of oleic acid–capped $Cd_{1-x}Zn_xSe_{1-y}S_y$ QDs brings the flowability to the resultant materials. A perfect dosage of α-CD is crucial for tuning their fluidic performance, whereas the presence of excess α-CD that can completely wrap the PEG chains could lead to a permanent solid material. Not only can the amount of α-CD but different PEG chains can also control their assembling structures. Interestingly, some of these QD nanofluids responded toward Cu^{2+} ion in fluorescence and contact angle [47b]. Very recently, using a similar QD core, chiral-responsive solvent-free liquid-like QDs was also reported [47c].

9.3 Conclusions

The physical robustness of nanocarbons/NPs was diminished by functionalizing them as solvent-free liquid-like materials through surface modification. Two different strategies were mainly adopted for producing solvent-free liquid-like

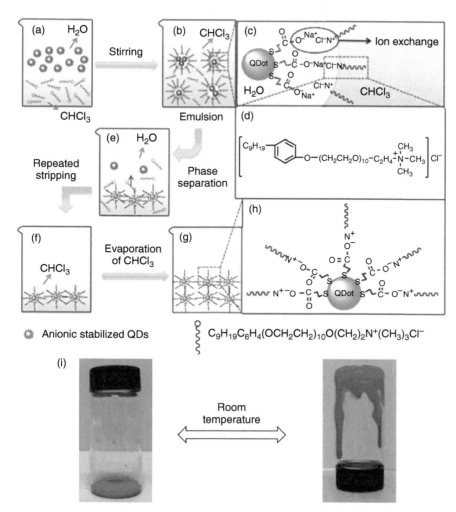

Figure 9.7 Schematic presentation showing the preparation of fluxible monodispersed liquid-like QDs. (a) QDs in water and NPEQ in CHCl$_3$ layers, (b) emulsion formed after stirring, (c) ion exchange between QDs and NPEQs, (d) chemical structure of NPEQ cationic surfactant, (e) water layer containing unreacted QDs and CHCl$_3$ layer containing NPEQ-functionalized QDs, (f) functionalized QDs in CHCl$_3$, (g) solvent-free liquid-like flexible QDs, (h) NPEQ-functionalized single QD. (i) Photo images of fluidity on reversing the vial at room temperature. Source: Reproduced with permission from Feng et al. [45]. Copyright 2010, John Wiley & Sons.

nanocarbons and metal NPs. The strategy of attaching alkyl chains to the C$_{60}$ surface was focused on for producing solvent-free alkylated-C$_{60}$ liquid derivatives. Substitution patterns as well as the chain lengths were important factors for maintaining their viscosity and inherent molecular properties. In another strategy, grafting of soft polymeric organic shells to the surface of nanocarbons/NPs through ionic/covalent interactions was used to transfer these solid nanomaterials into solvent-free nanofluids at RT. The observed functionalities of nanofluids are the combination of intrinsic physicochemical properties contributed from constituent composite components. Primary investigations revealed that the fluidity and mechanical properties of these nanofluids can be tuned by changing the grafting density of soft shells. NPs with various core functionalities such as conducting, magnetic, and luminescent, etc. were selectively chosen to bring those properties into their solvent-free liquid sates. Fluidity, high processability, excellent dispersity, and good thermal stability allow these hybrid NP fluids as promising potentials for a wide range of applications in electronic/optical inks, plasticizers, lubricant, energy transfer, imaging and therapy in drug delivery, magnetism, catalysis, and more. At this stage, the development of more facile methods and investigations into the effect of microscopic configurational interactions to the optical and physical properties of these liquid-like nanofluids are highly desired.

References

1 Su, D.S., Perathoner, S., and Centi, G. (2013). Nanocarbons for the development of advanced catalysts. *Chem. Rev.* 113: 5782–5816.
2 Li, Z., Liu, Z., Sun, H., and Gao, C. (2015). Superstructured assembly of nanocarbons: fullerenes, nanotubes, and graphene. *Chem. Rev.* 115: 7046–7117.
3 Georgakilas, V., Perman, J.A., Tucek, J., and Zboril, R. (2015). Broad family of carbon nanoallotropes: classification, chemistry, and applications of fullerenes, carbon dots, nanotubes, graphene, nanodiamonds, and combined superstructures. *Chem. Rev.* 115: 4744–4822.
4 Haley, M.M. and Tykwinski, R.R. (2006). *Carbon-Rich Compounds*. Weinheim: Wiley-VCH.
5 (a) Fukushima, T., Kosaka, A., Ishimura, Y. et al. (2003). Molecular ordering of organic molten salts triggered by single-walled carbon nanotubes. *Science* 300: 2072–2074. (b) Diederich, F. and Gómez-López, M. (1999). Supramolecular fullerene chemistry. *Chem. Soc. Rev.* 28: 263–277. (c) Sawamura, M., Kawai, K., Matsuo, Y. et al. (2002). Stacking of conical molecules with a fullerene apex into polar columns in crystals and liquid crystals. *Nature* 419: 702–705. (d) Nierengarten, J.-F. (2004). Chemical modification of C_{60} for materials science applications. *New J. Chem.* 28: 1177–1191. (e) Hirsch, A. and Brettreich, M. (2005). *Fullerenes: Chemistry and Reactions*. Weinheim: Wiley-VCH.
6 (a) Li, H., Babu, S.S., Turner, S.T. et al. (2013). Alkylated-C_{60} based soft materials: regulation of self-assembly and optoelectronic properties by chain branching. *J. Mater. Chem. C* 1: 1943–1951. (b) Michinobu, T., Okoshi, K., Murakami, Y. et al. (2013). Structural requirements for producing solvent-free room temperature liquid fullerenes. *Langmuir* 29: 5337–5344.
7 Li, Q., Dong, L., Fang, J., and Xiong, C. (2010). Property-structure relationship of nanoscale ionic materials based on multiwalled carbon nanotubes. *ACS Nano* 4: 5797–5806.
8 Tang, Z., Zhang, L., Zeng, C. et al. (2012). General route to graphene with liquid-like behavior by non-covalent modification. *Soft Matter* 8: 9214–9220.

9 Michinobu, T., Nakanishi, T., Hill, J.P. et al. (2006). Room temperature liquid fullerenes: an uncommon morphology of C_{60} derivatives. *J. Am. Chem. Soc.* 128: 10384–10385.
10 Hollamby, M.J. and Nakanishi, T. (2013). The power of branched chains: optimising functional molecular materials. *J. Mater. Chem. C* 1: 6178–6183.
11 Kramer, T.J., Babu, S.S., Saeki, A. et al. (2012). CdSe nanocrystal/C_{60}-liquid composite material with enhanced photoelectrochemical performance. *J. Mater. Chem.* 22: 22370–22373.
12 Hollamby, M.J., Karny, M., Bomans, P.H.H. et al. (2014). Directed assembly of optoelectronically active alkyl–π-conjugated molecules by adding *n*-alkanes or π-conjugated species. *Nat. Chem.* 6: 690–696.
13 Fernandes, N., Dallas, P., Rodriguez, R. et al. (2010). Fullerol ionic fluids. *Nanoscale* 2: 1653–1656.
14 (a) Bourlinos, A.B., Georgakilas, V., Tzitzios, V. et al. (2006). Functionalized carbon nanotubes: synthesis of meltable and amphiphilic derivatives. *Small* 2: 1188–1191. (b) Lei, Y., Xiong, C., Guo, H. et al. (2008). Controlled viscoelastic carbon nanotubes fluids. *J. Am. Chem. Soc.* 130: 3256–3257. (c) Gu, S.-Y., Liu, L.-L., and Yan, B. (2014). Effects of ionic solvent-free carbon nanotube nanofluid on the properties of polyurethane thermoplastic elastomer. *J. Polym. Res.* 21: 356.
15 Lei, Y., Xiong, C., Dong, L. et al. (2007). Ionic liquid of ultralong carbon nanotubes. *Small* 3: 1889–1893.
16 Bourlinos, A.B., Georgakilas, V., Boukos, N. et al. (2007). Silicone-functionalized carbon nanotubes for the production of new carbon-based fluids. *Carbon* 45: 1583–1585.
17 Zhang, J.-X., Zheng, Y.-P., Lan, L. et al. (2009). Direct synthesis of solvent-free multiwall carbon nanotubes/silica nonionic nanofluid hybrid material. *ACS Nano* 3: 2185–2190.
18 (a) Lan, L., Zheng, Y.P., Zhang, A.B. et al. (2012). Study of ionic solvent-free carbon nanotube nanofluids and its composites with epoxy matrix. *J. Nanopart. Res.* 14: 753. (b) Li, Q., Dong, L., Li, L. et al. (2012). The effect of the addition of carbon nanotube fluids to a polymeric matrix to produce simultaneous reinforcement and plasticization. *Carbon* 50: 2056–2060. (c) Wu, F., Zheng, Y., Qu, P. et al. (2013). A liquid-like multiwalled carbon nanotube derivative and its epoxy nanocomposites. *J. Appl. Polym. Sci.* 130: 2217–2224. (d) Zheng, Y., Yang, R., Wu, F. et al. (2014). A functional liquid-like multiwalled carbon nanotube derivative in the absence of solvent and its application in nanocomposites. *RSC Adv.* 4: 30004–30012. (e) Zhang, X., Zheng, Y.-P., Yang, R.-L., and Yang, H.-C. (2014). Nanocomposites with liquid-like multiwalled carbon nanotubes dispersed in epoxy resin without solvent process. *Int. J. Polym. Sci.* 2014: 712637. (f) Zhang, Q., Wu, J., Gao, L. et al. (2016). Influence of a liquid-like MWCNT reinforcement on interfacial and mechanical properties of carbon fiber filament winding composites. *Polymer* 90: 193–203. (g) Yang, Y.-K., Yu, L.-J., Peng, R.-G. et al. (2012). Incorporation of liquid-like multiwalled carbon nanotubes into an epoxy matrix by solvent-free processing. *Nanotechnology* 23: 225701.

19 (a) Li, P., Yang, R., Zheng, Y. et al. (2005). Effect of polyether amine canopy structure on carbon dioxide uptake of solvent-free nanofluids based on multiwalled carbon nanotubes. *Carbon* 95: 408–418. (b) Yang, R., Zheng, Y., Li, P. et al. (2016). Effects of acidification time of MWCNTs on carbon dioxide capture of liquid-like MWCNTs organic hybrid materials. *RSC Adv.* 6: 85970–85977.

20 Li, Q., Dong, L., Sun, F. et al. (2012). Self-unfolded graphene sheets. *Chem. Eur. J.* 18: 7055–7059.

21 Wu, L., Zhang, B., Li, H., and Liu, C.-Y. (2014). Nanoscale ionic materials based on hydroxyl-functionalized graphene. *J. Mater. Chem. A* 2: 1409–1417.

22 Li, P., Zheng, Y., Wu, Y. et al. (2014). Nanoscale ionic graphene material with liquid-like behavior in the absence of solvent. *Appl. Surf. Sci.* 314: 983–990.

23 Liu, X., Zeng, C., Tang, Z., and Guo, B. (2014). Liquefied graphene oxide with excellent amphiphilicity. *Chem. Lett.* 43: 222–224.

24 Li, P., Zheng, Y., Wu, Y. et al. (2014). A nanoscale liquid-like graphene@Fe_3O_4 hybrid with excellent amphiphilicity and electronic conductivity. *New J. Chem.* 38: 5043–5051.

25 (a) Li, P., Zheng, Y., Li, M. et al. (2016). Enhanced toughness and glass transition temperature of epoxy nanocomposites filled with solvent-free liquid-like nanocrystal-functionalized graphene oxide. *Mater. Des.* 89: 653–659. (b) Li, P., Zheng, Y., Shi, T. et al. (2016). A solvent-free graphene oxide nanoribbon colloid as filler phase for epoxy-matrix composites with enhanced mechanical, thermal and tribological performance. *Carbon* 96: 40–48.

26 Li, P., Shi, T., Yao, D. et al. (2016). Covalent nanocrystals-decorated solvent-free graphene oxide liquids. *Carbon* 110: 87–96.

27 Li, Q., Dong, L., Liu, Y. et al. (2011). A carbon black derivative with liquid behavior. *Carbon* 49: 1047–1051.

28 Segev-Bar, M. and Haick, H. (2013). Flexible sensors based on nanoparticles. *ACS Nano* 7: 8366–8378.

29 (a) Lohse, S.E. and Murphy, C.J. (2012). Applications of colloidal inorganic nanoparticles: from medicine to energy. *J. Am. Chem. Soc.* 134: 15607–15620. (b) Hodges, J.M., Morse, J.R., Fenton, J.L. et al. (2017). Insights into the seeded-growth synthesis of colloidal hybrid nanoparticles. *Chem. Mater.* 29: 106–119.

30 Kumar, S.K., Jouault, N., Benicewicz, B., and Neely, T. (2013). Nanocomposites with polymer grafted nanoparticles. *Macromolecules* 46: 3199–3214.

31 Bhattacharya, S. and Samanta, S.K. (2016). Soft-nanocomposites of nanoparticles and nanocarbons with supramolecular and polymer gels and their applications. *Chem. Rev.* 116: 11967–12028.

32 (a) Huxtable, S.T., Cahill, D.G., Shenogin, S. et al. (2003). Interfacial heat flow in carbon nanotube suspensions. *Nat. Mater.* 2: 731–734. (b) Wasan, D.T. and Nikolov, A.D. (2003). Spreading of nanofluids on solids. *Nature* 423: 156–159.

33 (a) Bourlinos, A.B., Raman, K., Herrera, R. et al. (2004). A liquid derivative of 12-tungstophosphoric acid with unusually high conductivity. *J. Am. Chem. Soc.* 126: 15358–15359. (b) Bourlinos, A.B., Chowdhury, S.R., Jiang, D.D. et al. (2005). Layered organosilicate nanoparticles with liquidlike behavior. *Small* 1: 80–82.

34 (a) Bourlinos, A.B., Herrera, R., Chalkias, N. et al. (2005). Surface-functionalized nanoparticles with liquid-like behavior. *Adv. Mater.* 17: 234–237. (b) Bourlinos, A.B., Giannelis, E.P., Zhang, Q. et al. (2006). Surface-functionalized nanoparticles with liquid-like behavior: the role of the constituent components. *Eur. Phys. J. E* 20: 109–117. (c) Rodriguez, R., Herrera, R., Archer, L.A., and Giannelis, E.P. (2008). Nanoscale ionic materials. *Adv. Mater.* 20: 4353–4358. (d) Bourlinos, A.B., Chowdhury, S.R., Herrera, R. et al. (2005). Functionalized nanostructures with liquid-like behavior: expanding the gallery of available nanostructures. *Adv. Funct. Mater.* 15: 1285–1290. (e) Bourlinos, A.B., Stassinopoulos, A., Anglos, D. et al. (2006). Functionalized ZnO nanoparticles with liquidlike behavior and their photoluminescence properties. *Small* 2: 513–516. (f) Fernandes, N.J., Wallin, T.J., Vaia, R.A. et al. (2014). Nanoscale ionic materials. *Chem. Mater.* 26: 84–96.

35 (a) Warren, S.C., Banholzer, M.J., Slaughter, L.S. et al. (2006). Generalized route to metal nanoparticles with liquid behavior. *J. Am. Chem. Soc.* 128: 12074–12075. (b) Zheng, Y., Zhang, J., Lan, L. et al. (2010). Preparation of solvent-free gold nanofluids with facile self-assembly technique. *ChemPhysChem* 11: 61–64.

36 (a) Agarwal, P., Qi, H., and Archer, L.A. (2010). The ages in a self-suspended nanoparticle liquid. *Nano Lett.* 10: 111–115. (b) McEwan, M. and Green, D. (2009). Rheological impacts of particle softness on wetted polymer-grafted silica nanoparticles in polymer melts. *Soft Matter* 5: 1705–1716. (c) McDonald, S., Wood, J.A., FitzGerald, P.A. et al. (2015). Interfacial and bulk nanostructure of liquid polymer nanocomposites. *Langmuir* 31: 3763–3770. (d) Nugent, J.L., Moganty, S.S., and Archer, L.A. (2010). Nanoscale organic hybrid electrolytes. *Adv. Mater.* 22: 3677–3680. (e) Texter, J., Qiu, Z.M., Crombez, R. et al. (2011). Nanofluid acrylate composite resins-initial preparation and characterization. *Polym. Chem.* 2: 1778–1788. (f) Rodriguez, R., Herrera, R., Bourlinos, A.B. et al. (2010). The synthesis and properties of nanoscale ionic materials. *Appl. Organometal. Chem.* 24: 581–589.

37 Ghosh, B., Ogawara, M., Sakka, Y., and Shirahata, N. (2012). White-light-emitting liquefiable silicon nanocrystals. *Chem. Lett.* 41: 1157–1159.

38 (a) Lin, K.-Y.A. and Park, A.-H.A. (2011). Effects of bonding types and functional groups on CO_2 capture using novel multiphase systems of liquid-like nanoparticle organic hybrid materials. *Environ. Sci. Technol.* 45: 6633–6639. (b) Park, Y., Shin, D., Jang, Y.N., and Park, A.-H.A. (2012). CO_2 Capture capacity and swelling measurements of liquid-like nanoparticle organic hybrid materials via attenuated total reflectance fourier transform infrared spectroscopy. *J. Chem. Eng. Data* 57: 40–45. (c) Petit, C., Bhatnagar, S., and Park, A.-H.A. (2013). Effect of water on the physical properties and carbon dioxide capture capacities of liquid-like Nanoparticle Organic Hybrid Materials and their corresponding polymers. *J. Colloid Interface Sci.* 407: 102–108. (d) Lin, K.-Y.A., Petit, C., and Park, A.-H.A. (2013). Effect of SO_2 on CO_2 capture using liquid-like nanoparticle organic hybrid materials. *Energy Fuels* 27: 4167–4174. (e) Petit, C., Lin, K.-Y.A., and Park, A.-H.A. (2013). Design and characterization of liquidlike POSS-based hybrid nanomaterials synthesized via ionic bonding and their interactions with CO_2. *Langmuir* 29: 12234–12242.

39 (a) Li, Q., Dong, L., Deng, W. et al. (2009). Solvent-free fluids based on rhombohedral nanoparticles of calcium carbonate. *J. Am. Chem. Soc.* 131: 9148–9149. (b) Huang, J., Li, Q., Li, D. et al. (2013). Fluxible nanoclusters of Fe_3O_4 nanocrystal-embedded polyaniline by macromolecule-induced self-assembly. *Langmuir* 29: 10223–10228. (c) Li, P., Zheng, Y., Yang, R. et al. (2015). Flexible nanoscale thread of $MnSn(OH)_6$ crystallite with liquid-like behavior and its application in nanocomposites. *ChemPhysChem* 16: 2524–2529. (d) Moganty, S.S., Jayaprakash, N., Nugent, J.L. et al. (2010). Ionic-liquid-tethered nanoparticles: hybrid electrolytes. *Angew. Chem. Int. Ed.* 49: 9158–9161. (e) Gu, S., Zhang, Y., and Yan, B. (2013). Solvent-free ionic molybdenum disulfide (MoS_2) nanofluids with self-healing lubricating behaviors. *Mater. Lett.* 97: 169–172. (f) Gu, S.-Y., Gao, X.-F., and Zhang, Y.-H. (2015). Synthesis and characterization of solvent-free ionic molybdenum disulphide (MoS_2) nanofluids. *Mater. Chem. Phys.* 149-150: 587–593.

40 Kim, Y., Kim, D., Kwon, I. et al. (2012). Solvent-free nanoparticle fluids with highly collective functionalities for layer-by-layer assembly. *J. Mater. Chem.* 22: 11488–11493.

41 Kim, Y. and Cho, J. (2013). Metal nanoparticle fluids with magnetically induced electrical switching properties. *Nanoscale* 5: 4917–4922.

42 Kim, D., Kim, Y., and Cho, J. (2013). Solvent-free nanocomposite colloidal fluids with highly integrated and tailored functionalities: rheological, ionic conduction, and magneto-optical properties. *Chem. Mater.* 25: 3834–3843.

43 Liu, D.-P., Li, G.-D., Su, Y., and Chen, J.-S. (2006). Highly luminescent ZnO nanocrystals stabilized by ionic-liquid components. *Angew. Chem. Int. Ed.* 45: 7370–7373.

44 Sun, L., Fang, J., Reed, J.C. et al. (2010). Lead–salt quantum-dot ionic liquids. *Small* 6: 638–641.

45 Feng, Q., Dong, L., Huang, J. et al. (2010). Fluxible monodisperse quantum dots with efficient luminescence. *Angew. Chem. Int. Ed.* 49: 9943–9946.

46 Zhou, J., Tian, D., and Li, H. (2011). Multi-emission CdTe quantum dot nanofluids. *J. Mater. Chem.* 21: 8521–8523.

47 (a) Zhou, J., Huang, J., Tian, D., and Li, H. (2012). Cyclodextrin modified quantum dots with tunable liquid-like behaviour. *Chem. Commun.* 48: 3596–3598. (b) Shi, F., Zhou, J., Zhang, L. et al. (2014). Cu^{2+} Ion responsive solvent-free quantum dots. *Small* 10: 3901–3906. (c) Zhang, J., Ma, J., Shi, F. et al. (2017). Chiral responsive liquid quantum dots. *Adv. Mater.* 29: 1700296.

48 (a) Nakashima, T. and Kawai, T. (2005). Quantum dots-ionic liquid hybrids: efficient extraction of cationic CdTe nanocrystals into an ionic liquid. *Chem. Commun.* 1643–1645. (b) Green, M., Rahman, P., and Boyle, D.S. (2007). Ionic liquid passivated CdSe nanocrystals. *Chem. Commun.* 574–576.

10

Solvent-Free Nanofluids and Reactive Nanofluids

John Texter

Coating Research Institute, School of Engineering Technology, Eastern Michigan University, Ypsilanti, MI 48197, USA

10.1 Introduction

It has been over a decade since solvent-free nanofluids (NFs) were introduced by the Giannelis group and other senior collaborators from Cornell including Archer and Wiesner [1–8]. Solvent-free nanofluids are nanoparticles that, because of their surface chemical functionality, exhibit classical liquidity and viscous flow in the absence of any added molecular or ionic solvent. Such nanofluids may or may not exhibit crystallization and melting, but all exhibit solidification and melting; and when in a melted state, they exhibit viscous flow. In addition to surface functionality that one way or another imparts liquidity and viscous flow in the absence of any conventional solvent, an inclusion of a minor surface component having reactive functionality defines another focus of this chapter. We begin by briefly and broadly reviewing solvent-free nanofluids, and then focus on reactive solvent-free nanofluids. Solvent-free nanofluids reported to date include zero- (0D), one- (1D), two- (2D), and three-dimensional (3D) examples.

Dilute dispersions of metal oxides and other inorganic phases having apparent synergistic increases in thermal conductivity [9] are not what we mean by the term nanofluids. This unfortunate usage of "nanofluids" in the realm of heat transfer has not provided sufficient hype to take such dilute dispersions into practical heat transfer applications. The best explanation for such apparent synergisms empirically appears to be related to nanoparticle aggregation effects [10–12].

The solvent-free nanofluids described and discussed here exhibit self-diffusion, conductivity, viscosity, and thermal properties that vary with morphology (Figure 10.1), size, and core–corona relative volume fractions. These interesting materials have promising applications as heat transfer liquids, CO_2 absorbing and storage matrices, lubricants, interfacial stabilizers, advanced thin films and coatings, and novel 3D materials.

Functional Organic Liquids, First Edition. Edited by Takashi Nakanishi.
© 2019 Wiley-VCH Verlag GmbH & Co. KGaA. Published 2019 by Wiley-VCH Verlag GmbH & Co. KGaA.

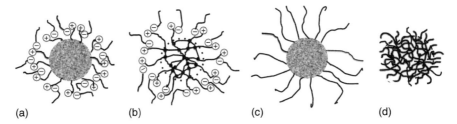

Figure 10.1 Types and classes of solvent-free nanofluids: (a) core–corona structure with grafted ionic liquid shell; (b) cross-linked supramolecular ionic liquid; (c) core/corona structure with tethered oligomers that melt; (d) cross-linked melting oligomers that compose nanogel particles.

10.1.1 Solvent-Free Nanofluids

The surface chemical functionality mentioned serves to define two classes of nanofluids. The first and most studied class is where this functionality can be described as a corona or shell of tethered ionic liquids (ILs). Ionic liquids are defined as organic salts that melt below 100 °C [13, 14]. Tethering or chemically linking such salts to nanoparticle surfaces or themselves (cross-linking) can produce solvent-free nanofluids that exhibit melting temperatures, T_m, or melting ranges, glass transitions, T_g, and viscous flow. This class can also be thought of as supramolecular ionic liquids (SILs) since its members are composed of a separate phase core or an auto-cross-linked core, as illustrated in Figure 10.1a,b. Only three examples corresponding to the morphology of Figure 10.1b have been reported [15–17]. The first two of these interesting materials have polyhedral oligomeric silsesquioxane (POSS) [18] cores. The third comprises 5- to 15-nm-diameter particles that are highly deformable and derive from autocondensation of a trimethoxysilane [17]. This last example is the first polydisperse particulate system to exhibit multiphase coexistence predicted some years earlier [19].

Another class of solvent-free nanofluids has surface functionality that can be described as tethered liquid polymers. When tethered liquid polymers provide a liquid-like shell, an ensemble of such particles also can exhibit viscous flow. Shown in Figure 10.1c,d are cases, respectively, where a core phase is surface functionalized with a polymeric brush that melts at some T_m, and where some oligomeric polymer that melts is cross-linked and may also be called a nanogel. Only one solvent-free nanofluid case corresponding to a nanogel, as pictured in Figure 10.1d, has been claimed [20]. It comprises 5- to 20-nm-diameter styrene/divinylbenzene nanoparticles that purportedly melt and flow in the 120–150 °C range [20]. This high-temperature flow domain limits this nanogel's range of practical liquidity. However, this example provides a novel and alternative mechanism for polymer transport by self-diffusion in comparison to reptation. There may be much lower melting examples discovered or synthesized in the future.

Solvent-free nanofluids of the type illustrated in Figure 10.1a were introduced in 2005, as mentioned earlier, and are a type of SIL. In an early paper [2], nanofluids based on silica cores and hematite cores were produced using an

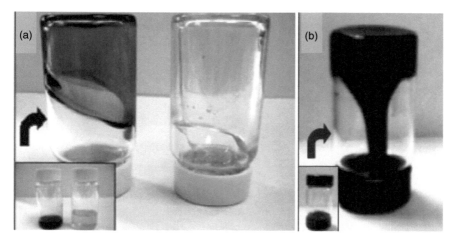

Figure 10.2 Silica (a) and hematite (b) nanoparticles surface modified with a methoxysilyl chloride, and then anion exchanged with a bulky sulfonate; (a) viscous silica-based nanofluid with (left) and without (right) methylene blue added; (b) ferrofluid nanofluid derived from hematite nanoparticles. Source: Adapted with permission from Bourlinos et al. [2]. Copyright 2005, Wiley.

ammonium silane and a soft sulfonate anion. This anion was exchanged for chloride counterions that accompanied the ammonium silane. A nanosilica core was provided (as Ludox SM), and the two inverted vials in Figure 10.2 show these solvent-free materials undergoing viscous flow. The darker example to the left in Figure 10.2a is an example with methylene blue dissolved in the solvent-free nanofluid. Instead of silica, hematite (γ-Fe_2O_3) forms the core of a solvent-free nanofluid pictured in Figure 10.2b. This hematite-based material is a ferrofluid. Such nanofluids are sometimes referred to as nanoionic materials (NIMs) [21, 22], and sometimes as nanoscale organic–inorganic hybrid materials (NOHMs) [23, 24]. Similar materials have been developed using anionic surface groups [21, 25] with various cationic counterions, where these materials also are well described as SILs. We can see that such core/shell types of nanofluids, in principle, may include a variety of nanoparticle cores offering diverse physical and chemical features for advanced material applications. We discuss some select applications later.

In addition to the seminal studies cited in our opening paragraph, the Giannelis and Archer groups have made continuing contributions to the synthesis, analysis, and application of solvent-free nanofluids [1–8, 21–33]. In addition to the feasibility demonstrated by making silica-core and hematite-core nanofluids [2] illustrated in Figure 10.2, many surface modification approaches to making examples of solvent-free nanofluids including those having cores made of the following have been undertaken: layered silica [3], nanosized titania [5], DNA [5], gold, platinum, palladium, and rhodium (surface modified using thiol chemistry) [8], ZnO [26, 34], carbon nanotubes [27], fullerols (hydroxylated fullerenes) [28], and PbS and PbSe quantum dots [29]. These and other nanofluid compositions are listed in Table 10.1.

Table 10.1 Solvent-free nanofluids.

Core	Core dimension (nm, nm^2)	Surface coupling chemistry	Surface species (reactive groups in corona)	Year	References
Au	1–11	Thiol	SPNMDO–NPEOPS	2006	[8]
Au	~20	HSCH$_2$CH$_2$SO$_3$H	SiPS–NM(C$_{8-10}$)$_3$	2007	[35]
Au	~20	Thiol	SES–NM$_3$C$_{10-12}$	2008	[36]
Au	4–20	Thiol	SC$_{10}$CO$_2$–Eth$_{18/25}$ [–OH]	2010	[37]
CaCO$_3$	20–50	Trimethoxysilyl	SiPNDMC$_{18}$–NPOEO$_{10}$S	2009	[38]
CaCO$_3$	60–70	Trimethoxysilyl	SiPNDMC$_{18}$–NPOEO$_{10}$S	2018	[39]
Carbon black	—	Trimethoxysilyl	SiPNDMC$_{18}$–NPOEO$_{10}$S	2011	[40]
CdS	—	Thiol	SECO$_2$–NPEO$_{10}$NM$_3$	2010	[41]
CdSe	—	Thiol	SECO$_2$–NPEO$_{10}$NM$_3$	2010	[41]
CdTe	10–20	HSCH$_2$CO$_2$H	SC$_1$CO$_2$–Eth$_{18/30}$	2011	[42]
CeO$_2$	6	Trimethoxysilyl triethoxysilyl	SiPNMDD–NPEOPS SiPA (acrylate)	2016	[43]
CPMV capsid (cowpea mosaic virus)	33	CO$_2$H amidation	CONHPNHM$_2$–NPEOPS	2012	[44]
DNA	2 × 12–23	O$_2$PO	PO–E$_3$NEO$_7$M	2001	[4]
DNA	2 × 45–140	O$_2$PO$^-$	PO–Eth$_{18/25}$	2005	[5]
γ-Fe$_2$O$_3$	9–11	Trimethoxysilyl	SiPNMDD–C$_{13-15}$OEO$_7$PS	2005	[2]
Fe$_3$O$_4$	2–3	Trimethoxysilyl	SiPNMR$_1$R$_2$–NPEOPS	2012	[45]
Ferritin	12	DMPA amidation	CONHPNHM$_2$–NPEOPS [NH$_2$]	2009	[46]
Fullerene	1	Pyrrolidine	C$_{60}$NP(OR)$_3$	2006	[47, 48]
Fullerene	1	Pyrrolidine	FB9ox	2008	[49]
Fullerene	1	Pyrrolidine	C$_{60}$NP(OR)(OR')(OR'')	2008	[48, 50]
Fullerene	1	Pyrrolidine	C$_{60}$NP-2,5-di(OR) C$_{60}$NP-3,5-di(OR)	2013	[51, 52]
Fullerol	1	–OH	O–NPO$_{35}$EO$_6$M [–OH]	2010	[28]
GO	50^2–200^2	NH$_2$C$_6$H$_4$SO$_3$H	S–NPO$_{35}$EO$_6$M	2012	[53]
GO	—	N$_3$CH$_2$CH$_2$OH trimethoxysilyl	SiPS–NH$_3$PO$_{10}$EO$_{31}$M	2014	[54]
GO	—	Trimethoxysilyl	SiPNMDD–NPEOPS	2014	[52]
GO@Fe$_3$O$_4$	—	Trimethoxysilyl	SiPNMDD–NPEOPS [–OH]	2016	[54, 55]
GO@ZnHOSn	—	Trimethoxysilyl	SiPOPOHNHPO$_{35}$EO$_6$M	2016	[56]
MWCNT	—	Glycidyl ether	OCH$_2$CH(OH)CH$_2$OE–PDMS	2007	[27]

(Continued)

Table 10.1 (Continued)

Core	Core dimension (nm, nm^2)	Surface coupling chemistry	Surface species (reactive groups in corona)	Year	References
MWCNT	$3 \times 0.5 \cdot 10^4$	Trimethoxysilyl	SiPNDMC$_{18}$–NPOEO$_{10}$S	2007	[57]
MWCNT	15×0.5–$3.4 \cdot 10^4$	–CO$_2$H	O–Eth$_{18/50}$	2008	[58]
MWCNT	20–$30 \times 3 \cdot 10^4$	–CO$_2$H, zwitterionization	O–HPAE [–OH]	2009	[59]
MWCNT	20–$30 \times 3 \cdot 10^4$	–CO$_2$H	F108	2009	[60]
MWCNT	$<8 \times 5$–$20 \cdot 10^4$	Trimethoxysilyl –CO$_2$H	SiPNDMC$_{18}$–NPEOPS O–Eth$_{18/z\ (z\ =\ 10,20,30)}$	2010	[60]
MWCNT	8–15×1–$3 \cdot 10^4$	Trimethoxysilyl –CO$_2$H, –OH	SiPNDMC$_{18}$–NPOEO$_{10}$S		[61]
MWCNT	—	–CO$_2$H, trimethoxysilyl	SiPS–Eth$_{18/25}$ [–OH]	2012	[62]
MWCNT	10–30×10^4	Trihydroxysilyl	SiPNMDD–NPEOPS	2012	[63]
MWCNT	11×10^4	–CO$_2$H, –OH	O–Eth$_{18/25}$	2014	[64]
MWCNT	—	–CO$_2$H	F108	2014	[65]
MWCNT	30–$50 \times 1.5 \cdot 10^4$	Trimethoxysilyl	SiPOPOHNHPO$_3$EO$_{19}$M SiPOPOHNHPO$_{10}$EO$_{31}$M	2015	[66]
MWCNT	2.5×10^4	–CO$_2$H	SiPS–Eth$_{18/25}$ F108	2015	[67]
Mg(OH)$_2$	—	Trimethoxysilyl	SiPNDMC$_{18}$–NPOEO$_{10}$S SiPA (acrylate)	2017	[68]
MnSn(OH)$_6$	—	Trihydroxysilyl	SiPS–NH$_3$PO$_{10}$EO$_{31}$M	2015	[69]
MoS$_2$	—	Trihydroxysilyl	SiPS–Eth$_{18/25}$	2013	[70]
MoS$_2$	—	Trihydroxysilyl	SiPS–Eth$_{18/25}$	2015	[71]
Myoglobin	5.3	CO$_2$H amidation	CONHPNHM$_2$–NPEOPS	2013	[72]
No core	4–15	Trimethoxysilyl autocondensation	SiPNMDD–NPEOPS	2013	[17]
POSS	1.2	—	SiPNHCOECO$_2$–C$_4$ImC$_1$	2010	[15]
POSS	1.2	Trimethoxysilyl	SiPOPOHNHEO$_6$PO$_{35}$M	2015	[16]
Pd	1.3–2.1	Thiol	SPNMDO–NPEOPS	2006	[7]
Phosphotungstic acid (H$_3$PW$_{12}$O$_{40}$)	1.2	O$^-$	O–Eth$_{18/25}$	2004	[1]
Pt	1.6–2.4	Thiol	SPNMDO–NPEOPS	2006	[8]
Pt	1.5–2	HSCH$_2$CH$_2$SO$_3$H	SiPS–NM(C$_{8-10}$)$_3$	2007	[35]
Rh	~2	Thiol	SPNMDO–NPEOPS	2006	[8]
Sepiolite		Trimethoxysilyl	SPNMDO–NPEOPS	2011	[73]

(Continued)

Table 10.1 (Continued)

Core	Core dimension (nm, nm²)	Surface coupling chemistry	Surface species (reactive groups in corona)	Year	References
SiO$_2$	9–11	Trimethoxysilyl	SiPNMDD–C$_{13-15}$OEO$_7$PS	2005	[2]
SiO$_2$	7, 12	Trimethoxysilyl	SiPOEO$_{6-9}$M	2005	[6]
SiO$_2$	~7	Trimethoxysilyl	SiPNMDD–NPEOPS SiPNMDD-i-stearate	2006	[7]
SiO$_2$	13–25	Trihydroxysilyl	SiPS–Eth$_{18/25}$	2008	[21]
SiO$_2$	9–11	Trimethoxysilyl	SiPNMDD–NPEOPS	2009	[74, 75]
SiO$_2$	9–11	Trimethoxysilyl	SiPMImC$_{10}$–NPEOPS	2009	[74, 76]
SiO$_2$	9–11	Triethoxysilyl	SiPOEO$_9$C$_{15}$	2009	[76]
SiO$_2$	9–11	Triethoxysilyl	SiPOEO$_9$PC$_9$	2009	[76]
SiO$_2$	18	Trihydroxysilyl	SiPS–Eth$_{18/25}$ [–OH]	2010	[22]
SiO$_2$	18	–OH	SiPOEO$_{6-9}$M	2010	[23]
SiO$_2$	18	Trimethoxysilyl	SiPS–Eth$_{18/25}$	2010	[25]
SiO$_2$	18	Trimethoxysilyl	SiPS–Eth$_{18/25}$	2010	[23]
SiO$_2$	9–11	Trimethoxysilyl triethoxysilyl	SiPNMDD–NPEOPS SiPA (acrylate)	2009, 2010, 2011	[75–78]
SiO$_2$	9–11	Trimethoxysilyl triethoxysilyl	SiPNMDD–NPEOPS SiP–NCS	2009, 2010	[75–77, 79]
SiO$_2$	9–11	Trimethoxysilyl triethoxysilyl	SiPNMDD–NPEOPS SiP–NH$_2$	2009, 2010, 2017	[75–78]
SiO$_2$	10	Trihydroxysilyl	SiPS–NH$_3$EO$_n$H SiPS–NH$_3$PDMSNH$_2$	2011	[80]
SiO$_2$	10–15	Trihydroxysilyl	SiPS–NH$_3$PO$_{10}$EO$_{31}$M	2011	[81]
SiO$_2$	10–15	Trihydroxysilyl	SiPS–NH$_3$PO$_{29}$EO$_6$M	2011	[81]
SiO$_2$	10–15	Trihydroxysilyl	SiPS–PEI	2011	[81]
SiO$_2$	10–15	Trimethoxysilyl	SiPOPOHNHEO$_{31}$PO$_6$H	2011	[81]
SiO$_2$	10–15	Trimethoxysilyl	SiPOPOHNHEO$_6$PO$_{29}$H	2011	[81]
SiO$_2$	7–22	Trihydroxysilyl	SiPS–NH$_3$PO$_6$EOM SiPS–NH$_3$PO$_{35}$EO$_6$M	2011	[82]
SiO$_2$	7–22	Trihydroxysilyl	SiPS–NH$_3$PO$_{35}$EO$_6$M	2012	[83]
SiO$_2$	12	Trimethoxysilyl	SiPNMDD–NPEOPS	2015	[84]
SiO$_2$	12–17	Trihydroxysilyl	SiPS–Eth$_{18/10-60}$	2017	[85]
SiO$_2$	7–22	Trimethoxysilyl	SiPOPOHNB$_3$–NPEOPS	2017	[86]
SiO$_2$/MWCNT	—	–O$^-$	F108	2009	[87]
SiO$_x$ layers	—	Trichlorosilane	SiC$_{18}$H$_{37}$	2005	[3]
SiO$_2$ hollow	20–40	Trimethoxysilyl	SiPNMDD–NPEOPS	2015	[88]

(Continued)

Table 10.1 (Continued)

Core	Core dimension (nm, nm²)	Surface coupling chemistry	Surface species (reactive groups in corona)	Year	References
SiO$_2$ hollow	10–50	Trihydroxysilyl	SiPS–NH$_3$PO$_{10}$EO$_{31}$M	2018	[89]
SnO$_2$	2–3	–OH	SiPOEO$_{6-9}$M	2006	[51]
TiO$_2$	5	–OH	SiPNMDD–NPEOPS	2005	[5]
TiO$_2$	15	–OH	SiPOEO$_{6-9}$M	2010	[23]
TiO$_2$	8	–OH	SiPOEO$_{6-9}$M	2006	[90]
ZnO	3–7	Trimethoxysilyl	SiPNMDD–NPEOPS	2006	[26]
ZnO	—	Thiol	SECO$_2$–NPEO$_{10}$NM$_3$	2010	[41]
ZrO$_2$	86	Trimethoxysilyl	SiC$_{11}$ImC$_4$–TFSI	2010	[32]
ZrO$_2$	45	Trimethoxysilyl	SiPNDMC$_{18}$–NPOEO$_{10}$S	2015	[91]

Early and continuing contributions have also been made by the groups of Xiong and coworkers [38–41, 57, 58, 60, 61, 85, 91], Zheng and coworkers [37, 45, 54–56, 59, 62, 63, 65–67, 69, 73, 87, 92–95], Nakanishi and coworkers [47, 48, 50–52], and others [34, 36, 42, 46, 49, 90]. Multiwalled carbon nanotubes (MWCNTs) have been used as cores and liquefied using polydimethylsiloxane (PDMS), polyethyleneglycol (PEG) sulfonates, ammonium PO$_y$EO$_x$ oligomers, and Pluronic-type triblock copolymers. They represent the largest class of 1D solvent-free nanofluids, with other members including "cores" derived from DNA and MnSn(OH)$_6$. 2D "cores" used to produce nanofluids include the abovementioned layered silica, graphene oxide (GO), MoS$_2$, and a clay mineral, sepiolite (Mg$_4$Si$_6$O$_{15}$(OH)$_2$·6H$_2$O). Special mention should be made of zero-dimensional systems derived from cores based on phosphotungstic acid [1], POSS [15, 16], fullerene [36, 47, 49, 50, 52], fullerol [28], and ferritin [46], in addition to many very small quantum dot solvent-free nanofluids. A listing of these studies is given in Table 10.1. Structures for many of the surface-modifying species listed in Table 10.1 are presented in Scheme 10.1.

Metal oxides such as SnO$_2$ and TiO$_2$ have also been stabilized with corona of 350 Da PEGMe to produce nanofluids [90]. Relatively large (40–60 nm) nanocrystals of CaCO$_3$ were used to produce very viscous ionic nanofluids [38]. Gold nanoparticles (20 nm) surface modified with sulfoethylmercaptan and trialkyl(decyl-dodecyl) methylammonium cations to produce nanofluids have been demonstrated to be useful microelectromechanical (MEMS) switch-contact lubricants. Mercaptoacetic acid was used to synthesize CdTe quantum dots; subsequent reaction with a tertiary amine yielded CdTe nanofluids in the Zhou Laboratory [42]. Cölfen and coworkers have reported a nanofluid exhibiting a liquid crystalline phase and a liquid phase of a cationic ferritin protein, a solvent-free liquid protein [46]. Lin and Park introduced nanosilica-templated nanofluids, ionic and nonionic, that exhibit CO$_2$ sequestration activity [81]. Yu and Koch recently introduced a theoretical treatment of nanofluids cores attached to a corona of oligomers and demonstrated derivation

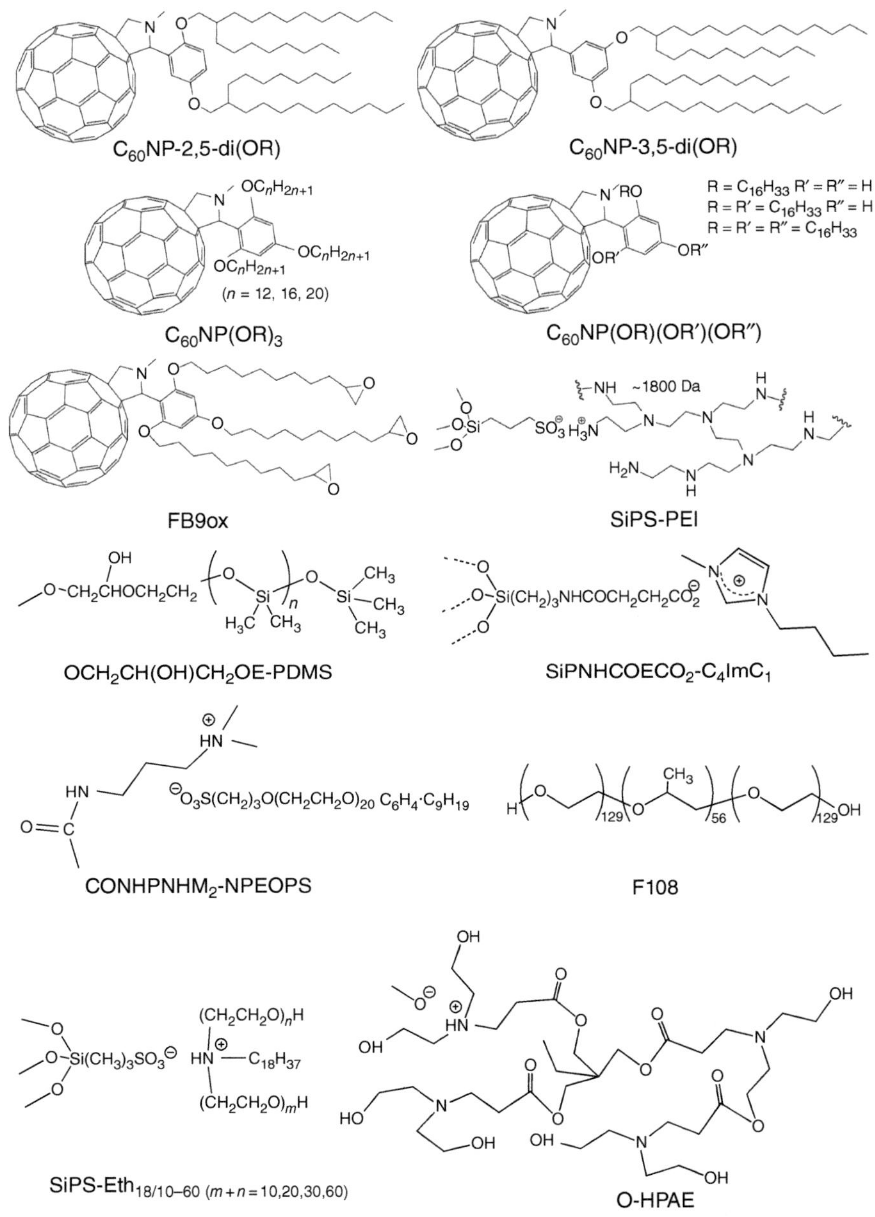

Scheme 10.1 Surface corona/cap and corona group structures.

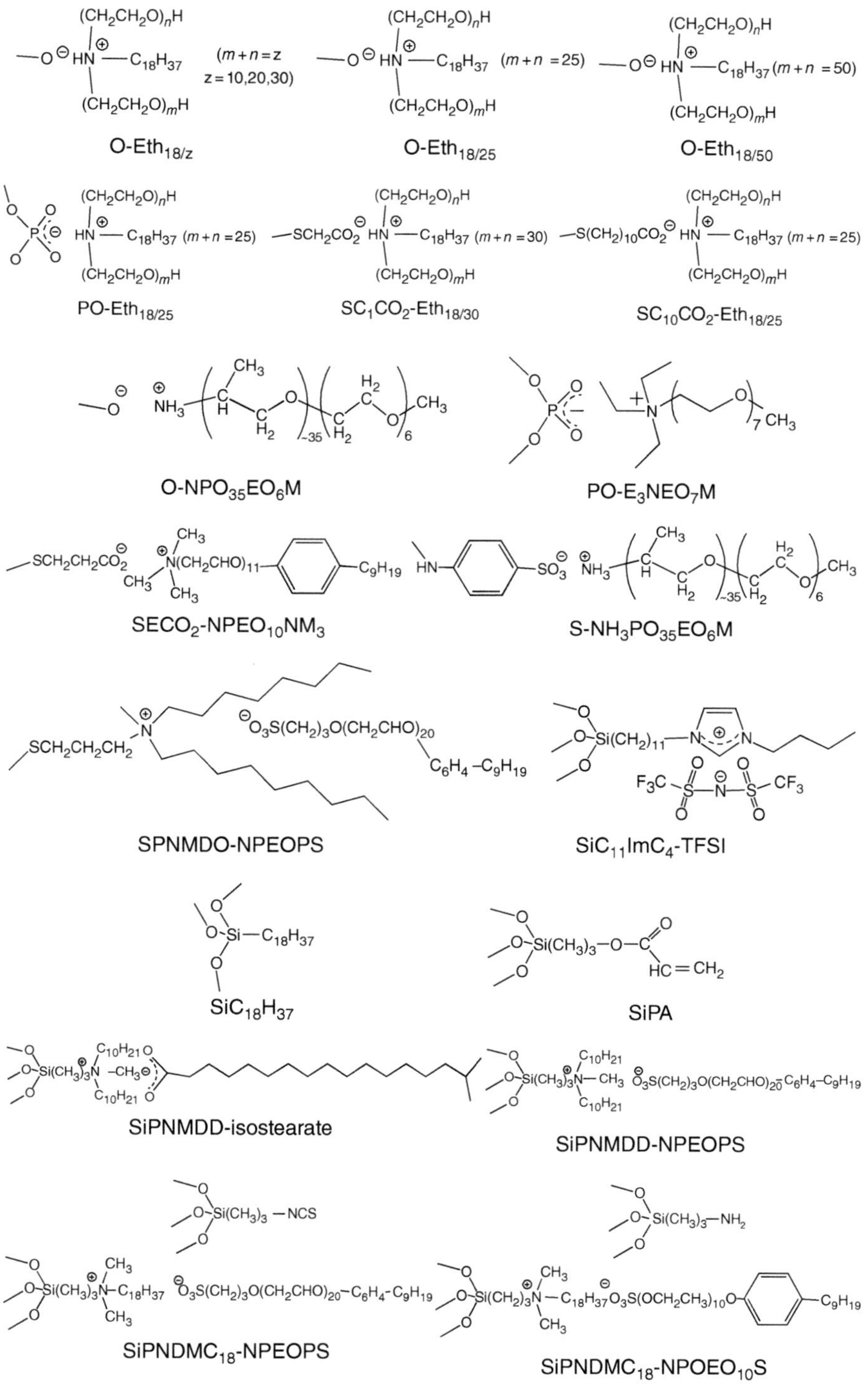

Scheme 10.1 (Continued)

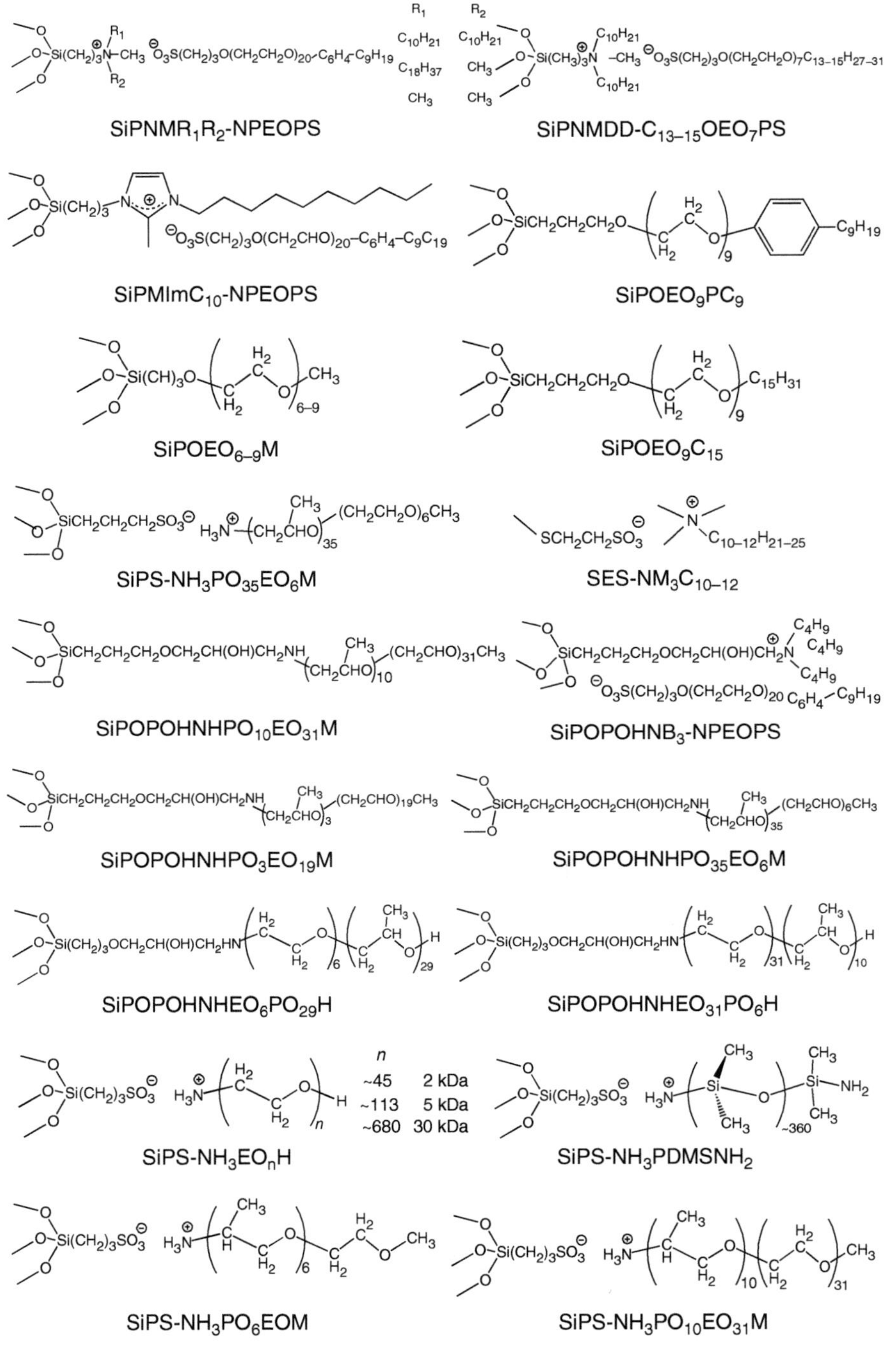

Scheme 10.1 (Continued)

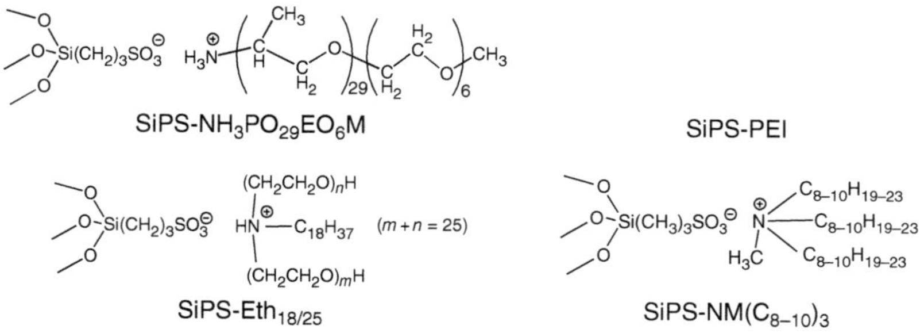

Scheme 10.1 (Continued)

of radial distribution functions among other results. This report appears to be the first theoretical treatment of inorganic–polymer hybrid nanofluids liquid structure [96] and is discussed further.

These collective results have provided a bridge between molecular liquids and ionic liquids and what we term supramolecular ionic and nonionic liquids. Zero-dimensional (0D) examples based on POSS [15, 16], fullerene [36, 39, 49–52], phosphotungstic acid [1], and ferritin [46] may be considered as having molecular cores, and as having nanoparticle cores, and, in this sense, we see a nanoscale link between molecular and particulate liquids. Various nanometals and quantum-dot-based nanofluids (Table 10.1) also bridge these limits as they grow from ~1 nm to tens of nanometers. The growth of such very small 0D nanofluids into larger 3D examples tens of nanometers in diameter to 50–90 nm in diameter as for $CaCO_3$ [38] and ZrO_2 [22, 91], respectively, defines a dimensional state of the art for spheroidal nanofluid nanoparticles. These 0D cores exemplify organic (protein), hybrid organic–inorganic, and inorganic (phosphotungstic acid, carbon, Au) materials.

One-dimensional cores span a similar range of material types exemplified by duplex oligomeric DNA (10–30 kDa, 15–45 base pairs, 2 nm in diameter and 45–140 nm in length) [5], MWCNT (<8 nm in diameter and 50–200 nm in length [60]; "ultralong" at up to 500 nm) [59, 60, 62, 63, 65–67, 92], and $MnSn(OH)_6$ threads (10- to 40-nm long) [69].

Two-dimensional cores to date that yield nanofluids are all inorganic: GO [53, 55, 93, 97]; silica [3]; and MoS_2 [70, 71]. The number of such 2D-core nanofluids is likely to grow as 2D nanosheet materials grow in number and application [98, 99]. They likely will define exciting new nonaqueous liquids for catalysis.

A significant question that persists is how large may a solvent-free nanofluid particle be and still exhibit liquidity, for example, at temperatures below 100 °C? This size most likely varies with core diameter and with molecular type and length of surface-modifying species that impart liquidity. For example, in the case of DNA where oligomeric duplex DNA < 30 kDa was liquefied with a particular ammonium counterion, longer DNA (2000 base pairs) modified similarly yielded a sticky solid that softened but did not melt below 120 °C [5]. The MWCNT nanofluids modified by complexing surface $-CO_2H$ with methyl, stearyl, PO_xEO_y amines ($Eth_{18/n\,=\,x+y}$) with core diameters <8 nm had corona–canopy shells 2- to 5-nm thick and were 50- to 200-nm long [60]. Similar surface modification yielded MWCNT-based nanofluids with 500-nm-long nanotubes [57]. These nanofluids become less viscous as PO_xEO_y diblock length increases.

Hollow-sphere-based nanofluids have been reported with an aim of using them in gas separation and storage applications [98]. Hollow silica spheres served as "hollow cores" for these nanofluids and were synthesized in aqueous triblock copolymer solutions of TEOS (tetraethylorthosilicate) or TMOS (tetramethylorthosilicate) [88]. These nanofluids were prepared by two routes (i) surface modification with a cationic didecylmethyl ammonium propyl trimethoxysilane and pairing with a large sulfonate anion [88]; (ii) surface modification with a sulfopropyl trihydroxysilane and pairing with a diblock ammonium cation [89]. A pronounced gas CO_2 permeability behavior for membrane-supported nanofluids of hollow silica was reported, and an enhancement mechanism based on rapid diffusion through sphere interiors was supported with solid-sphere-based nanofluid control data [88]. This approach was extended by making similar hollow-silica spheres surface modified with an anionic silylpropylsulfonate paired with an ammonium moiety. Hollow spheres of varying diameter produced nanofluids that decreased in viscosity with increasing diameter and increased CO_2 absorption capacity with increasing diameter [89].

10.1.2 Simulation and Theoretical Modeling

An early study examining the role of surface-modifying components [7] showed that nanofluids with stearate and bulky sulfonate anions had their glass transition temperatures and local mobility controlled by the anion and that interactions between the corona groups tuned the particle–particle spatial correlation. Nuclear magnetic resonance (NMR) analyses of solvent-free nanofluids [30] have also shown that these supramolecular objects do not diffuse too slowly for such analyses.

Modeling studies have provided some important insights. A molecular dynamics (MD) study by Chremos and Panagiotopoulos of oligomer-grafted nanoparticles showed that particle (Figure 10.1c) density variations could result in both solid and liquid behaviors [100]. Interestingly, solid behavior emerged at lower particle densities. Solid–solid transitions in a core-free solvent-free nanofluid have been reported with temperature lowering and having accompanying density lowering [17]; this example (Figure 10.1b) is morphologically different from that of Figure 10.1c. These solid–liquid differences were attributed to subtle effects of apparent particle size with varying grafting density [100]. Liquid-to-anisotropic aggregate transitions and anisotropic-to-isotropic transitions were observed with variations in temperature and with variations in particle (core) size relative to oligomer bead size (oligomers formed strings of beads) [100]. Similar MD studies augmented with density functional theory calculations [101] found that oligomer-grafted nanoparticle behaviors could be explained at low temperatures by the oligomers filling empty interstices between the (core) nanoparticles. This space filling leads to an entropic attraction between particles. Increasing temperature leads to decreasing viscosity and increasing free volume. In a related MD study focusing on dynamics [102], longer chains for solvent-free nanofluids of the type in Figure 10.1c exhibited higher diffusivities in bulk than did nanofluids with shorter chains. Shorter chain NFs appeared to crystallize at a volume fraction of about 0.2, and longer chain moieties behaved like hyperbranched and melted star polymers [102].

POSS and silica nanoparticles grafted with oligomeric poly(ethylene oxide) (PEO) were examined by MD simulations where POSS was functionalized with eight dodecameric oligomers, and 2-nm-diameter silica was grafted with 12 dodecameric oligomers and with 12 hexameric oligomers [103]. Simulations were done to estimate transport properties of these theoretical nanofluids and to contrast these properties with those of similar nanoparticle cores dispersed in nongrafted oligomeric PEO. At higher temperatures, these nanofluids had higher viscosities and lower self-diffusion coefficients than comparison non-grafted particle–oligomer mixtures. Relative dynamics increased with temperature lowering. At lower temperatures, lower viscosities and lower self-diffusion relative to non-grafted mixtures were observed [103, 104]. Coarse-grained MD simulations showed that core–core interaction potentials significantly affected radial distribution functions.

Particles of the type illustrated in Figure 10.1c and composed of semi-hollow spheres comprising uniformly distributed surface beads bonded to their nearest neighbors and a center (of sphere) bead, where these surface beads serve as tethering points, and where GD represents the percentage of surface beads tethered and N_m represents the number of bead units per oligomer, were studied using MD and a bead-spring model [81, 105]. Calculated radial core–core distribution functions having the same core volume fraction showed similar behavior; particles with shorter chains exhibited higher and more narrow peaks. Longer chains produced distribution functions that were similar to particles blended with nontethered oligomers. Increasing corona thickness produced increasing first peak positions and intensity in core–core radial distributions. Diffusivities were observed to increase with increasing oligomer length. Shear-stress simulations indicated these fluids were shear thinning [105].

A density functional theory [96] analysis of a Case 3 system comprising 10-nm-diameter cores and tethered uncharged polymer chains have provided a corroborating understanding of nanofluid self-diffusion and linear viscoelasticity [106]. They conclude that transport properties in these nanofluids are constrained by the need for the tethering oligomers to occupy inter-core interstices (and minimize free-volume).

Charged solvent-free nanofluids of the corona–cap type illustrated in Figure 10.1a have also been studied by MD simulations [107, 108]. An initial study examined negatively charged semi-featureless 1.2-nm-diameter nanoparticles with negative charges fixed on the respective particle surfaces reminiscent of an ionized hydroxyl group (Case 1) and with sulfonate charges fixed at nanoparticle surface sites (Case 2). Uncharged particles of the same size with uncharged tethered oligomers were also studied [107]. Case 1 simulations also used linear ammonium oligomers as counterions, and Case 2 simulations used tertiary ammonium ions as more sterically hindered counterions. Simulations showed that linear counterions exhibited much faster surface diffusion, hopping from one anionic site to another, than did the bulkier tri-oligooxyethylene ions in Case 2. However, calculated radial distribution functions showed preferential dimerization in Case 1, where two core particles shared in close contact a single linear ammonium cation. Case 2 systems exhibited similar core–core radial distribution functions. The tri-oligooxyethylene cations also exhibited faster jumping from one core to another. These and other observations suggested that Case 2 nanofluids exhibit lower temperature liquidity than do Case 1 nanofluids. The uncharged system exhibited greater liquidity and faster self-diffusion (1.5×10^{-8} cm^2 s^{-1}) than Cases 1 and 2 ($2 \pm 1 \times 10^{-9}$ cm^2 s^{-1}) [107].

A full extension of this Case 1 system relying on linear PEO ammonium counterions examined diffusivities, viscosities, conductivities, and chain migration kinetics on core surfaces and from one core to another [108]. Diffusivities were calculated from mean-squared displacements over 300–400 K for cation chain lengths of 14 units. Self-diffusion coefficients of about 9×10^{-10} cm^2 s^{-1} to about 6×10^{-8} cm^2 s^{-1} were calculated for the particles (cores) over this range, and the counterion self-diffusion coefficients ranged from about 2×10^{-8} to about 4×10^{-7} cm^2 s^{-1}. Observed Arrhenius behavior yielded a core-diffusion activation energy of 57 kJ mol^{-1} and a counterion-diffusion activation energy of 37 kJ mol^{-1}. Effects of counterion chain length on self-diffusion showed that core diffusion reached a lower plateau as chain length increased, and corresponded to that of a grafted nanoparticle in an excess of the grafted polymer. The counterion chain self-diffusion behavior agrees with that of a Rouse model [108].

Calculated viscosities also exhibited Arrhenius behavior with activation energy of about 35 kJ mol^{-1}. This value is similar to experimentally obtained nanofluid activation energies from viscosity measurements, such as 12–38 kJ mol^{-1} reported [17] for a nanogel of the type pictured in Figure 10.1b, wherein the counterion is a single-chain alkylaryloligoethylene oxide sulfonate. A minimum viscosity was calculated for a chain length of 14 oxyethylene units [108]. Viscosity increased almost 10-fold as chain length was reduced to four units. This effect was rationalized by the accompanying increase in core volume fraction. It was explained that increasing chain length above 14–29 units increases viscosity because the molecular weight of the oligomeric counterions becomes dominant in determining viscosity.

An experimental Case 1 system having a silica core and a cationic corona with linear anionic capping showed that viscosity decreased as grafted surface density of corona substituents increased [76, 108]. This same system showed that melting points decreased as surface grafting density increased [109]. These observations are also consistent with the theoretical studies discussed for uncharged core–corona systems.

Conductivities were estimated by computing the time derivative of the mean-squared displacements of the collective translational dipole moments of all ions. The self-diffusion of the counterions dominated this quantity. A conductivity activation energy of about 43 kJ mol^{-1} was obtained [108] and was compared to a similarly sized phosphotungstic acid ($H_3PW_{12}O_{40}$) IL [1] that exhibited activation energies for a conductivity of 44.6 kJ mol^{-1} at 144 °C and 90.6 kJ mol^{-1} at 25 °C.

It was mentioned that linear counterions exhibited relatively rapid diffusion from one surface anionic site to another [108], and this same qualitative feature was observed in this more detailed Case 1 study [109]. Besides, it was shown that transfer of counterions from one core to another is best explained as happening during near-approach collisions of cores. These simulations indicated that isolated counterions in interstices were only rarely observed [109]. While it is easy to rationalize charges not wanting to occupy free volumes that are virtual vacuum pores, the diffuseness of surface charges and counterion charges would be expected to affect such counterion transport, particularly if the chain units were good "solvents" for their respective cations.

10.1.3 Reactive Solvent-Free Nanofluids

A liquid fullerene nanofluid surface modified with alkyl glycidyl ether groups was engineered to allow photopolymerization of coatings of such nanofluids [49]. Photoinitiation resulted in cationic polymerization of the glycidyl ethers to form a cross-linked film containing 52% (w/w) fullerene. This solvent-free nanofluid had a viscosity of about 8200 Pa s at room temperature, which would appear to most observers to be solid. An Arrhenius analysis suggested that this material should have a viscosity of 1 Pa s at 114 °C. Examples of solvent-free reactive nanofluids were next reported at the international symposium on "Solvent-Free and Almost Solvent-Free Nanofluids" held at the 237th National ACS Meeting in Salt Lake City [75, 109] and the 2009 European Coatings Congress [76]. These papers and follow-up papers [43, 77–79, 110] examine reactive nanofluids based on silica cores having reactive acrylate, amino, isothiocyanate, and glycidyl ether reactive groups.

These reactive nanofluids have been used to produce new classes of resins and coatings. Amine surface functionalization produces nanofluids that can be used to produce polyurea coatings and resins. When isothiocyanate-functionalized nanofluids are made and mixed with other diisocyanate prepolymers, low-volatile organic compound (VOC) coatings that air cure into protective polyurea clearcoats can be obtained, wherein air moisture converts isocyanate (isothiocyanate) groups to amines that react with proximal isocyanates (isothiocyanates) to produce urea linkages and polyurea resins.

When such nanofluids are functionalized with acrylate groups and mixed with other acrylate monomers, zero VOC coating formulations that can be used to make clearcoats curable by ultraviolet (UV) are obtained [43, 78, 79]. Such clearcoats showed that these nanofluids are distinguishable from the same nanosilica surface modified with acrylate groups but not as a nanofluid. These nanofluids offer the possibility of increasing toughness without increasing brittleness. They also preferentially segregate to the interface and may serve as lubricants for polymer surfaces undergoing frictional shear [78]. We discuss these systems later.

Although these examples [43, 78, 79, 110] are the only purposely designed reactive nanofluids, we stress that quite a few of the nanofluids already reported have this potential intrinsically. This is particularly true of the GO-based materials that retain unreacted hydroxyl and glycidyl ether reactivity (for coupling with isocyanates, carboxylates, acid chlorides, amides, and amines). Similarly, the hydroxylated metal oxides and fullerols have this intrinsic reactivity. Other cases include surface modifications that add hydroxyl moieties in the canopy [59, 62]. More details on these reactive nanofluids are given later, with emphasis on future applications.

10.2 Syntheses of Nanofluids

Syntheses of select examples are provided to demonstrate representative approaches in the preparation of solvent-free nanofluids. These syntheses illustrate core/corona/cap structures of the type in Figure 10.1a, a corresponding core-free corona/cap structure in Figure 10.1b, and core/corona structures in Figure 10.1c. The synthesis [20] of the one example reported to date of the nanogel in Figure 10.1d is rather convoluted and not discussed further.

10.2.1 Core–Corona–Cap Nanofluid

The preparation of the seminal silica-based nanofluid illustrated in Figure 10.1 serves to illustrate the relative simplicity and accessibility of these materials. Ludox SM (30% by weight), 3.5 ml, is diluted with 20 ml of deionized (DI) water. To this suspension is added 5 ml of a 40% methanol solution of $(CH_3O)_3Si(CH_2)_3N^+(CH_3)(C_{10}H_{21})_2Cl^-$ (Gelest) [2]. A resulting white precipitate is aged 24 hours with gentle shaking. These solids are then washed three times with water, twice with ethanol, resuspended in ethanol, and dried at 70 °C. To 1 g of this intermediate chloride salt is added 15 ml of a 10.5% w/v aqueous solution of $R(OCH_2CH_2)_7O(CH_2)_3SO_3K$, where R is a mixture of C_{13} to C_{15} alkyl chains. This mixture is heated at 70 °C for 24 hours, after which the aqueous phase is removed, followed by several aqueous washes and drying at 70 °C. Differential scanning calorimetry (DSC) measurements indicate [2] a T_m at about −18 °C. The corona/cap structure, SiPDD-A_{15}OES, illustrated in Scheme 10.1 is perhaps a little more complicated than depicted. This complication is because one generally does not get exclusively silylalkoxy coupling with surface hydroxyl groups, but also obtains oligomerization of silylalkoxy groups with each other producing a hairy brush structure.

A large percentage of nanofluids of the type illustrated in Figure 10.1a have had cationic groups tethered to the core, and a detached anionic cap. The tethering of anionic salts to core surfaces with cationic caps is a useful alternative synthetic approach, and as a practical matter provides greater synthetic and design flexibility. We illustrate a synthesis of such an example which also is based on a nanosilica core. Ludox HS 30 (3 g) is slowly added to a solution of 3-(trihydroxysilyl)-1-propane sulfonic acid (4 g) and is diluted with 20 ml of DI water with vigorous stirring [12]. The pH is then adjusted to 5 using 1 M NaOH dropwise, and this suspension is heated to 70 °C for 24 hours with stirring. This suspension is then dialyzed for 48 hours against DI water (SnakeSkin dialysis tubing, 10 kDa molecular weight cutoff, MWCO), and then is passed through a column (Dowex HCR-W2 ion exchange resin) to ensure sulfonate protonation. An equivalent of Ethomeen$_{18/25}$ [$(C_{18}H_{37})N(CH_2COH_2O)_nH(CH_2CH_2O)_mH$, $n+m = 25$] is then prepared at 10% by weight in DI water and is then added dropwise to the sulfonated suspension to the equivalence point. The solvent is removed *in vacuo* at 35 °C [12] to produce a corona/cap structure, SiPS–Eth$_{18/25}$, illustrated in Scheme 10.1.

A reactive analog of the silica-based nanofluid illustrated in Figure 10.1 is obtained using two surface-modifying groups, one to impart nanofluidity and the other to impart corona reactivity for coupling to materials external to the corona. An aqueous suspension of Ludox SM (7 ml, 3.22 g SiO_2) is prepared by diluting with 40 ml DI H_2O. Then the acrylate reagent, $(CH_3O)_3Si(CH_2)_3CO_2C_2H_3$ (0.185 g), is added to a methanol solution of $(CH_3O)_3Si(CH_2)_3N(CH_3)(C_{10}H_{21})_2Cl$ (42%, 9.5 ml, 3.99 g reagent) [78]. This solution is then added slowly to the aqueous Ludox SM suspension. A white precipitate forms immediately, and this suspension is stirred at 25 °C for 24 hours, and is then washed with DI H_2O. This chloride salt (1.0 g) is then reacted with 1.5 g (20% excess) NPEOPSK (4-nonylphenyleicosa (oxyethylene)-3-sulfopropyloxy potassium) in 10-ml $CHCl_3$. This mixture is stirred at 25 °C for 24 hours. The acrylate nanofluid product is added to methanol and DI water to remove KCl and excess sulfonate salt, and dried in a vacuum oven at 25 °C. The corona/cap comprises a mixture of SiPNMDD–NPEOPS and SiPA structures (Scheme 10.1). The characterization and application of this acrylate nanofluid are discussed later (Scheme 10.2).

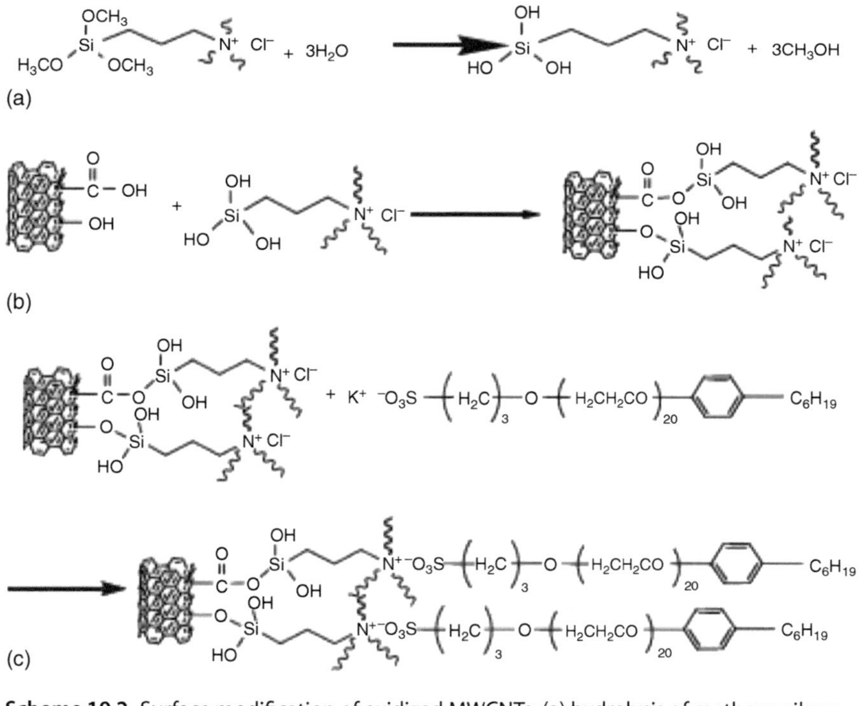

Scheme 10.2 Surface modification of oxidized MWCNTs; (a) hydrolysis of methoxy-silane coupling agent to produce hydroxy silane coupling agent; (b) condensation of a coupling agent with active surface hydroxyl groups to link silane coupling agent to surface; (c) anion exchange of soft sulfonate for chloride resulting in a solvent-free nanofluid. Source: Reproduced with permission from Ref. [67]; Copyright 2012 Springer.

Another important class of nanofluids is based on metal nanoparticle cores, and such cores are particularly suited for thiol-anchoring chemistry. Reactive nanogold solvent-free nanofluids that also have reactive hydroxyl groups in the canopy are an interesting group of nanofluids [87] yet to be anchored in resins or used as a reagent using their reactive surface hydroxyls. Their synthesis uses sodium citrate to reduce $HAuCl_4$ in the presence of 11-mercaptoundecanoic acid (MUA) as reported by Natan and coworkers [111]. After aging for 24 hours, an MUA-stabilized nanogold dispersion is reacted with Ethomeen (18/25) [$(C_{18}H_{37})N(CH_2COH_2O)_nH(CH_2CH_2O)_mH$, $n + m = 25$] to produce an anionic core ($-CO_2^-$) stabilized by a bulky ammonium cation, $H(CH_2CH_2)_mH(OCH_2CH_2)_n(C_{18}H_{37})NH^+$ where $m + n = 25$ [37]. The resulting corona/cap structure, $SC_{10}CO_2$–$Eth_{18/25}$, is illustrated in Scheme 10.1, and exhibits a T_g (−61 °C) and a T_m (1 °C); the T_m is close to that for the Ethomeen tertiary amine (4.2 °C) [37].

10.2.2 Core-Free Corona–Cap Nanofluid

Only one example of this class of nanofluids [17] (Figure 10.1b) has been reported to date. Its accessibility was postulated when an analytical ultracentrifugation analysis of a core/corona/cap material, SiPNMDD–NPEOPS (Scheme 10.1), revealed the existence of a very high molecular weight material of lower density. To a 40-ml vial is added 9.10 g of N,N-didecyl-N-methyl-N-(3-trimethoxysilylpropyl)ammonium chloride in methanol, 0.209 g of Amberlyst A26 resin, and 1.283 g water. The vial is capped and mixed for two days using a roller mill or a wrist action shaker. A yellow liquid is separated from the ion exchange resin by decanting and is used as is for anion exchange. The potassium salt, NPEOPSK, is converted to the silver salt, NPEOPSAg, by adding 4.5 g (0.0265 mol) $AgNO_3$ to 100 ml CH_2Cl_2 in a 250-ml Erlenmeyer flask wrapped with aluminum foil. NPEOPSK (26 g, 0.0206 mol) is added and stirring is continued for 24 hours. Centrifugation sediments the KNO_3 precipitate, and the supernatant is decanted and taken to dryness *in vacuo* overnight at room temperature. Yield (typically 95%) is checked by silver ion potentiometry using a silver-ion-specific electrode. The previously described chloride salt (5.03 g) is dissolved in 20 ml CH_2Cl_2 and 200 ml CH_3OH. This solution is then titrated with 0.100 N NPEOPSAg in CH_3OH to a potential corresponding, approximately, to the equivalence point. Centrifugation removes the precipitated AgCl. The product containing supernatant is decanted and dialyzed against DI water using 10 kDa MWCO SnakeSkin® tubing for seven days. The product solution is lyophilized and then dried *in vacuo* at 95 °C to obtain a pale-yellow product nanofluid (Figure 10.3).

10.2.3 Core–Corona Nanofluid

Examples of the type illustrated in Figure 10.1c that are not SILs may utilize alkyloligo(oxyethylene) and alkylaryloligo(oxyethylene) to compose coronas for core/corona structures. Suitable surface coupling agents may be prepared by condensing hydroxyl-terminated silanes with chloropropyltriethoxysilane to make the corresponding ether. For example, pentyldecylnona(oxyethylene)hydroxide ($C_{15}H_{31}(OCH_2CH_2)_9OH$; $C_{15}E_9$) when condensed with chloropropyltriethoxysilane yields a new liquid coupling agent. This coupling agent is then condensed on Ludox SM (9- to 11-nm diameter, by adding 7.0 ml of this coupling agent

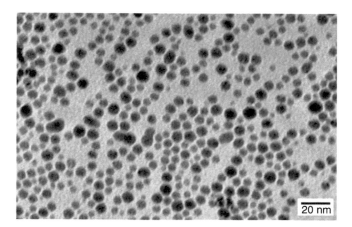

Figure 10.3 TEM of core-free nanofluids derived from autocondensation of sol–gel coupling agent, SiPNMDD; scale bar corresponds to 20 nm.

dropwise into 3.0 ml of Ludox SM (30%) suspension diluted with 15 ml of H_2O followed by stirring at 60 °C for 24 hours. The product is washed and dialyzed against DI water, lyophilized, and dried *in vacuo* at 50 °C [76]. The resulting solvent-free nanofluid has a T_g of −78 °C and a T_m of −19 °C.

10.3 UV Reactive Nanofluids

Several examples of solvent-free nanofluids susceptible to curing by UV irradiation have been reported using fullerene epoxy materials (Scheme 10.3) [49] and mixtures of SiPNMDD–NPEOPS and SiPA as corona/cap surface-modifying groups. The first one used UV-initiated cationic polymerization of epoxy groups. The second example used nanosilica as a core [77, 78] to demonstrate feasibility, and the third used nanoceria [43]. Similar surface modification, in principle, can be achieved with any nanometal-oxide core material as well as with any nano-object exhibiting significant number densities of hydroxyl groups. Acrylate and other types of vinyl surface modification are also suitable for thick films and plates using thermally initiated polymerization to produce various kinds of bulk materials, although photoinitiated UV polymerization is particularly useful for thin films and coatings.

10.3.1 Model Coatings and Thermomechanical Characterization

The reactive fullerene-epoxy NF [49] discussed earlier, in chloroform solution, was spin coated along with a UV photoinitiator, *p*-(octyloxyphenyl) phenyliodonium hexafluoroantimonate, for cationic polymerization. Very thin films exhibited hardness modulus values as a function of depth, determined by nanoindentation, ranging from about 0.32 to 0.22 GPa from 100- to 200-nm penetration depth, respectively. Unfortunately, control films obtained with the same epoxy material without fullerene, B9ox (Scheme 10.3), were not discussed.

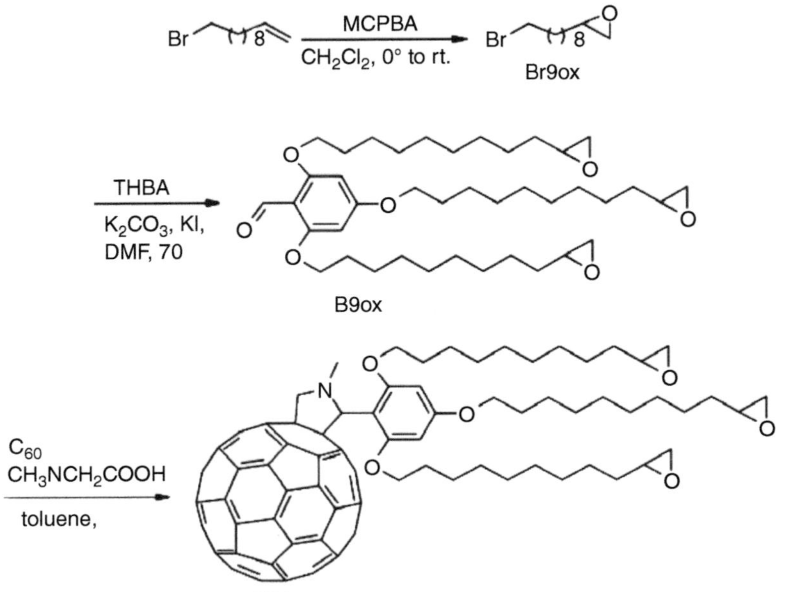

Scheme 10.3 Synthesis of alkyl glycidyl ether–functionalized fullerene-based solvent-free nanofluid (FB9ox). Source: Adapted with permission from Ref. [49]; Copyright 2008 Wiley.

It was mentioned that comparable values of hardness were reported for other epoxy resins [112], but those involved primary and secondary amine hardening of bisphenol A epoxy prepolymers, differing markedly in composition and hardening mechanism.

Nanofluids based on nanosilica with a tetraalkylsilyl ammonium surface group and a soft sulfonate counterion (SiPNMDD–NPEOPS) [75, 78] and silylpropylacrylate (SiPA) surface groups were combined with varying amounts of tetraacrylate monomer, THPETA (Scheme 10.4), to formulate a series of UV-cured coatings that showed several significant value-added features in this early report of reactive nanofluids [78].

Coatings on glass slides with 0–80% by weight nanofluids were made in combination with THPETA and cured using UV irradiation and photoinitiation with

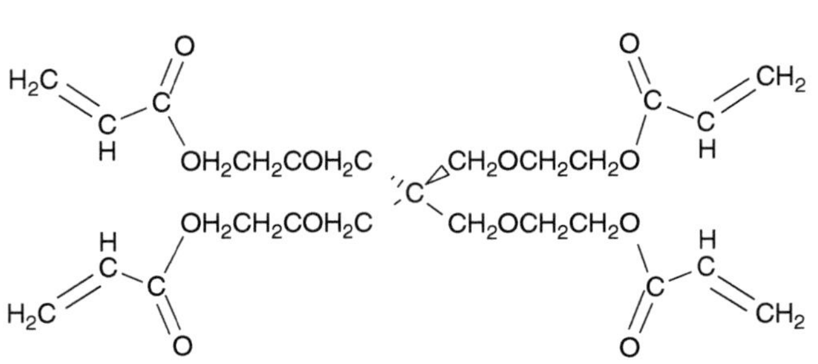

Scheme 10.4 THPETA (tetrahydroxyethylpentaerythritol tetraacrylate).

DMPA (2,2-dimethoxy-2-phenylacetophenone). Additional control coatings were also made using the same core silica with only acrylate surface modification, $(SiO_2)_{PA}$ in combination with THPETA. Nanoindenting, atomic force microscopy (AFM), and DSC were used to characterize several material effects.

At low loading, 0.5–5% NF, and only several weeks equilibration at ambient after curing, minor increases in storage modulus (up to 12% relative to a 3.2 GPa value of the THPETA matrix control) were reported [78] at various probe penetration depths (1000–3000 nm). After equilibration at ambient for 10 months, deviations at these same penetration depths from the matrix control were not experimentally significant. Such added NF increases toughness, and the significance of this aspect is discussed at greater length in Section 10.5.

At very high loadings of 50–80% by weight, it was observed that both storage modulus and hardness decreased significantly with increasing NF loading. Here, it was pointed out [78] that inserting such a liquid nanofiller into a matrix can result in material softening. Softening can be understood when a matrix-continuous phase is harder than the filler phase. This relationship can be expressed by a simple model [78, 113, 114]:

$$E_{res} = E_{mat} + (\chi_p E_p - E_{mat})\varphi_p$$

where the subscripts "res, mat, and p" refer, respectively, to final resin or composite, matrix material, and nanofiller particle material; E refers to storage or hardness modulus; χ_p is a parameter between 0 and 1; and φ_p is volume fraction of nanofiller. This equation fits the experimental data at high loading almost quantitatively when the small modulus of this nanofluid is discounted (~0), as it is orders of magnitude less than E_{mat}.

DSC of the very highly nanofluid-loaded coatings (50–80% NF) showed two trends with increasing NF loading: (i) T_g (not evident in matrix control) shifts from −62 to −55 °C (−64 °C in neat NF) and the magnitude of the endothermic heat flow also increases with loading; (ii) a small and endothermic T_m (not present in the control matrix) shifts from 12 °C to about 30 °C. This nanofluid has a T_g of about −64 °C and a T_m at about 13 °C. These T_g and T_m in the coatings, therefore, most likely emanate from the corresponding thermal transitions in the neat nanofluid. Such thermal effects suggest that these composite films could serve as thermally switchable barriers to small molecule permeation and transport. It was suggested that the T_m endotherms indicate that some of the nanofluid particles form attractive clusters that melt on warming, and above melting a similarly positive effect on permeability likely would be realized.

An unexpected result was reported for such coatings at various contents of nanofluids. AFM studies of elevation and phase at the composite/air interface for a resin loaded with 2% by weight NF is illustrated in Figure 10.4. Both approaches (elevation and phase) clearly illustrate the same particle shapes at the interface, and the elevation data show that these NF nanoparticles protrude almost 50% out of the resin, or 50% above the resin/air interface. A detailed analysis indicated that the surface concentration of such nanoparticles greatly exceeded that expected for a uniform distribution of particles in the matrix and that it shows marked preferential surface segregation. This segregation is a phenomenon well known for nanoparticle stabilizers at liquid/liquid interfaces in Pickering

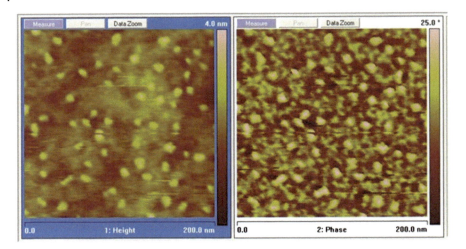

Figure 10.4 AFM tapping mode images (left, elevation, and right phase) of 2% (w/w) nanofluid in THPETA after UV curing. Source: Reproduced with permission of Texter et al. [78]. Copyright 2011, Royal Society of Chemistry.

emulsions [115, 116], but has never previously been documented by AFM at liquid/air or composite/air interfaces. In addition, because of the well-known lubricity of some ionic liquids [117, 118] and for certain solvent-free nanofluids [3, 25, 36], it was proposed [78] that such surface-segregated nanofluid particles in composites may serve as useful lubricants for composite surface/solid interfaces. Tribology support for such conjecture has not yet been reported.

Preferential surface segregation of the nanofluid illustrated in the AFM images in Figure 10.4, and softness of these particles (arising from the ionic liquid-like shell), suggests such interfaces may be self-lubricating. This possible application is in line with suggestions [3, 25, 119] that bulk nanofluid (NIM) may serve as a lubricant. Such nanofluids may be formulated over a wide range in intrinsic viscosity, and their surface activity can be tuned for various applications by modifying counterion type and the ionic liquid type of surface-modifying groups. Hardness data for these nanofluid-containing samples have shown that the surface is softer than the bulk. This aspect is undoubtedly due to an incomplete reaction of acrylate groups close to the surface. Nanofluid shell ionic liquid presumably adds to this softness, and may, therefore, be a useful variable in formulating polymer coatings for tribological applications that may be subject to high shear. Having nanofluid particles partially protruding from surfaces may, in any case, make lubrication with similar nanofluids, as earlier proposed [3, 119], a synergistic effect.

A 2D layered material, $Mg(OH)_2$, was transformed into a reactive solvent-free nanofluid by means similar to those described for the example discussed (Figure 10.2). The $Mg(OH)_2$ cores were 1500–2500 nm in diameter and 50–100 nm thick. UV reactivity was imparted by coating with SiPA and a cationic octadecyldimethyl(3-trimethoxysilylpropyl) ammonium chloride surface group was matched (chloride exchanged) with the soft sulfonate counterion ($SiPNDMC_{18}$-$NPOEO_{10}S$) to promote viscous liquidity [68]. Polyethyleneglycol diacrylate (PEGDA; 600 Da) was used to form a host matrix in which this Mg(OH)$_2$ nanofluid was dispersed and polymerized by UV photoinitiation.

It was noted that, when cured, films of these nanofluids in PEGDA exhibited many particles oriented parallel to composite–air interfaces at composite surfaces. This preferential segregation is a consequence of a Pickering stabilization phenomenon, now well known for graphene and GO [116]. PEGDA solutions of these nanofluids were prepared at nanofluid levels 0%, 10%, 40%, 50%, and 60% by weight nanofluid. Cotton composites were prepared by coating these fluids on 5 cm × 5 cm swatches, and then exposing to UV to induce curing. The 40% and 50% nanofluid composite coatings yielded increased mechanical properties relative to the matrix polymer control in tensile strength and Young's modulus [68].

10.3.2 UV Protective Coatings

An example, among many other possibilities, is the formulation of another solvent-free nanofluid of the type illustrated in Figure 10.1a with a cerium oxide (ceria, CeO$_2$) core. Films and nanoparticles of ceria are of interest because of ceria's intrinsic UV absorbance and its photocatalytic and other properties [120–122]. A particularly interesting property is that when used as an additive in diesel fuel [123], the amounts of nitroxides and soot produced by combustion are decreased concomitantly with increased combustion efficiency [124, 125]. Such a nanoceria was used by Maniglia et al. [43] to produce a solvent-free nanofluid using the same types of surface modification used in the reactive acrylate system discussed in the preceding section. A nominally 6-nm-diameter nanoceria was used, and quite a large corona of cross-linked SiPNMDD–NPEOPS and SiPA groups (cross-linked through sol–gel condensation of trialkoxy silanes) was obtained; the corona volume fraction was 0.87, and the corresponding weight fraction was 0.69.

UV-cured coatings of ceria-based nanofluid in combination with EGDMA (ethylene glycol dimethacrylate) and with a nominally 400 Da polyoxyethylene diacrylate yielded highly cross-linked films with absorption coefficients $\alpha_{350nm} = 6.6 \pm 0.8 \text{ cm}^2 \text{ mg}^{-1}$ and $\alpha_{300nm} = 24.5 \pm 3.5 \text{ cm}^2 \text{ mg}^{-1}$ [43]. Average near-UV protection over 300–350 nm of 1–3 optical density units can be obtained with 0.065–0.19 mg cm^{-2} of such a ceria-based nanofluid; representative absorbance spectra are illustrated in Figure 10.5. A practical application of such clearcoats is to lengthen the service life of plastic materials in an exterior ambient application since such UV protection mitigates against radical formation and bond cleavage in epoxy-, polyurethane/polyurea-, and polyester-based materials emanating from phenyl groups and heteroatom lone pair electrons.

10.4 Polyurethane and Polyurea Coupling of Nanofluids

Polyurethane and polyurea coupling are essential classes of condensation chemistries. Polyurethanes usually are formed by addition condensation coupling of isocyanates with hydroxyl functionalities. Amines readily react with isocyanates to form urea linkages, and isocyanates and isothiocyanates

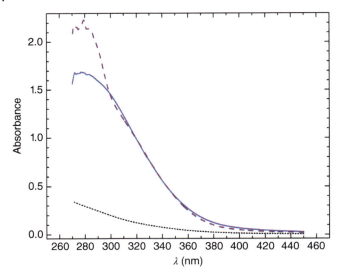

Figure 10.5 UV absorption spectra of solvent-free nanoceria nanofluid coatings: 10.5% ceria nanofluid, photoinitiated with DMPA (0.6%) using H-bulb UV (_____); 9.7% ceria nanofluid, photoinitiated with Darocur 1173 (7.3%)/DMPA (1.1%) mixture using D-bulb UV (– – – –) and of the control coating (·········) of THPETA resin (Scheme 10.4) without any nanofluid. Source: Adapted with permission from Maniglia et al. [43]. Copyright 2017, Elsevier.

slowly react with water molecules to lose CO_2 or COS, respectively, to form primary amines. These relatively simple chemistries are the basis for air-cured polyurethane/polyuria coatings. Here, we show a couple of approaches to making such air-cured coatings using reactive nanofluids in combination with polyurethane prepolymers and with a popular aliphatic diisocyanate monomer, IPDI (isophorone diisocyanate).

10.4.1 Air-Cured Polyurethane Coupling with Isothiocyanate Nanofluid

Air-cured (humid air) clearcoats involving polyurethane prepolymers and polyurea coupling were reported [77, 79] using nanosilica-based isothiocyanate nanofluids in combination with a prepolymer, polypropylene oxide-isophorone diisocyante, PPG-di-IPDI (Scheme 10.5). The reactive isothiocyanate groups of this nanofluid and isocyanate groups from the prepolymer have no facile

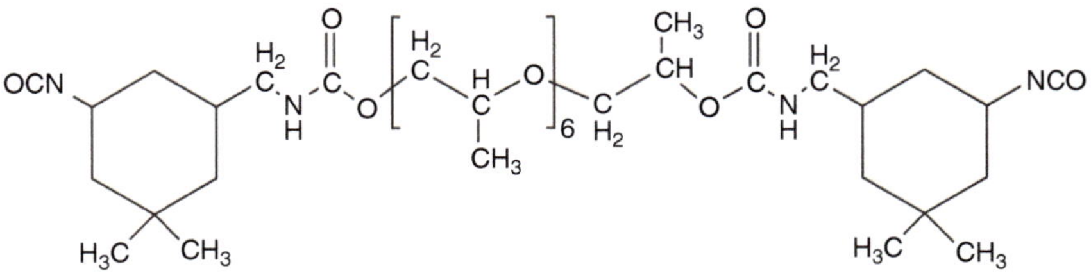

Scheme 10.5 PPG-di-IPDI.

chemistry with each other, but both react slowly with water adsorbed from the ambient humid air. This slow reaction triggers loss of COS or CO_2 and yields reactive primary amines that react quickly with proximal isothiocyanates or isocyanates to form urea linkages. These chemistries, along with other secondary coupling reactions [79], include primary chain extension and cross-linking urea linkage formation by addition reaction of amines with isocyanates or isothiocyanates. A reaction of nanofluid-NCS with nanofluid-amine or with isophorone (prepolymer) amine can yield a thiourea linkage. Similarly, a reaction of nanofluid-NCS with a urea-NH can form a thiobiuret, and a reaction of nanofluid-NCS with a urethane NH can form a thioallophanate [79]. Thus, there are four reactions leading to chain extension or cross-linking and another four less facile cross-linking reactions forming biuret and allophonate linkages.

Initial coatings obtained by mixing isothiocyanate-functionalized nanofluids with hexamethylene diisocyanate produced hard air-cured coatings that were opaque [79]. This opacity resulted from such solvent-free coatings being too viscous and thick (~12 μm), so that when air curing occurred, too much of the CO_2 and COS gas produced on reaction with water molecules was trapped within the coating as highly light scattering bubbles. A skin effect at the coating/air interface initiated this effect. These bubbles accounted for the opacity observed.

Liquid and miscible mixtures of $(SiO_2)_{NSO_3,NCS}$ nanofluid, and PPG-di-IPDI were mixed with some MEK (methyl ethyl ketone) to facilitate making thinner coatings, and these were air cured successfully to yield transparent films [79]. Photographs of a successfully cured coating are shown in Figure 10.6; curing conditions were 24 °C and 44% RH (relative humidity). The photograph in Figure 10.6a was taken with the coating (on a glass slide) on top of black paper to accentuate any light scattering. One can see only a little scattering from the right and left edges of the coating, attesting to the clarity of the coating. Figure 10.6b illustrates a coating photographed over text from the seminal paper [2], and we see the yellow color of these coatings. This yellow comes from the isothiocyanate groups of the $(SiO_2)_{NSO_3,NCS}$ nanofluids after predominantly forming thiourea linkages.

Various formulations were coated to explore nanofluid to PPG-di-IPDI prepolymer weight ratios of 0–16% by weight. Coatings were made on slide covers to facilitate nanoindenter mechanical analysis, and these coatings were cured at 50% RH for seven days at 25 °C. The storage modulus of the PPG-di-IPDI cross-linked matrix was about 2.1–2.2 GPa. Nanofluid weight fractions of 0.01 and 0.02 yielded moduli of 2.4–2.7 GPa. At 0.04 weight fraction, the modulus is intermediate to these values; at 0.08 weight fraction, it falls below the control value, 1.8–2.2 GPa; and at 0.16 weight fraction, it falls significantly to 1.2–2 GPa. A similar weight fraction series of hardness moduli exhibit similar values relative to the control. So we again see that adding nanofluids, at sufficiently large weight fractions, decreases the storage moduli and hardness moduli and leads eventually to softening.

Photographs of the indentations made by a Berkovich pyramidal tip illustrated in Figure 10.7 show that the PPG-di-IPDI control coating undergoes viscoelastic recovery. We see that the illustrated pyramidal indentation in Figure 10.7a has nearly completely recovered. However, an indentation in a coating with 4% weight

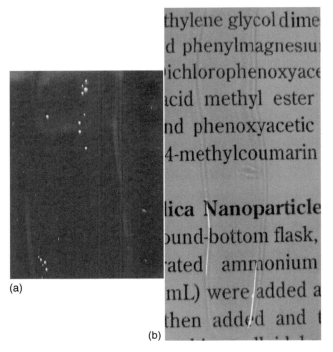

Figure 10.6 Photographs of resin composed of PPG-di-IPDI and $(SiO_2)_{NSO_3,NCS}$ nanofluid (equimolar in NCO and NCS) after curing; (a) clearcoat-coated slide with black background showing slight visibility of light being scattered from coating edges; (b) clearcoat on top of text from seminal Giannelis and coworkers paper [2]. Source: Reproduced with permission from Texter et al. [79]. © 2013, Wiley.

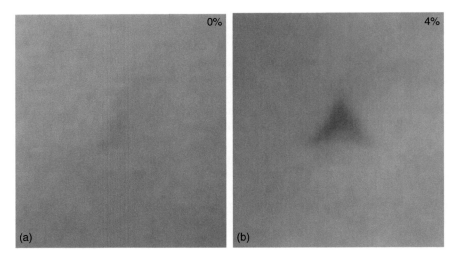

Figure 10.7 Comparison of nanoindentations made in (a) a control coating of PPG-di-IPDI and in (b) a 4% $(SiO_2)_{NSO_3,NCS}$/PPG-di-IPDI coating; magnification of the 0% NF coating (a) is twice that of the NF-containing coating (b). Source: Adapted with permission from Texter et al. [79]. Copyright 2013, Wiley.

fraction NF shows no sign of viscoelastic recovery. It was suggested that such indentations exceeded the linear viscoelastic regime, as the extant cross-linking in these slightly harder coatings results in greater thermoset-like behavior.

10.4.2 Air-Cured TDI Coupling with Amino Nanofluid

An amine-functionalized nanofluid was mixed with tolyldiisocyanate (TDI), the most widely used monomer for making polyurethanes, and allowed to air cure at 44% RH and 25 °C [77, 110]. We expect primary amines on such nanofluid to react with TDI on mixing to convert their reactive corona from an amine to isocyanates, by forming a urea linkage with one of the isocyanate groups on TDI. During ambient curing in humid air, water molecules trigger CO_2 ejection from both types of isocyanates and yield amines highly reactive for forming urea linkages with proximal isocyanate groups.

Coatings made by mixing amino-nanofluid and TDI at weight fractions of 0, 0.3, 0.4, 0.5, 0.6, 0.7, 0.8, and 0.9 were incubated at 25 °C and 44% RH. Coatings from 0 to 0.8 weight fraction nanofluid yielded an interesting variation in optical and physical properties. The zero weight fraction control produced a granular powder, and a slightly more cohesive and brittle powder agglomerate was obtained at 0.3 weight fraction. At 0.4 weight fraction nanofluid, the coating was hard but opaque due to entrapped CO_2; the polymerized material at the air/coating interface was too impermeable for CO_2 to escape, although not for water to diffuse in. At higher nanofluid weight fractions, the cured coatings become transparent. From 0.4 to 0.8 weight fraction nanofluid, the coatings also were hard and tack-free. At contents higher than 0.8 weight fraction, tacky coatings were obtained.

Storage (elastic) modulus data for these coatings are given in Figure 10.9 for measurements made by nanoindenting at various penetration depths. The storage modulus in these coatings appears insensitive to penetration depth. The modulus decreases steadily with increasing nanofluid content, and drops over an order of magnitude from 0.3 to 0.8 weight fraction $(SiO_2)_{NSO_3,NH_2}$ (5.3–0.21 GPa). On increasing nanofluid content to 0.9 weight fraction, this modulus decreases another order of magnitude (0.21–0.013 GPa).

Rahman and Brazel studied conventional ionic liquids as plasticizers for polyvinyl chloride [126]. Ionic liquids are novel plasticizers, but they are expected to exhibit leaching problems similar to those already well known from dialkyl phthalates. Using nanofluids as plasticizers may offer advantages in some applications, but leaching must still be expected in long service life applications. We can expect that such nanofluids, SILs, will exhibit nearly zero vapor pressure, which is an environmental plus.

Reactive nanofluids are bound to their host matrix and cannot leach. This property makes them superior to extant plasticizers. The data in Figure 10.8 demonstrate that such plasticization can be tuned over different ranges in modulus. The acrylate nanofluids discussed earlier may be ideally suited for certain applications, such as those involving PVC (polyvinylchloride) and PMMA (polymethylmethacrylate), particularly for making new types of materials that are very flexible. We have seen that weight fraction can dramatically affect

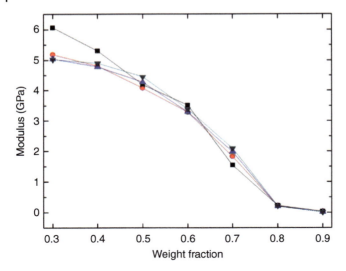

Figure 10.8 Storage modulus (from nanoindenting experiments) for TDI–(SiO$_2$)$_{NSO_3,NH_2}$ polyurea resins at different penetration depths (■–100 nm; ●–1000 nm; ▲–2000 nm; ▼–3000 nm) as a function of (SiO$_2$)$_{NSO_3,NH_2}$ weight fraction.

modulus and associated plasticization. It is also possible to dramatically modify these properties by varying counterion type. Such reactive nanofluids represent a new and useful class of SIL plasticizers.

10.4.3 Polyurethane Shape-Memory Materials

Nanofluids comprising a corona derived by coupling 3-(HO$_3$S) propyltrihydroxysilane with 10-nm-diameter silica were complexed with α,ω-diOH-PEO (poly[ethylene oxide]) oligomers of 2, 5, and 30 kDa molecular weight. These liquid-like nanofluids were then reacted with hexamethylene diisocyanate (HMDI) in chloroform to form viscoelastic films by curing these mixtures (with excess HMDI) in Teflon molds at 70 °C [80]. DSC, elastic modulus (E'), and shape-memory behavior were examined. A comparison film derived from the 5 kDa capping α-NH$_2$,ω-OH-PEO and reacted with an excess of HMDI (no nanofluid) exhibited a pronounced melting endotherm at 54 °C. This melting point agrees well with melting points of 5.1 kDa PEO blocks in diblock copolymers of poly(butadiene-b-ethylene oxide) (nearly equal mole ratio of monomer units) [127]. Pure PEO has a glass transition between −55 and −48 °C and a melting point at about 65–68 °C [128]. Glass transitions were not visible in the DSC data provided for nanofluid volume fraction of resulting resins (shape-memory polymers) at 0%, 10%, 13%, and 16%. However, only data above 0 °C were provided, and these transitions would be expected to be seen at much lower temperatures.

Elastic modulus (E') data over 10–16% volume fraction exhibited low-temperature (−100 °C) values increasing from 2 to 5 GPa with increasing nanoparticle volume fraction. Each composition exhibited significant exponential decreases to 700 MPa (16%), 300 MPa (13%), and 200 MPa (10%) as temperature increased to about −5, 1, and 19 °C, respectively. As temperature increased further, more dramatic decreases in E' were reported to about 15 MPa at 10 °C (16%), about 5 MPa at 23 °C (13%), and to about 1.6 MPa at 28 °C (10%). It was inferred that these last transitions corresponded to melting of PEO oligomers [80], although there was little to no overlap with the DSC endotherms. Disparities may be due to scan rate effects, although changes in the E' data indicate more complex thermal behaviors than evident in the DSC data.

Elastic modulus data were also reported for films prepared with varying PEG molecular weights of 2, 5, and 30 kDa with nanoparticle volume fractions of 18.1%, 10%, and 12.2%, respectively. Low-temperature (−100 to −70 °C) values increased with increasing volume fraction from 2 to 2.2 GPa, to about 6 GPa. These materials also exhibited a two-stage decreasing E' behavior with the second stage being more steep than the first. It was pointed out that the PEO molecular weight could be used to tailor the transition temperature of the most dramatic change in modulus [80]; the same can be said of tailoring with nanoparticle volume fraction.

Shape-memory effects were illustrated by folding a square sheet of film in half and then again in half with a second fold perpendicular to the first fold and fixing this material in place by cooling to −10 °C [80]. This folded sheet was seen to unfold essentially completely by raising its temperature to about 60 °C.

10.4.4 PDMS-Amino Nanofluids Coupling with HMDI

A nanofluid similar to that discussed in the previous section having a sulfonic acid corona and a cap composed of α,ω-diamino-PDMS oligomers of 27 kDa molecular weight was reported to make highly elastic polyuria films [80]. The amino group proximal to the corona accepts the sulfonic acid proton to become a linear ammonium moiety. This solvent-free nanofluid was mixed with excess HMDI in chloroform and cured in a Teflon mold by heating at 70 °C to remove the solvent. Silica core content after curing was 18% by weight. While such films purportedly exemplify a shape-memory polymeric system, supporting data for such behavior was not included. However, the elastic modulus for such a film was reported over the −100 to 50 °C interval. The elastic modulus, E', at −100 °C was slightly greater than 1 GPa. As temperature increased, E' decreased exponentially to about 500 MPa at −50 °C. After that, E' plunged precipitously to about 2.5 MPa at −30 °C and rose a bit to 2.8 MPa through 45 °C. Such large decreases in E' are typically accompanied by a peak in loss modulus generally assigned to a glass transition, but DSC data were not provided.

Studies of α,ω-diOH-PDMS cross-linked with tetraethoxysilane report this transition at about −110 °C [80]. We would expect this α,ω-diamino-PDMS system, when cross-linked with HMDI, to be nonexhaustively cross-linked because of differences in coupling of isocyanates with a primary amine and with urea-linkage NH-groups [77, 79] where a biuret cross-linked product can be obtained on coupling with an additional isocyanate. Cross-linked α,ω-diOH-PDMS systems all exhibit a pronounced melting endotherm at about −50 °C [80], and they exhibit E' of about 10 GPa below their glass transition.

Therefore, we suspect that this transition observed with this cross-linked α,ω-diamino-PDMS system is a melting transition. Measurements of both elastic and loss moduli to temperatures as low as −140 °C and DSC measurements should resolve these uncertainties. Another chemical pathway not previously discussed involves a reaction of HMDI with the surface-proximal ends of the diamine caps. Those sulfonate-proximal ammonium groups would be susceptible to the same coupling chemistry exhibited by the distal amino groups. Such new materials do offer intriguing new material possibilities.

10.4.5 Polyurethane Coupling with Hydroxyl Nanofluid

The same type of nanosilica discussed earlier was surface modified with glycidyl ether groups using 3-glycidyloxypropyl-trimethoxysilane (GPTMS) [86]. This glycidyl ether group was then converted by reacting it with tributylamine with HCl to produce an ammonium group with a proximal hydroxyl group for polyurethane coupling. The soft sulfonate NPEOPS anion was substituted as a counterion to make these nanosilica core materials room-temperature solvent-free nanofluids. This nanofluid was used to modify a matrix polyurethane derived from MDI (4,4′-diphenyl methane diisocyanates; 5 mequiv) and polytetramethylene ether glycol (PTMEG 1000; 2.5 mequiv), by extending this reaction with 2.25 mequiv ethylene glycol and varying amounts of nanofluid, amounting to 0%, 2%, 4%, 6%, and 8% by weight relative to the matrix monomers, the ethylene glycol, and this nanofluid.

This nanofluid behaves as an exotic cross-linking agent because of its hydroxyl functionality. Mechanical tests showed that tensile strength increased with added nanofluid to 4% and then decreased to the control matrix level at 6% and fell below this level at 8%. This behavior is similar to the impact of added nanofluid discussed earlier in Section 10.4.1 and illustrated in Figures 10.6 and 10.7. These variations in tensile strength also appeared reflective in apparent Young's modulus evident from their stress–strain curves [86]. Surface resistivity measurements also showed that the partial ionic character of these nanofluids produced films that exhibited sufficient ionic conductivity to serve as antistatic agents.

10.5 Epoxy Coupling with Amino Nanofluid

The previously discussed amino nanofluid having a tetra-alkylammonium and amino corona and sulfonate cap was used as a variable additive to make some amine-hardened epoxy resins [109, 129]. The epoxy moiety used was poly(propylene glycol) diglycidyl ether with a number average molecular weight of 640 Da. An epoxy-to-amine mole ratio of 1.5 was maintained using an auxiliary amine, poly(propylene glycol) bis(2-aminopropyl ether) with a number average molecular weight of 230 Da, and varying amounts of amino nanofluid to make the formulations examined 0%, 4%, 16%, and 32% by weight amino nanofluid. Reaction storage modulus was examined using parallel plate rheometry (at 1 Hz) as a function of reaction time, and an initial induction time was observed to increase with increasing amino nanofluid content. About the same storage modulus plateau of 0.1 MPa was observed in each system at a reaction temperature of 90 °C.

Two qualitative effects were seen: (i) An increasing induction time was observed with increasing amounts of incorporated amino nanofluid. (ii) The final storage modulus value decreased from about 0.15 MPa at 0% nanofluid to about 0.07 MPa with 32% nanofluid. It was hypothesized [129] that the reactive amine in the nanofluid corona may have been buried among a larger grafted density of didecylmethylammonium groups that sterically obstructed access to the nanofluid amines by the glycidyl ether moieties, thereby causing an increase in induction time. A somewhat different and somewhat equivalent effect may be put forward as a partitioning effect, wherein the glycidyl ether moiety, based on poly(propylene oxide), is more compatible with the auxiliary amine moiety also based on poly(propylene oxide). The softening seen in the longtime decrease in final storage modulus has been seen in several examples discussed early in this chapter, where the nanofluid is fundamentally softer than the reference amine matrix.

10.6 Using Nanofluids to Make Composites Tougher

The impact of nanofillers on composite moduli is usually to increase storage modulus (and Young's modulus) and to increase brittleness. When a nanofiller modulus is greater than a host matrix modulus, as is usually the case, the composite modulus increases with nanofiller loading and toughness decreases (e.g. as measured by the area under an engineering stress–strain curve). The equation describing volume fraction effects on composite modulus given in Section 10.3.1 provides loading effects whether the nanofiller modulus is higher than that of the matrix or less than that of the matrix. The use of reactive nanofluids in composites appears to offer one the ability to increase modulus or to maintain modulus while increasing toughness.

10.6.1 Nanosilica Polyacrylate Nanocomposites

Surface-modified reactive nanoparticles (nanofillers) are believed to be the best way to incorporate nanofillers into composites. The chemical ties between matrix and nanoparticle make stress transfer from the continuous phase to the discontinuous phase more efficient. Reactive nanofluids offer similarly strong coupling between phases; but in the case of reactive nanofluids, these interactions mitigate against brittleness and lead to increased toughness. A comparison of these two types of nanofillers is given in the hardness moduli exhibited in Figure 10.9. One set of data are for a case where a nanosilica is prepared with its surface modified with acrylate groups, SiPA, but no SIL component. The other set uses the same nanosilica nanofiller, but the surface modification uses both SiPA functionalization as well as SiPNMDD–NPEOPS functionalization. Separately, we see in Figure 10.9 that the acrylate-modified silica increases modulus with loading and that the acrylate nanofluids decreases modulus with increasing loading. Together, they would maintain modulus while increasing toughness.

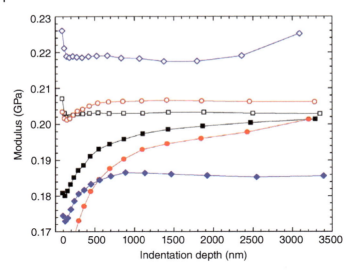

Figure 10.9 Hardness modulus as a function of penetration depth for 1 (■,□), 2 (●,○), and 5% (w/w) (◆,◇) acrylate sulfonate nanofluid (■,●,◆) in acrylate resin and for acrylate-modified nanosilica (□,○,◇) in acrylate resin. Source: Reproduced with permission from Texter et al. [79]; Copyright 2011, Royal Chemical Society.

10.6.2 MWCNT Polyamide Nanocomposites

(MWCNT)$_{SiODDMN-NPEOS}$ nanofluid as prepared by Lei et al. [57], was used to toughen polyamide nanocomposites by melt blending and compared with control MWCNT/polyamide blends produced by melt blending [61]. Nanocomposite blends were prepared at MWCNT weight concentrations of 0–2% MWCNT for both nanofluids and regular MWCNT nanocomposites. Tensile strength and percent elongation (at break) measurements were obtained using a universal testing machine. The control materials increased to a tensile strength at 0.5% loading of about 150 MPa and decreased to about 135 to 130 MPa as loading increased to 1–2%. The brittleness accompanying this increase in modulus dropped from an elongation percent of about 230% at 0% loading to about 50% at 0.5% loading to about 30% at 2% loading. This is a typical type of brittleness onset seen with direct incorporation of nanofillers into polymeric matrices. The (MWCNT)$_{SiN}$ (SiN = SiPNDMC$_{18}$–NPOEO$_{10}$S) nanofluid-loaded nanocomposites exhibited interesting alternative behaviors. The tensile strength at 0% loading, 110 MPa, increased to about 133 MPa at 0.5% loading with a concomitant increase to 284% elongation at break. This shows an increase in modulus with a simultaneous increase in toughness, opposite of the effect seen with the non-nanofluid MWCNT control. Further nanofluids loading results in further tensile strength increase to about 134 MPa and then a small decrease to about 130 MPa at 2% loading [61]. The elongation at break gradually decreases with loading to about 140% at 2% loading, more than four times that obtained with the control formulation.

10.6.3 MnSn(OH)$_6$ Thread Epoxy Nanocomposites

The threadlike nanofluids developed by Li et al. [69] were used to toughen epoxy resins based on a bisphenol A prepolymer epoxy (CYD-128) having an average equivalent weight of 189 Da, methyl tetrahydrophthalic anhydride (MeTHPA) as curing agent, and 2,4,6-tris(dimethylaminomethyl)phenol (DMP-30) as an accelerator. MnSn(OH)$_6$ threads with an SiPS–H$_2$NPO$_{10}$EO$_{31}$M cap and canopy were formulated with the abovementioned resin-forming materials at weight concentrations of 0–1.6% and cured at a sequence of temperatures from 90 to 140 °C for 30- to 90-minutes heating periods. A control material was formulated with the same epoxy resin-forming materials, and unmodified crystalline MnSn(OH)$_6$ was used as nanofiller over the same loading interval. Flexural strength and impact toughness were measured for MnSn(OH)$_{6[SiNP]}$ (SiNP = SiPS-H$_2$NPO$_{10}$EO$_{31}$M) epoxy nanocomposites and for MnSn(OH)$_6$ epoxy nanocomposite control materials.

Flexural strength and impact toughness for the control materials decreased approximately linearly with increasing MnSn(OH)$_6$ loading. Solvent-free nanofluid-loaded nanocomposite material increased in flexural strength and impact toughness up to 1% nanofluid loading and then decreased with further loading [69]. These data also showed that it is possible to increase flexural strength and toughness simultaneously with small loading. DSC measurements also showed that the T_g of this nanofluids-loaded epoxy increased from 98 to 128 °C over 0–1% loading and then decreased to 110 °C at 1.5% nanofluid.

10.6.4 Graphene Oxide Epoxy Nanocomposites

GO nanoribbons were oxidatively prepared from MWCNT [95] and then surface functionalized with a diblock-copolymer-modified trimethoxysilane. The resulting GO nanoribbon colloids were not presented as solvent-free nanofluids, although loading these materials into epoxy resins exhibited some properties similar to those discussed earlier. We would expect such colloids to melt at a reasonable temperature, but a thermal characterization was not reported [95].

GO solvent-free nanofluids modified with nano-iron oxide particles were used as nanofillers for epoxy resins [55]. Epoxy resins very similar to those discussed earlier for MnSn(OH)$_6$ were prepared and analyzed. Tensile strength or flexural strength was not reported. Impact toughness, however, was dramatically improved in comparison with control materials substituting graphite for GO nanofluid. Resins prepared with graphite filler steadily decreased in impact toughness with up to 2% loading of graphite. Addition of GO@Fe$_3$O$_4$ nanofluid, however, increased impact toughness 50% at 0.5% by weight loading, 150% at 1% loading, 100% at 1.5% loading, and 35% at 2% loading [55].

10.7 Summary and Future Prospects

Solvent-free 0D, 1D, 2D, and 3D nanofluids based on enveloping a core-phase or a cross-linked core with a corona or corona–canopy shell imparting liquidity have been synthesized on length scales as small as 1 nm in the 0D limit and on lengths approaching 100 nm in diameter for 3D materials. A bridge between molecular ILs and SIL appears continuous, with fullerene-based, POSS-based, and Keggin-ion-based ILs establishing a 0D limit, and larger POSS-related 3D SIL demonstrated in the 5- to 15-nm range (Figure 10.3). Such soft particles have theoretical and practical potentialities. Theoretical opportunities extend from the already demonstrated effects of polydispersity on frustrating crystallization and include tailoring peripheral functionalities to explore their impact on interparticle potentials and effects on crystallization. Practical applications will stem from other surface modifications that turn such particles into reactive nanofluids that can be used to make new 1D, 2D, and 3D cross-linked materials exhibiting high elastic limits. Simulation advances and theoretical treatments are providing a molecular-scale understanding of how various potential components emanating from structural aspects of nanofluids affect a variety of physical properties. We now have a reasonably good understanding of how a transition from 0D to 3D in nanofluid nanoparticle structure transforms from properties of the core material to those of the corona material, particularly in the case of nonionic SIL of the type shown in Figure 10.1c.

Threadlike 1D materials derived from DNA, carbon nanotubes, and other threadlike phases have been liquefied with lengths on the order of hundreds of nanometers. GO and MoS_2 core materials have provided 2D nanofluids with particulate areas approaching 10^4 nm^2. Such materials and many other related 2D materials are intensively being examined as catalysts for energy storage and generation reactions. We expect, therefore, that 1D and 2D SIMs derived from such materials will provide new venues for heterogeneous catalysis in a pseudomolecular phase that is homogeneous in a supramolecular sense. The nonvolatility of SIM and related supramolecular nanofluids may offer new thermal ranges for catalysis.

Addition of reactive nanofluids to a resin can result in marked *softening*, rather than the conventionally seen hardening upon addition of nanofillers. A simple mixture model shows that moduli in such resins can be tuned by varying nanofluid volume fraction, corona–canopy type, and molecular weight. It has also been shown that elastic and storage moduli can be maintained while increasing toughness. Such reactive nanofluids can be used as plasticizers that are not subject to phase separation and leaching, providing greener avenues for formulating many extant materials. Similarly, such plasticized resins may also serve as thermal phase change materials at high reactive nanofluid loading. Because of a great many different features now available in nanoparticles and nanosheets, and the variety of chemical approaches available to transform nanoparticles into solvent-free nanofluids, we can expect to see many new and interesting materials being derived in the future based on such nanofluids. We expect such 2D systems to provide new avenues of tuned heterogeneous catalysis, wherein the nano-dimension of these supports make such catalysis essentially homogeneous.

Acknowledgments

I thank my students and research assistants, postdoctoral research associates, and collaborators Joseph Byrom, Rafael Maniglia, Danial Chojnowski, Zhiming Qiu, Kejian Bian, Rene Crombez, Ken Reed, and Weidian Shen, whose collaborations were invaluable. Financial support from United States ONR grant award N00014-04-1-0763, United States Army Research Laboratory contract W911QX-06-C-0102, WPAFB UTC prime contract FA8650-05-D-5807, task order BH subcontract agreement 06-S568-BH-C1, and United States AFOSR grant FA9550-08-1-0431 is gratefully acknowledged.

References

1 Bourlinos, A.B., Raman, K., Herrera, R. et al. (2004). A liquid derivative of 12-tungstophosphoric acid with unusually high conductivity. *J. Am. Chem. Soc.* 126: 15358–15359.
2 Bourlinos, A.B., Herrera, R., Chalkias, N. et al. (2005). Surface-functionalized nanoparticles with liquid-like behavior. *Adv. Mater.* 17: 234–237.
3 Bourlinos, A.B., Chowdhury, S.R., Jiang, D.D. et al. (2005). Layered organosilicate nanoparticles with liquid like behavior. *Small* 1: 80–82.
4 Leone, A.M., Weatherly, S.C., Williams, M.E. et al. (2001). An ionic liquid form of DNA: redox active molten salts of nucleic acids. *J. Am. Chem. Soc.* 123: 218–222.
5 Bourlinos, A.B., Chowdhury, S.R., Herrera, R. et al. (2005). Functionalized nanostructures with liquid-like behavior: expanding the gallery of available nanostructures. *Adv. Funct. Mater.* 15: 1285–1290.
6 Bourlinos, A.B., Chowdhury, R., Jiang, D.D., and Zhang, Q. (2005). Weakly solvated PEG-functionalized silica nanoparticles with liquid-like behavior. *J. Mater. Sci.* 40: 5095–5097.
7 Bourlinos, A.B., Giannelis, E.P., Zhang, Q. et al. (2006). Surface-functionalized nanoparticles with liquid-like behavior: the role of the constituent components. *Eur. Phys. J. E* 20: 109–117.
8 Warren, S.C., Banholzer, M.J., Slaughter, L.S. et al. (2006). Generalized route to metal nanoparticles with liquid behavior. *J. Am. Chem. Soc.* 128: 12074–12075.
9 Choi, S.U.S. (2009). Nanofluids: from vision to reality through research. *J. Heat Transfer* 131: 033106. (9 pp).
10 Wen, D.S., Lin, G.P., Vafaei, S., and Zhang, K. (2009). Review of nanofluids for heat transfer applications. *Particuology* 7: 141–150.
11 Prasher, R., Phelan, P.E., and Bhattacharya, P. (2006). Effect of aggregation kinetics on the thermal conductivity of nanoscale colloidal solutions (nanofluid). *Nano Lett.* 6: 1529–1534.

12 Evans, W., Prasher, R., Fish, J. et al. (2008). Effect of aggregation and interfacial thermal resistance on thermal conductivity of nanocomposite and colloidal nanofluids. *Int. J. Heat Mass Transfer* 51: 1431–1438.
13 Welton, T. (1999). Room-temperature ionic liquids. Solvents for synthesis and catalysis. *Chem. Rev.* 99: 2071–2084.
14 Dupont, J., de Souza, R.F., and Suarez, P.A.Z. (2002). Ionic liquid (molten salt) phase organometallic catalysis. *Chem. Rev.* 102: 3667–3692.
15 Tanaka, K., Ishiguro, F., and Chujo, Y. (2010). POSS ionic liquid. *J. Am. Chem. Soc.* 132: 17649–17651.
16 Bai, H.P., Zheng, Y.P., Li, P.P., and Zhang, A.B. (2015). Synthesis of liquid-like trisilanol isobutyl-POSS NOHM and its application in capturing CO_2. *Chem. Res. Chin. Univ.* 31: 484–488.
17 Texter, J., Bian, K., Chojnowski, D., and Byrom, J. (2013). Organosiloxane supramolecular liquids – surface energy driven phase transitions. *Angew. Chem. Int. Ed.* 52: 2511–2515.
18 Schwab, J.J. and Lichtenhan, H.D. (1998). Polyhedral oligomeric silsesquioxane (POSS) – based polymers. *Appl. Organomet. Chem.* 12: 707–713.
19 Fasolo, M. and Sollich, P. (2003). Equilibrium phase behavior of polydisperse hard spheres. *Phys. Rev. Lett.* 91: 068301. (6 pp).
20 Antonietti, A., Pakula, T., and Bremser, W. (1995). Rheology of small spherical polystyrene microgels: a direct proof for a new transport mechanism in bulk polymers besides reptation. *Macromolecules* 28: 4227–4233.
21 Rodriguez, R., Herrera, R., Archer, L.A., and Giannelis, E.P. (2008). Nanoscale ionic materials. *Adv. Mater.* 20: 4353–4358.
22 Jespersen, M.L., Mirau, P.A., von Meerwall, E. et al. (2010). Canopy dynamics in nanoscale ionic material. *ACS Nano* 4: 3735–3742.
23 Nugent, J.L., Moganty, S.S., and Archer, L.A. (2010). Nanoscale organic hybrid electrolytes. *Adv. Mater.* 22: 3677–3680.
24 Lu, T.Y., Das, S.K., Moganty, S.S., and Archer, L.A. (2012). Nanoscale organic hybrid electrolytes. *Adv. Mater.* 24: 4430–4435.
25 Rodriguez, R., Herrera, R., Bourlinos, A.B. et al. (2010). The synthesis and properties of nanoscale ionic materials. *Appl. Organomet. Chem.* 24: 581–589.
26 Bourlinos, A.B., Stassinopoulos, A., Anglos, D. et al. (2006). Functionalized ZnO nanoparticles with liquidlike behavior and their photoluminescence properties. *Small* 2: 513–516.
27 Bourlinos, A.B., Georgakilas, V., Boukos, N. et al. (2007). Silicone-functionalized nanotubes for the production of new carbon-based fluids. *Carbon* 45: 1583–1595.
28 Fernandes, N., Dallas, P., Rodriguez, R. et al. (2010). Fullerol ionic fluids. *Nanoscale* 2010: 1653–1656.
29 Sun, L., Fang, J., Reed, J.C. et al. (2010). Lead-salt quantum-dot ionic liquids. *Small* 6: 638–641.
30 Oommen, J.M., Hussain, M.M., Emwas, A.-H.M. et al. (2010). Nuclear magnetic resonance study of nanoscale ionic materials. *Electrochem. Solid-State Lett.* 13: K87–K88.

31 Agarwal, P., Qi, H.B., and Archer, L.A. (2010). The ages in a self-suspended nanoparticle liquid. *Nano Lett.* 10: 111–115.
32 Moganty, S.S., Jayaprakash, N., Nugent, J.L. et al. (2010). Ionic-liquid-tethered nanoparticles: hybrid electrolytes. *Angew. Chem. Int. Ed.* 49: 9158–9161.
33 Kim, D. and Archer, L.A. (2011). Nanoscale organic–inorganic hybrid lubricants. *Langmuir* 27: 3083–3094.
34 Liu, D.-P., Li, G.-D., Su, Y., and Chen, J.-S. (2006). Highly luminescent ZnO nanocrystals stabilized by ionic-liquid components. *Angew. Chem. Int. Ed.* 45: 7370–7373.
35 Voevodin, A.A., Vaia, R.A., Patton, S.T. et al. (2007). Nanoparticle-wetted surfaces for relays and energy transmission contacts. *Small* 3: 1857–1963.
36 Patton, S.T., Voevodin, A.A., Vaia, R.A. et al. (2008). Nanoparticle liquids for surface modification and lubrication of MEMS switch contacts. *J. Micromech. Syst.* 17: 741–746.
37 Zheng, Y.P., Zhang, J., Lan, L. et al. (2010). Preparation of solvent-free gold nanofluids with facile self-assembly technique. *ChemPhysChem* 11: 61–64.
38 Li, Q., Dong, L., Deng, W. et al. (2009). Solvent-free fluids based on rhombohedral nanoparticles of calcium carbonate. *J. Am. Chem. Soc.* 131: 9148–9149.
39 Wang, X.S., Shi, L., Zhang, J.Y. et al. (2018). In situ formation of surface-functionalized ionic calcium carbonate nanoparticles with liquid-like behaviours and their electrical properties. *R. Soc. Open Sci.* 5: 170732.
40 Li, Q., Dong, L.J., Liu, Y. et al. (2011). A carbon black derivative with liquid behavior. *Carbon* 49: 1033–1051.
41 Feng, Q.S., Dong, L.J., Huang, J. et al. (2010). Fluxible monodisperse quantum dots with efficient luminescence. *Angew. Chem. Int. Ed.* 49: 9943–9946.
42 Zhou, J., Tian, D.M., and Li, H.B. (2011). Multi-emission CdTe quantum dot nanofluids. *J. Mater. Chem.* 21: 8521–8523.
43 Maniglia, R., Reed, K., and Texter, J. (2017). Reactive CeO_2 nanofluids for UV protective films. *J. Colloid Interface Sci.* 506: 346–354.
44 Patil, A.J., McGrath, N., Barclay, J.E. et al. (2012). Liquid viruses by nanoscale engineering of capsid surfaces. *Adv. Mater.* 24: 4557–4563.
45 Tan, Y., Zheng, Y.P., Wang, N., and Zhang, A.B. (2012). Controlling the properties of solvent-free Fe_3O_4 nanofluids by corona structure. *Nano-Micro Lett.* 4: 208–214.
46 Perriman, A.W., Cölfen, H., Hughes, R.W. et al. (2009). Solvent-free protein liquids and liquid crystals. *Angew. Chem. Int. Ed.* 48: 6242–6246.
47 Michinobu, T., Nakanishi, T., Hill, J.P. et al. (2006). Room temperature liquid fullerenes: an uncommon morphology of C_{60} derivatives. *J. Am. Chem. Soc.* 128: 10384–10385.
48 Nakanishi, T. (2010). Supramolecular soft and hard materials based on self-assembly algorithms of alkyl-conjugated fullerenes. *Chem. Commun.* 46: 3425–3436.
49 Lintinen, K., Efimov, A., Hietala, S. et al. (2008). Cationic photopolymerization of liquid fullerene derivative. *J. Polym. Sci., Part A: Polym. Chem.* 46: 5194–5201.

50 Nakanishi, T., Takahashi, H., Michinobu, T. et al. (2008). Fine-tuning supramolecular assemblies of fullerenes bearing long alkyl chains. *Thin Solid Films* 516: 2401–2406.

51 Li, H., Babu, S.S., Turner, S.T. et al. (2013). Alkylated-C_{60} based soft materials: regulation of self-assembly and optoelectronic properties by chain branching. *J. Mater. Chem. C* 1: 1943–1951.

52 Babu, S.S. and Nakanishi, T. (2013). Nonvolatile functional molecular liquids. *Chem. Commun.* 49: 9372–9382.

53 Zeng, C.F., Tang, Z.H., Guo, B.C., and Zhang, L.Q. (2012). Supramolecular ionic liquid based on graphene oxide. *Phys. Chem. Chem. Phys.* 14: 9838–9845.

54 Li, P.P., Shi, T., Yao, D.D. et al. (2016). Covalent nanocrystals-decorated solvent-free graphene oxide liquids. *Carbon* 110: 87–96.

55 Li, P.P., Zheng, Y.P., Li, M.Z. et al. (2016). Enhanced toughness and glass transition temperature of epoxy nanocomposites filled with solvent-free liquid-like nanocrystal-functionalized graphene oxide. *Mater. Des.* 89: 653–659.

56 Li, P.P., Zheng, Y.P., Li, M.Z. et al. (2016). Enhanced flame-retardant property of epoxy composites filled with solvent-free and liquid-like graphene organic hybrid material decorated by zinc hydroxystannate boxes. *Composites A* 81: 172–181.

57 Lei, Y.A., Xiong, C.X., Dong, L.J. et al. (2007). Ionic liquid of ultralong carbon nanotubes. *Small* 3: 1889–1893.

58 Lei, Y.A., Xiong, C.X., Guo, H. et al. (2008). Controlled viscoelastic carbon nanotube fluids. *J. Am. Chem. Soc.* 130: 3256–3257.

59 Zhang, J.X., Zheng, Y.P., Yu, P.Y. et al. (2009). The synthesis of functionalized carbon nanotubes by hyperbranched poly(amine-ester) with liquid-like behavior at room temperature. *Polymer* 50: 2953–2957.

60 Li, Q., Dong, L.J., Fang, J.F., and Xiong, C.X. (2010). Property–structure relationship of nanoscale ionic materials based on multiwalled carbon nanotubes. *ACS Nano* 4: 5797–5806.

61 Li, Q., Dong, L.J., Li, L.B. et al. (2012). The effect of the addition of carbon nanotube fluids to a polymeric matrix to produce simultaneous reinforcement and plasticization. *Carbon* 50: 2045–2060.

62 Zhang, X., Zheng, Y.-P., Lan, L., and Yang, H.-C. (2012). Synthesis and properties of a solvent-free MWCNT-based nanofluids. *New Carbon Mater.* 29: 193–202.

63 Lan, L., Zheng, Y.P., Zhang, A.B. et al. (2012). Study of ionic solvent-free carbon nanotube nanofluids and its composites with epoxy matrix. *J. Nanopart. Res.* 14: 753. (10pp).

64 Gu, S.Y., Liu, L.L., and Yan, B.B. (2014). Effects of ionic solvent-free carbon nanotube nanofluids on the properties of polyurethane thermoplastic elastomer. *J. Polym. Res.* 21: 356. (10 pp).

65 Zhang, X., Zheng, Y.-P., Yang, R.-L., and Yang, H.-C. (2014). Nanocomposites with liquid-like multiwalled carbon nanotubes dispersed in epoxy resin without solvent process. *Int. J. Polym. Sci.* 712637. (6pp). doi: https://doi.org/10.1155/2014/712637.

66 Qu, P., Zheng, Y.P., Yang, R.L. et al. (2015). Effect of canopy structures on CO_2 capture capacity and properties of NONMs. *Colloid. Polym. Sci.* 293: 1623–1634.

67 Li, P.P., Yang, R.L., Zheng, Y.P. et al. (2015). Effect of polyether amine canopy structure on carbon dioxide uptake of solvent-free multiwalled carbon nanotubes. *Carbon* 95: 408–418.

68 Weng, P.X., Yin, X.Z., Yang, S.W. et al. (2018). Functionalized magnesium hydroxide fluids/acrylate-coated hybrid cotton fabric with enhanced mechanical, flame retardant and shape-memory properties. *Cellulose* 25: 1425–1436.

69 Li, P.P., Zheng, Y.P., Yang, R.L. et al. (2015). Flexible nanoscale thread of $MnSn(OH)_6$ crystallite with liquid-like behavior and its application in nanocomposites. *ChemPhysChem* 16: 2524–2529.

70 Gu, S.Y., Zhang, Y.H., and Yan, B.B. (2013). Solvent-free ionic molybdenum disulfide (MoS_2) nanofluids with self-healing lubricating behaviors. *Mater. Lett.* 97: 169–172.

71 Gu, S.Y., Gao, X.F., and Zhang, Y.H. (2015). Synthesis and characterization of solvent-free ionic molybdenum disulphide (MoS_2) nanofluids. *Mater. Lett.* 149–150: 587–593.

72 Sharma, K.P., Bradley, K., Brogan, A.P.S. et al. (2013). Redox transitions in an electrolyte-free myoglobin fluid. *J. Am. Chem. Soc.* 135: 18311–18314.

73 Zheng, Y.P., Zhang, J.X., Lan, L., and Yu, P.Y. (2011). Sepiolite nanofluids with liquid-like behavior. *Appl. Surf. Sci.* 257: 6171–6174.

74 Clemans, D., Amad, R., Benhamida, B. et al. (2009). Bactericidal activity of solvent-free nanofluids. Abstracts of Papers, 237th ACS National Meeting, Salt Lake City, UT, USA, 22–26 March 2009; COLL-017, Presented 22 March 2009.

75 Shen, W., Crombez, R., Qiu, Z., and Texter, J. (2009). Nanoindenting studies of nanofluid-based coatings. Abstracts of Papers. 237th ACS National Meeting, Salt Lake City, UT, USA, 22–26 March 2009; COLL-020, Presented 22 March 2009.

76 Texter, J., Qiu, Z., and Chojnowski, D. (2009). Solvent-free supramolecular liquids for novel resins and applications. European Coatings Congress, Nuremberg, Germany, Session 2. Nanotechnology, Paper 2.4, Presented 30 March 2009.

77 Texter, J., Qiu, Z., Crombez, R. et al. (2010). Solvent-free supramolecular liquids for novel resins & applications. *China Coat. J.* 38–42: 44–48.

78 Texter, J., Qiu, Z., Crombez, R. et al. (2011). Nanofluid acrylate composite resins – initial preparation and characterization. *Polym. Chem.* 2: 1778–1787.

79 Texter, J., Qiu, Z., Crombez, R., and Shen, W. (2013). Nanofluid polyurethane/polyurea resins – thin films and clearcoats. *J. Polym. Sci., Part A: Polym. Sci.* 51: 3439–3448.

80 Agarwal, P., Chopra, M., and Archer, L.A. (2011). Nanoparticle netpoints for shape-memory polymers. *Angew. Chem. Int. Ed.* 50: 8670–8673.

81 Lin, K.-Y.A. and Park, A.-H.A. (2011). Effects of bonding types and functional groups on CO captures using novel multiphase systems of liquid-like nanoparticle organic hybrid materials. *Environ. Sci. Technol.* 45: 6633–6639.

82 Park, Y.J., Decatur, J., Lin, K.-Y.A., and Park, A.-H.A. (2011). Investigation of CO_2 capture mechanisms of liquid-like nanoparticles organic hybrid materials vis structural characterization. *Phys. Chem. Chem. Phys.* 13: 18115–18122.

83 Park, Y.J., Shin, D., Jang, Y.N., and Park, A.-H.A. (2012). CO_2 capture capacity and swelling measurements of liquid-like nanoparticle organic hybrid materials via attenuated total reflectance Fourier transform infrared spectroscopy. *J. Chem. Eng. Data* 57: 40–45.

84 Xu, Y.T., Zheng, Q., and Song, Y.H. (2015). Comparison studies of rheological and thermal behaviors of ionic liquids and nanoparticle ionic liquids. *Phys. Chem. Chem. Phys.* 17: 19815–19819.

85 Yang, S.W., Tan, Y.Q., Yin, X.Z. et al. (2017). Preparation and characterization of monodisperse solvent-free silica nanofluids. *J. Disp. Sci. Technol.* 38: 425–431.

86 He, H.G., Yan, Y.R., Qiu, Z.M., and Tan, X. (2017). A novel antistatic polyurethane hybrid based on nanoscale ionic material. *Prog. Org. Coat.* 113: 110–116.

87 Zhang, J.X., Zheng, Y.P., Lan, L. et al. (2009). Direct synthesis of solvent-free multiwall carbon nanotubes/silica nonionic nanofluids hybrid material. *ACS Nano* 3: 2185–2190.

88 Zhang, J.S., Chai, S.H., Qiao, Z.A. et al. (2015). Porous liquids: a promising class of media for gas separation. *Angew. Chem. Int. Ed.* 54: 932–936.

89 Shi, T., Zheng, Y.P., Wang, T.Y. et al. (2018). Effect of pore size on the carbon dioxide adsorption behavior of porous liquids based on hollow silica. *ChemPhysChem* 19: 130–137.

90 Xiong, H.M., Shen, W.Z., Wang, Z.-D. et al. (2006). Liquid polymer nanocomposites PEGME–SnO_2 and PEGME–TiO_2 prepared through solvothermal methods. *Chem. Mater.* 18: 3850–3854.

91 Li, H.R., Jiang, M., Hu, D.B. et al. (2015). Solvent-free zirconia nanofluids/silica single-layer multifunctional hybrid coatings. *Colloids Surf., A* 464: 26–32.

92 Zhang, J.X., Zheng, Y.P., Lan, L. et al. (2009). Modified carbon nanotubes with liquid-like behavior at 45 °C. *Carbon* 47: 2776–2781.

93 Li, P.P., Zheng, Y.P., Wu, Y.W. et al. (2014). Nanoscale ionic graphene material with liquid-like behavior in the absence of solvent. *Appl. Surf. Sci.* 314: 983–990.

94 Zhang, J.-X., Zheng, Y.-P., Lan, L. et al. (2013). The preparation of a silica nanoparticle hybrid ionic nanomaterial and its electrical properties. *RSC Adv.* 3: 16714–16719.

95 Li, P.P., Zheng, Y.P., Shi, T. et al. (2016). A solvent-free graphene oxide nanoribbon colloid as a filler phase for epoxy-matrix composites with enhanced mechanical, thermal and tribological performance. *Carbon* 96: 40–48.

96 Yu, H.-Y. and Koch, D.L. (2010). Structure of solvent-free nanoparticle-organic hybrid materials. *Langmuir* 26: 16801–16811.

97 Wu, L.S., Zhang, B.Q., Lu, H., and Liu, C.-Y. (2014). Nanoscale ionic materials based on hydroxyl-functionalized graphene. *J. Mater. Chem. A* 2: 1409–1417.
98 Sun, Y.F., Gao, S., Lei, F.C., and Xie, Y. (2015). Atomically-thin two-dimensional sheets for understanding active sites in catalysis. *Chem. Soc. Rev.* 44: 623–636.
99 Tao, H.C., Zhang, Y.Q., Gao, Y.N. et al. (2017). Scalable exfoliation and dispersion of two-dimensional materials – an update. *Phys. Chem. Chem. Phys.* 19: 921–960.
100 Chremos, A. and Panagiotopoulos, A.Z. (2011). Structural transitions of solvent-free oligomer-grafted nanoparticles. *Phys. Rev. Lett.* 107 (5): 105503.
101 Chremos, A., Panagiotopolous, A.Z., Yu, H.-Y., and Koch, D.L. (2012). Structure of solvent-free grafted nanoparticles: molecular dynamics and density functional theory. *J. Chem. Phys.* 135: 114901. (12 pp).
102 Chremos, A., Panagiotopolous, A.Z., and Koch, D.L. (2012). Dynamics of solvent-free grafted nanoparticles. *J. Chem. Phys.* 136: 144092. (12 pp).
103 Hong, B.B., Chremos, A., and Panagiotopolous, A.Z. (2012). Molecular dynamics simulations of silica nanoparticles grafted with poly(ethylene oxide) oligomer chains. *J. Phys. Chem. B* 116: 2385–2395.
104 Hong, B.B., Chremos, A., and Panagiotopolous, A.Z. (2013). Simulations of the structure and dynamics of nanoparticle-based ionic liquids. *Faraday Discuss.* 136: 204904. (9 pp).
105 Goyal, S. and Escobedo, F.A. (2011). Structure and transport properties of polymer grafted nanoparticles. *J. Chem. Phys.* 135: 184902. (12 pp).
106 Yu, H.-Y. and Koch, D.L. (2014). Self-diffusion and linear viscoelasticity of solvent-free nanoparticle organic hybrid materials. *J. Rheol.* 58: 369–394.
107 Hong, B.B., Chremos, A., and Panagiotopolous, A.Z. (2012). Simulations of the structure and dynamics of nanoparticle-based ionic liquids. *Faraday Discuss.* 154: 29–40.
108 Hong, B.B. and Panagiotopolous, A.Z. (2012). Diffusivities, viscosities, and conductivities of solvent-free ionically grafted nanoparticles. *Small* 9: 6091–6102.
109 Chojnowski, D., Qiu, Z., and Texter, J. (2009). Nanofluid rheology—thermal transitions and reactive curing. Abstracts of Papers, 237th ACS National Meeting, Salt Lake City, UT, USA, March 22–26, 2009; COLL-017, Presented 22 March 2009.
110 Texter, J., Qiu, Z., Crombez, R., and Byrom, J. (2019). Reactive nanofluids for tuning resing hardness. *MRS Adv.* (submitted for publication).
111 Graber, K.C., Freeman, R.G., Homer, M.B., and Natan, M.J. (1995). Preparation and characterization of Au colloid monolayers. *Anal. Chem.* 67: 735–739.
112 Lam, C.C.L. and Chong, A.C.M. (2000). Effect of cross-link density on strain gradient plasticity in epoxy. *Mater. Sci. Eng., A* 281: 156–161.
113 Fu, S.Y., Xu, G., and Mai, Y.-W. (2002). On the elastic modulus of hybrid. Particle/short fiber/polymer composites. *Composite Part B* 33: 291–299.

114 Fu, S.Y., Feng, X.-Q., Lauke, B., and Mai, Y.-W. (2008). On the elastic modulus of hybrid. Particle/short fiber/polymer composites. *Composite Part B* 39: 933–961.

115 Binks, B.P. (2002). Particles as surfactants similarities and differences. *Curr. Opin. Colloid Interface Sci.* 7: 21–41.

116 Texter, J. (2015). Graphene oxide and graphene flakes as stabilizers and dispersing aids. *Curr. Opin. Colloid Interface Sci.* 20: 454–464.

117 Liu, B.Y. and Jin, N.X. (2016). The application of ionic liquid as functional material: a review. *Curr. Org. Chem.* 20: 2109–2116.

118 Zhou, F., Liang, Y.M., and Liu, W.M. (2009). Ionic liquid lubricants: designed chemistry for engineering applications. *Chem. Soc. Rev.* 9: 2590–2599.

119 Archer, L.A., Olenick, L.L., and Nugent, J.L. (2010). Nanoparticle hybrid materials (NOHMS). WO 2010/083041 A1, 22 July 2010.

120 Hailstone, R.K., DiFrancesco, A.G., Leong, J.G. et al. (2009). A study of lattice expansion in CeO_2 nanoparticles by transmission electron microscopy. *J. Phys. Chem. C* 113: 15155–15159.

121 Li, X.D., Li, J.G., Huo, D. et al. (2009). Facile synthesis under near-atmospheric conditions and physicochemical properties of hairy CeO_2 nanocrystals. *J. Phys. Chem. C* 113: 1806–1811.

122 Reed, K., Cormack, A., Kulkarni, A. et al. (2014). Exploring the properties and applications of nanoceria: is there still plenty of room at the bottom? *Environ. Sci. Nano* 1: 390–405.

123 DiFrancesco, A.G., Hailstone, R.K., Reed, K.J., and Prok, G.R. (2014). Cerium-containing nanoparticles. US Patent 8,883,865 B2.

124 Annamalai, M., Dhinesh, B., Nanthagopal, K. et al. (2016). An assessment on performance, combustion and emission behavior of a diesel engine powered by ceria nanoparticle blended emulsified biofuel. *Energy Convers. Manage.* 123: 372–380.

125 Zhang, J.F., Lee, K.B., He, L.C. et al. (2016). Effects of a nanoceria fuel additive on the physicochemical properties of diesel exhaust particles. *Environ. Sci. Proc. Impacts* 18: 1333–1342.

126 Rahman, M. and Brazel, C.S. (2006). Ionic liquids: new generation stable plasticizers for poly(vinyl chloride). *Polym. Degrad. Stab.* 91: 3371–3382.

127 Gao, W.-P., Bai, Y., Chen, E.-Q., and Hou, Q.-F. (2005). Crystallization and melting of poly(ethylene oxide) confined in nanostructured particles with cross-linked shells of polybutadiene. *Chin. J. Polym. Sci.* 23: 275–284.

128 Polu, A.R. and Rhee, H.-W. (2017). Effect of organic–inorganic hybrid nanoparticles (POSS–PEG(n = 4)) on thermal, mechanical, and electrical properties of PEO-based solid polymer electrolytes. *Adv. Polym. Tech.* 36: 145–151.

129 Chojnowski, D. (2011). Rheological characteristics of solvent-free nanofluids: thermal transitions and reactive curing. Thesis. Eastern Michigan University. http://commons.emich.edu/cgi/viewcontent.cgi?article=1847&context=theses (downloaded 11 December 2016).

11

Solvent-Free Liquids and Liquid Crystals from Biomacromolecules

Kai Liu[1], Chao Ma[2], and Andreas Herrmann[2,3,4]

[1] Chinese Academy of Sciences, Changchun Institute of Applied Chemistry, State Key Laboratory of Rare Earth Resource Utilization, 130022, Changchun, China
[2] University of Groningen, Zernike Institute for Advanced Materials, Nijenborgh 4, 9747 AG Groningen, The Netherlands
[3] DWI–Leibniz Institute for Interactive Materials, Forckenbeckstr. 50, 52056, Aachen, Germany & Institute of Technical and Macromolecular Chemistry, RWTH Aachen University, Worringerweg 2, 52074, Aachen, Germany
[4] RWTH Aachen University, Germany

11.1 Introduction

Biological macromolecules, such as DNA, RNA, and proteins, operate as persistent nanoscale objects within aqueous environments, and as a consequence the ubiquity of water as a solvent is a basic requirement for biomacromolecule self-organization and activity [1]. Beyond the biological context, biomacromolecular components are of increasing interest for integration into technological systems [2–6]. However, the processing and engineering of these biological components is currently limited to methods based primarily on aqueous media due to the insolubility and potential destabilization of the folded structure in organic solvents. Moreover, the biological materials as freeze-dried powders face general difficulty and safety concerns associated with storage and manipulation. Therefore, considering the many technologies that are incompatible with solvent systems (e.g. high- and low-temperature applications), investigation of function of biomacromolecular liquids in a solvent-free environment would expand the usefulness of biomacromolecules outside of the set of conditions dictated by biology. The ability to prepare solvent-free liquids comprising extremely high concentrations of structurally and functionally intact biomacromolecules should have a significant impact on advancing the bioinspired design and processing of biologically derived nanostructures. Biomacromolecular liquids might replace conventional polymeric ionic liquids [7] as the dispersion medium for organic and inorganic moieties if biocompatibility and biodegradability are desired. They could also be used as injectable depots in the field of drug delivery if the presence of high concentration of bioactive compounds, such as in the design of barrier dressings for wound healing and artificial skin, are needed [8, 9]. Such materials may offer opportunities for the development of flexible and printable bioelectronic components, where water is detrimental for device performance.

Functional Organic Liquids, First Edition. Edited by Takashi Nakanishi.
© 2019 Wiley-VCH Verlag GmbH & Co. KGaA. Published 2019 by Wiley-VCH Verlag GmbH & Co. KGaA.

From a fundamental perspective, this type of biomacromolecular liquids allows for the investigation of their structural stability and functions in the absence of any solvent, which should be significantly distinct from typical biological aqueous systems. Therefore, the development of solvent-free biomacromolecular liquids is an attractive goal both from basic science and application perspectives.

Recently, it was found that solvent-free liquids can be made from biomacromolecules by electrostatic complexation with surfactants containing flexible alkyl tails, followed by dehydration [10–13]. This simple and generic method enabled the formation of solvent-free liquids from biosystems ranging from nucleic acids [10–12] and proteins [13–15] to even whole viruses [16, 17], spanning a size range from only a few nanometers to one micrometer. Interestingly, in some reports, solvent-free liquid crystals (LCs) of biomcromolecules were also fabricated, where fluidity and ordering are introduced by ionic self-assembly [17–20]. The biomacromolecules in these biofluids adopted structures close to their native states and retained biological function [21]. In this context, the flexibility of processing various biological building blocks into water-free liquids combined with their stability enabled several applications in biocatalysis [22], bioelectronics [23, 24], and potentially in biomedicine.

In this chapter, we outline seminal and recent research in this exciting field. We start from DNA liquids and their application in electronics. We then enter a discussion of solvent-free protein liquids and their utilization for enzyme catalysis and electrochemical device fabrication. Finally, examples of virus-based liquids are introduced. In the end, the mechanism for the formation of biomacromolecular liquids in the absence of solvent is discussed.

11.2 Solvent-Free Nucleic Acid Liquids

11.2.1 Fabrication of Solvent-Free Nucleic Acid Liquids

For the preparation of nucleic acid liquids, an oligonucleotide (15–100 bp) and a cationic surfactant containing poly(ethylene glycol) (PEG) tails (Figure 11.1a,b) were electrostatically complexed in a very simple procedure, including a final dehydration step [10, 11]. The obtained water-free DNA liquids were viscous, optically transparent, and easily flowing above 60 °C. A variety of characterization methods including FTIR spectroscopy, ultraviolet (UV) spectroscopy, and circular dichroism (CD) indicated that the double-stranded DNA was maintained in the solvent-free melts. Rheological investigations manifested the DNA liquid-like behavior by a higher loss modulus G'' than a storage modulus G'. The viscosity decreased with increasing temperature with values commensurate to their liquid-like states. Besides the fabrication of DNA liquids by exchange replacement of the sodium counterions by PEGylated quaternary ammonium surfactants, an alternative toward a meltable DNA derivative comprised the direct neutralization of the proton form of high-molecular-weight acidized DNA (2000 bp) by a tertiary-PEGylated amine (Figure 11.1c). In this way, a waxy complex that reversibly melts at ~40 °C was obtained [11]. Metal coordinated cationic complexes employed as surfactants (Figure 11.1d) also

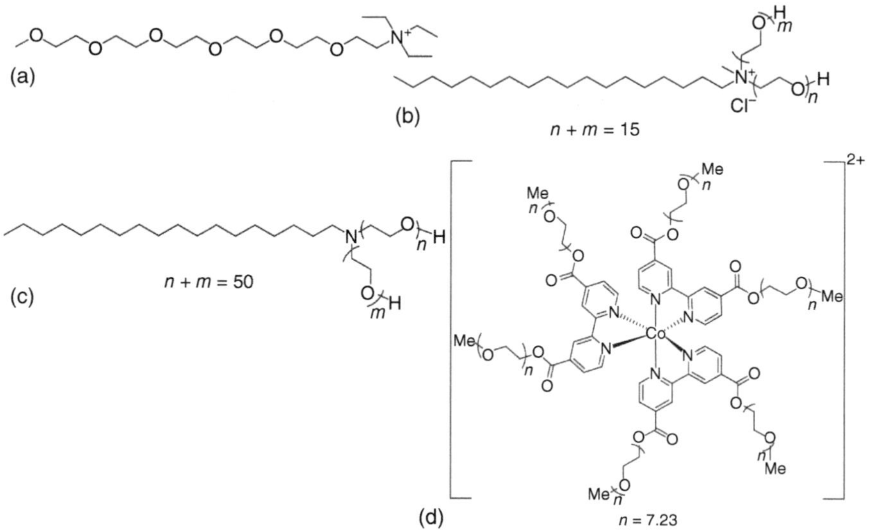

Figure 11.1 Overview of surfactants containing poly(ethylene glycol) (PEG) tails for the fabrication of solvent-free DNA liquids. (a, b) Cationic surfactants with PEG tails [10, 11]. (c) Amine surfactant to complex with acidized DNA by direct proton exchange [11]. (d) Polypyridyl complex of Co ion decorated with polyether chains as cationic surfactants [10]. Source: Liu et al. 2017 [25]. Reproduced with permission from ACS.

enabled room-temperature DNA melts [10]. Furthermore, it was feasible to blend the DNA liquids with hydrophobic molecules without affecting their fluidic properties. For instance, methylene blue and laser dyes derived from coumarin and rhodamine 6G can be blended with the DNA liquids, giving colored samples that may find applications as DNA-based optical materials [11].

Very recently, our group reported that the combination of DNA and cationic surfactants is a generic scheme for the production of a series of DNA fluids, including smectic LCs and isotropic liquids (Figure 11.2) [12, 17]. Solvent-free DNA–surfactant melts were prepared by electrostatic complexation of single-stranded oligonucleotides (6, 14, 22, 50, and 110 mer) with cationic surfactants containing linear, flexible alkyl chains. Three surfactants with two aliphatic chains of variable lengths, i.e. dioctyldimethylammonium bromide (DOAB), didecyldimethylammonium bromide (DEAB), and didodecyldimethylammonium bromide (DDAB), were used to prepare the stable DNA–surfactant LC mesophases and liquid phases. Polarized optical microscopy (POM) images showed that the DNA LCs exhibited typical focal-conic textures characteristic of smectic lamellar structures (Figure 11.2a,b). After heating above the clearing temperatures, the DNA–surfactant complexes transformed into the disordered liquid state (Figure 11.2c) wherein the oligonucleotide–surfactant hybrids showed no ordering. Upon heating over the clearing temperature, the birefringent focal-conic textures disappear completely (Figure 11.2d), resulting in a transparent isotropic fluid. Small-angle X-ray scattering (SAXS) measurements indicated long-range ordered smectic layers of the DNA–surfactant LCs. In this lamellar phase, DNA sublayers are alternatively intercalated between aliphatic

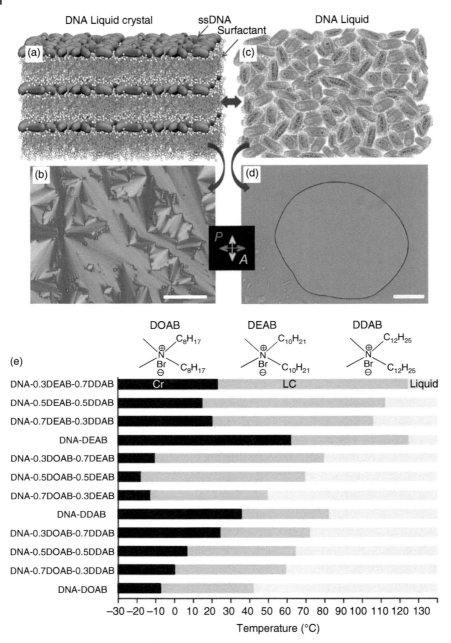

Figure 11.2 Solvent-free liquid crystals and liquids of DNA–surfactant complexes [12, 17]. (a) Proposed lamellar structure in the liquid crystalline phase. (b) Typical polarized optical microscopy (POM) image of the DNA–surfactant mesophases, showing well-defined focal-conic textures of smectic layers. (c) Schematic of disordered DNA–surfactant complex in the isotropic liquid phase and (d) corresponding POM image of the isotropic liquid not showing any birefringent textures. Both POM images were acquired with an inserted one-quarter wave plate. The scale bar is 100 μm. (e) Overview of phase-transition temperatures (melting/clearing points) of binary and ternary DNA–surfactant complexes from crystalline (Cr) to liquid crystalline (LC) and then to isotropic liquid, which depend strongly on the specific length of the aliphatic chains of the surfactants. Source: Liu et al. 2017 [25]. Reproduced with permission from ACS.

hydrocarbon sublayers (Figure 11.2a). Each repeating layer consists of a cationic surfactant bilayer that electrostatically interacts with an anionic oligonucleotide sublayer. Within the DNA sublayer, the oligonucleotide chains of single-stranded DNA (ssDNA) are randomly packed, without any positional or orientational order. The topography of the lamellar structure was visualized directly by freeze-fracture transmission electron microscopy (FF-TEM). Furthermore, SAXS analysis of the DNA–surfactant melts in the isotropic liquid states showed no clear diffraction, with only one very broad halo associated with disordered DNA-surfactant scattering. In addition, liquid crystals and liquids from RNA were fabricated following the same complexation procedure.

The DNA–surfactant melts were thermally stable over a wide range of temperatures until decomposition occurred at approximately 200 °C. The phase transitions (crystal-liquid crystal-isotropic liquid) temperatures can be controlled over an extremely broad temperature range (Figure 11.2e) [12]. Fluid DNA was achieved at temperatures as low as −20 °C and the transition from the crystal to the LC can be adjusted in a temperature window of 65 °C. The transition from LC to isotropic liquid can be tuned between 41 and 130 °C. Interestingly, we found a correlation between the phase-transition temperatures and the length of the surfactant alkyl chains. When the aliphatic chain length of the surfactant was increased in the binary (or ternary) complexes, the melting and clearing points were generally increased. Furthermore, rheology measurements indicated their fluid behaviors. There was a clear correlation between the viscosity of the DNA–surfactant fluids and the alkyl chain lengths of the surfactant alkyl chains. The viscosity increased with longer DNA or alkyl chains. These new solvent-free DNA–surfactant melts have negligible volatility, exhibit high DNA content, and their high thermal stability might make them suitable for many technologies that are incompatible with aqueous systems.

11.2.2 Electrical Applications Based on Solvent-Free Nucleic Acid Liquids

The nature of the solvent-free DNA liquids rendered these materials useful for incorporation into microelectronic circuits that utilize DNA for both self-assembly and electronic connections [10, 26]. For instance, when the Co^{2+}-containing DNA liquid (Figure 11.1d) was interrogated electrochemically by a microelectrode, the neat melt exhibited an electrochemical signal due to the $Co^{(2+/3+)}$ oxidation reaction. This behavior indicated electron hopping at the Co center and ionic diffusion in the undiluted liquid. The obtained faradaic current was very low since the DNA counterion induces a qualitatively higher viscosity compared to the pristine surfactant system and the rigid DNA helices impede transport of the metal-containing surfactant to the electrode. When mixed metal surfactants (Co^{2+} and Fe^{2+}) were complexed with DNA for electrochemical investigations, additional oxidation of the guanine base in DNA was observed besides the two electrochemical signals from the oxidation of $Co^{(2+/3+)}$ and $Fe^{(2+/3+)}$. This indicated that electrogenerated Fe^{3+} was a sufficiently strong oxidant to oxidize guanine [27]. Further investigations demonstrated that in pure Co–DNA melts, DNA could suppress the net electron transfer rate in

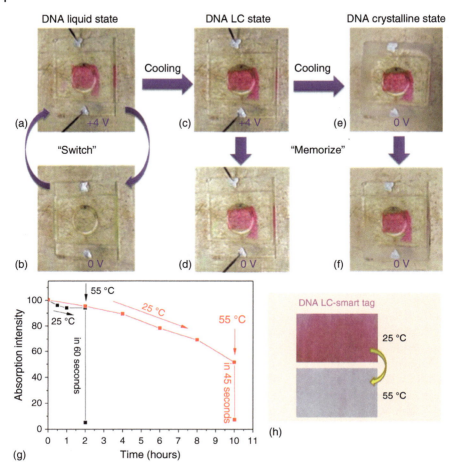

Figure 11.3 Phase-dependent electrochromic device based on solvent-free DNA–surfactant complexes [24]. (a, b) Switchable electrochromism between the colored (magenta) and colorless states in the isotropic liquid phase. (c, d) Remarkable optical memory of the liquid crystal can be observed as a persistent colored state. (e, f) Cooling the colored DNA liquid crystalline phase to the crystalline phase can further increase the relaxation time of the color impression. (g, h) The activated DNA electrochromic device demonstrated functionality as combined time and temperature indicator. Source: Liu et al. 2017 [25]. Reproduced with permission from ACS.

the reduction process of $Co^{(2+/+)}$ due to the very low mobility of the anionic phosphate groups of the DNA counterion [26].

Solvent-free DNA liquids not only acted as a scaffold for metal redox reactions but it was found that the nucleobases of DNA can be reversibly oxidized in pristine DNA–surfactant fluids and phase-dependent electrochromism was studied (Figure 11.3) [24]. In the isotropic liquid phase, electric-field-induced coloration and bleaching had a switching time of seconds. The magenta color was due to radical cation formation of nucleobases [28, 29]. Upon transition to the smectic LC phase, a remarkable optical memory of the written state was observed without color decay for many hours in the absence of applied voltage. Therefore, the volatility of the optoelectronic state can be controlled simply by changing the phase of the DNA–surfactant fluid material.

As shown in Figures 11.3a,b, an application of double potential steps of 0 and 4 V caused DNA–surfactant complexes to change color between transparent and magenta reversibly in the isotropic liquid phase with a switch time of seconds. The electrochromic switch time of these liquid materials correlated with the length of the DNA used, suggesting that the rate of DNA oxidation was limited by the rate of mass transport to the electrode. When the DNA–surfactant liquid was cooled to the smectic LC phase while an applied voltage of 4 V was maintained, the magenta color was conserved (Figure 11.3c). Even after the cell voltage was returned to 0 V, the coloration state was temporally preserved in the mesophase and completely bleached within 24 hours (Figure 11.3d). Further cooling the DNA–LC material in the magenta color state from the liquid crystal to the crystalline phase in the absence of applied voltage extended the persistence time of the magenta state (Figures 11.3e,f). The complete recovery of the colorless state in the crystalline phase was observed within about 30 hours. POM was used to investigate the mechanism of optical memory in the DNA–surfactant material. It was found that reorientation of the oxidized DNA–surfactant smectic layers took place due to the applied voltage during the transition process from the isotropic to the mesophase. In the parallel aligned oxidized DNA–surfactant complex, the surfactant sublayers may act as insulating barrier and prevent electron hopping. Therefore, the reduction process of DNA radical cations was slowed down.

In the electrochromic DNA–surfactant liquid devices, the phase-dependent radical cation reduction in the LC and crystalline phases can be regarded as a clock function. The largely variable clearing points due to utilization of different surfactant mixtures can be exploited in the context of ceiling temperature indicator (Figure 11.3g,h). Therefore, this type of DNA melts offers great opportunities for developing smart tag applications for packaging of perishable food or temperature-sensitive medical products and drugs.

11.3 Solvent-Free Protein Liquids

11.3.1 Fabrication of Solvent-Free Protein Liquids

Inspired by previous work on nanoparticle liquids [30, 31], the Mann group reported the synthesis of solvent-free protein liquids, including three fundamental steps (Figure 11.4a-c) [13, 14, 21].: (i) cationization of globular proteins (e.g. ferritin, myoglobin [Mb]) via carbodiimide-mediated coupling of the surface-accessible carboxylic acid side chains to N,N'-dimethyl-1,3-propanediamine; (ii) electrostatically induced complexation between cationized proteins and anionic surfactants; and (iii) lyophilization of the resultant complexes to yield thermally induced protein liquids. The produced ferritin–surfactant and myoglobin–surfactant complexes exhibit melting temperatures around 25 °C, and the formed liquids are viscous (Figure 11.4d-f). They are thermally

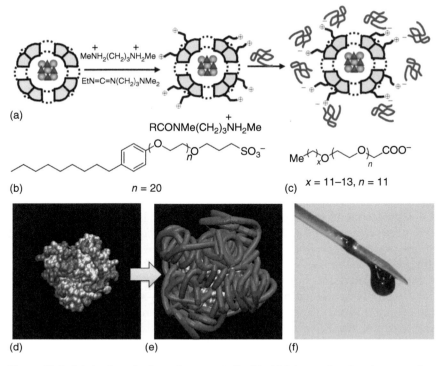

Figure 11.4 Fabrication of solvent-free protein liquids. (a) Scheme showing the general route for the preparation of protein liquids. The first step involves the N-(3-dimethylaminopropyl)-N-ethylcarbodiimide hydrochloride (EDC)-initiated coupling of N,N-dimethyl-1,3-propanediamine to carboxylic acid surface residues of proteins to produce a cationized protein. The second step involves electrostatic complexation of cationized protein with anionic surfactants to form a protein–surfactant hybrid. (b, c) Anionic surfactants that were electrostatically bound to cationized proteins. (d, e) Graphic representation of electrostatic binding of the cationized protein with anionic surfactants. (f) Photograph showing a gravity-induced flow of a solvent-free protein–surfactant liquid. Source: Liu et al. 2017 [25]. Reproduced with permission from ACS.

stable, degradation only occurs after 400 °C. Thermogravimetric analysis (TGA) indicated a typical water content of Mb of less than 0.23% [21]. This corresponds to only six water molecules per protein–surfactant complex, which is far less than required to cover the solvent accessible surface (i.e. 526 H_2O per Mb) [32]. It is also considerably less than the number of site-specific structural water molecules (c. 36 H_2O per Mb molecule), and an order of magnitude lower than that required for protein motion and function (60 H_2O per protein) [33–35].

Besides the liquid phase, the ferritin–surfactant exhibited viscoelastic and smectic LC behavior [13]. POM investigation of the ferritin melt showed temperature-dependent focal-conic textures that persisted in a very small temperature window up to 37 °C. This phase transition was confirmed by an endothermic peak observed by differential scanning calorimetry (DSC). Significantly, rheological studies on the optically anisotropic ferritin melt showed an abrupt shear thinning to a limiting shear viscosity at 32 °C, which transformed to a Newtonian fluid at 50 °C. In addition, SAXS experiments indicated a lamellar structure with layer spacing of 13 nm, which is similar to the external diameter of the ferritin molecule. With increasing temperature, SAXS profiles showed no evidence of LC ordering, indicating the isotropic liquid phase of the ferritin–surfactant melt. These results are intriguing as ferritin is a spherical nanoparticle, which is not expected to exhibit anisotropic assembly behavior. However, the cationization of ferritin and subsequent complexation with surfactant could result in the transformation from a spherical native protein into an ellipsoidal complex [13]. This new type of shape anisotropy of the ferritin–surfactant complex could promote LC formation.

Regarding myoglobin–surfactant liquids, they exhibited a high level of structural integrity as confirmed by the studies of attenuated total reflectance–Fourier transform infrared (ATR–FTIR) and CD spectroscopies [14]. At room temperature, myoglobin molecules in the solvent-free liquid state retained their near-native architecture with minimal perturbation of their α-helical secondary structure. Interestingly, the viscous deoxy–myoglobin melts showed the ability to reversibly bind dioxygen or other molecules (CO, SO_2) when exposed to a dry atmosphere of the gases. Due to their high viscosity, the melts took a few minutes rather than milliseconds for the dioxygen molecules to bind under equilibrium conditions. The binding experiments showed classical behavior with regard to the oxygen binding affinity and an absence of cooperativity. Importantly, the observations were almost identical to those obtained for deoxy–myoglobin under physiological conditions. This indicated that the structure and function of the myoglobin were preserved in the absence of any solvent. Next, temperature-dependent structure and refolding behavior of myoglobin liquids were investigated by high-resolution synchrotron radiation as well as CD and UV–vis spectroscopies [21]. Thermal denaturation experiments showed that unfolding occurred in the solvent-free liquid state. But the half denaturation temperature of myoglobin (160 °C) was around 90 °C higher than that for aqueous myoglobin. This implied that the surfactant-assisted solvent-free environment significantly stabilized the proteins. Further investigations confirmed that the thermophilic behavior was primarily due to the entropic contributions associated with dehydration and the corresponding decrease in dielectric constant of the protein interior. This leads to additional interactions (hydrogen bonding, electrostatic, and van der Waals forces) within the protein–surfactant complex, which allows maintaining the folded structure of the polypeptide chain at higher temperatures than in the aqueous phase. Moreover, the restriction in conformational freedom due to molecular crowding of proteins and the lack of translational mobility of the complexed surfactant may also contribute to the thermal stability. The protein molecules underwent reversible refolding from a temperature of 155 °C, presumably because the surfactant shell of the complex had sufficient configurational flexibility and level of molecular interactivity to facilitate the reversible transfer between different secondary structure domains, as well as regulating order-to-disorder equilibrium transitions. Furthermore, incoherent neutron scattering (INS) and specific deuterium labeling of surfactants were performed to separately study protein and surfactant dynamics and probe atomic motions on the nanoseconds to picoseconds time scales and

Å-length scales in the myoglobin–surfactant liquids [36]. It was found that the dynamics of myoglobin within the complex closely resemble those of hydrated myoglobin. The surfactant shell fulfills a function similar to that of the water hydration layer that is required for protein chain mobility and activity. Those results demonstrate that the surfactant coating can plasticize protein structures in a similar way as water hydration.

Besides ferritin and myoglobin, solvent-free lysozyme–surfactant liquids were prepared and studied by synchrotron radiation CD spectroscopy [37]. The high thermal stability of the solvent-free constructs can be exploited to trap an intermediate unfolding state that is normally highly reactive toward aggregation in aqueous environment. The initial intermediate was in equilibrium with the native state below 78 °C, but transformed into an irreversible β-sheet-enriched state at higher temperature that cannot be refolded on cooling. It eventually transformed into the fully denatured state at 178 °C. This behavior was ascribed to the decreased stability of the native state in the absence of any appreciable hydrophobic effect. On the other hand, molecular constriction in the solvent-free melt could contribute toward the entropically derived stabilization of the intermediate. Consequently, stabilization of the β-sheet intermediate state within a combined surfactant coronal layer suggested that solvent-free protein melts could be used to isolate new structural intermediates of protein unfolding and lead to a greater understanding of transient analogues in aqueous environment. To better understand the atomistic structure of the protein–surfactant complexes and illustrate the influence of surfactant corona on the phase behavior and properties of these solvent-free liquids, molecular dynamics simulations were carried out on the systems [38]. It was demonstrated that the derived structural parameters were highly consistent with experimental values. The coronal layer structure of surfactant was responsive to the dielectric constant of the medium and the mobility of the surfactant molecules was significantly hindered in the solvent-free state. This provides a basis for the origins of retained protein dynamics in these novel biofluids. Recently, an anisotropic glucose oxidase–surfactant complex was synthesized and shown to exhibit temperature-dependent phase behavior in the solvent-free state [39]. At close to room temperature, the complex crystallized with an expanded interlayer spacing of ~12 nm and interchain correlation lengths consistent with alkyl tail–tail and PEG–PEG ordering. The complex displayed a birefringent spherulitic texture and melted at 40 °C to produce a solvent-free LC phase with the preserved enzyme secondary structure. It was found that the melt exhibited a birefringent dendritic texture below the conformation transition temperature (T_c) of glucose oxidase (58 °C) and retained interchain PEG–PEG ordering. These results indicated that the shape anisotropy of the protein–surfactant building block played a key role in the formation of ordered structure of the complex. This self-organization behavior was also associated with restrictions in the intramolecular motions of the protein core of the complex.

In addition to the fabrication of liquids from globular proteins, rodlike polypeptides were also used to form solvent-free melts [19, 20, 40]. For instance, the complexation of the cationic poly(L-lysine) (PLL) or H-shaped hexapeptide with lecithin resulted in a stable and lamellar ordered liquid crystalline and liquid phases [19, 20]. Recently, the Ikkala group developed a ternary fluid system based on the electrostatic interaction of PLL with dodecylbenzenesulfonate (DBS) and additional complexation of dodecylbenzenesulfonic acid (DBSA) by hydrogen bonding [40]. When the stoichiometric ratio of PLL and DBSA was between 1.5 and 2.0, a fluid-like liquid crystalline material with an α-helical secondary structure of PLL and hexagonal cylindrical self-assembly was achieved. Heating above 120–140 °C led to a partial change into lamellar β-sheet secondary structures. The heating thus produced a mixed structure, whereby the complete conversion was inhibited probably by topological constraints. For PLL/DBSA ratio of 3.0, the heating resulted in disordered structures with a random coil conformation. All transitions were irreversible on the time scale of the measurements.

Very recently, solvent-free LCs and liquids involving unfolded polypeptides were investigated as well (Figure 11.5) [15, 17]. A series of supercharged polypeptides (SUPs) with the pentapeptide repeat motif $(VPGEG)_n$ were produced by genetic engineering (Figure 11.5a). At position four of this peptide motif, there is a glutamic acid that after multimerization by recursive directional ligation leads to unfolded peptide chains with negative charges ranging from 9, over 18, 36, 72, to 144. These SUPs were complexed with cationic surfactants (DOAB, DEAB, and DDAB). After dehydration, the anhydrous SUP–surfactant complexes exhibited non-Newtonian (smectic LC, Figure 11.5b) and Newtonian (isotropic

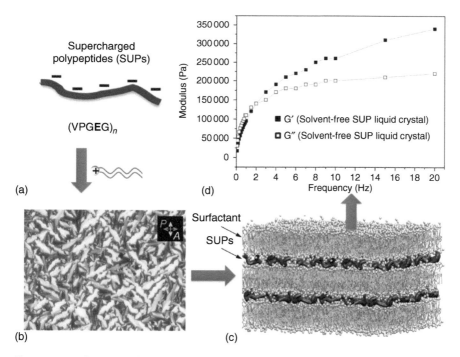

Figure 11.5 Fabrication of solvent-free fluids based on supercharged polypeptides (SUPs) [15]. (a) Negatively charged SUPs are produced by genetic engineering and combined with cationic surfactants. (b) POM image of the SUP–surfactant smectic liquid crystal. (c) Proposed lamellar bilayer structure of the liquid crystalline phase. (d) Rheological investigation of the solvent-free SUP–surfactant fluids, indicating high elasticity of the SUP liquid crystals. Source: Liu et al. 2017 [25]. Reproduced with permission from ACS.

liquid) fluid behaviors. The fluids were thermally stable over a wide range of temperatures until decomposition occurred at around 200 °C. DSC results showed two endothermic peaks during heating corresponding to the phase transitions of crystal-LC and LC-isotropic liquid, respectively.

Typical focal-conic textures that are characteristic for smectic layer structures were found by POM. The intrinsic long-range lamellar-ordered structure of SUP–surfactant fluids were confirmed by SAXS, where each repeating layer consists of tail-to-tail interacting cationic surfactants, which electrostatically interact with the anionic SUPs (Figure 11.5c). The rheological measurements indicate that the solvent-free SUP fluids underwent a thermal transformation from the LC state with viscoelastic properties to the isotropic liquid state with Newtonian behavior. Importantly, the elastic moduli of the SUP–surfactant LCs were of MPa magnitude (Figure 11.5d), confirming their remarkable elasticity. Indeed, repeatable rheological behavior in the smectic phase was obtained after thermal treatment in the isotropic phase, confirming the recoverable mechanical properties of the present fluid system. Such a strong and recoverable elasticity was achieved neither by the single SUP components, which are brittle and inextensible, nor by cationic surfactants, nor by SUP–surfactant amorphous complexes. Thus, the high-elastic mechanical behavior originated from the lamellar spatially segregated SUP–surfactant structures. Due to the modular assembly of the SUP–surfactant complexes, their mechanical properties can be easily tuned by varying the lengths of the alkyl chains of the surfactants or the molecular weight of the SUP backbone. In addition, green fluorescent protein (GFP) was fused to the SUPs. The characteristic fluorescent property of GFP was maintained in the solvent-free SUP–surfactant fluid systems, indicating that the folded protein was not denatured by the surfactant environment.

11.3.2 Electrochemical Applications Based on Solvent-Free Protein Liquids

As discussed, the myoglobin liquids are of remarkable thermal stability. They retain biological function at an exceptionally high protein concentration. These features make the liquids very suitable for bioelectrochemistry applications. Recently, the Mann group reported on an entirely new approach to study the redox activity of myoglobin in a highly crowded anhydrous myoglobin–surfactant liquid without any external electrolyte solution [41]. A three-electrode configuration was fabricated to record electrochemical properties (Figure 11.6a,b). The myoglobin liquid was deposited onto a highly oriented pyrolytic graphite (HOPG) electrode, and Pt (counter) as well as Ag (pseudoreference) wires were then directly inserted into the protein droplet. Cyclic voltammetric responses indicated the quasi-reversible one-electron oxidation of the heme moiety. The temperature-dependent experiments suggested that the kinetics of the electrochemical responses were controlled by planar diffusion to the electrode surface. Further analysis of the electrochemical data combined with the structural features of the liquid obtained from diffuse reflectance UV–vis, SAXS, and rheology measurements revealed that charge transport in the melt occurs via electron hopping from heme to heme coupled to the migration of mobile ionic species. However, the diffusion coefficients for charge transport in these highly viscous melts were approximately 4×10^{-12} and 5×10^{-11} cm^2 s^{-1} at 25 and 60 °C, respectively, which were several orders of magnitude lower than that reported for myoglobin dispersed in hydrated polyelectrolytes [42].

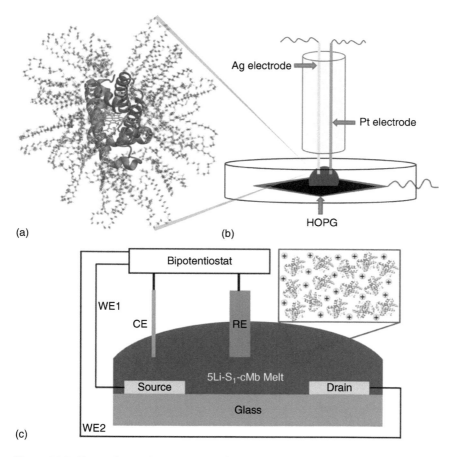

Figure 11.6 Electrochemical investigation of the solvent-free myoglobin–surfactant liquids [41]. (a) Molecular model of the myoglobin–surfactant complex. (b) Schematic diagram showing the three electrode cell configuration, where Pt (counter) and Ag (pseudoreference) wires were inserted directly into a drop of the myoglobin–surfactant melt, which was in contact with a highly oriented pyrolitic graphite (HOPG) electrode. (c) Schematic showing the structure of electrochemical field-effect transistor that was used for conductivity measurements [23]. The setup included immersion of a counter electrode (CE) and reference electrode (RE) into a drop of myoglobin–surfactant melt in which LiPF$_6$ was dispersed in the myoglobin liquid. The drop of myoglobin melt was uniformly spread over the two working electrodes (WE1 and WE2), acting as source and drain, respectively. (d) The corresponding conductivity measurements for the myoglobin–surfactant melt blended with LiPF$_6$ (gray curve) and the pristine myoglobin–surfactant melt (black curve) at 30 °C. Conductivity plots at 150 °C are shown in the inset. Source: Liu et al. 2017 [25]. Reproduced with permission from ACS.

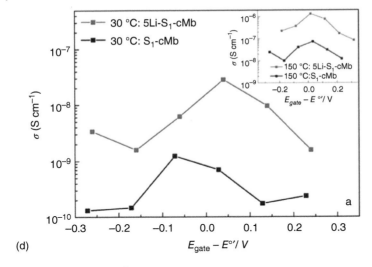

Figure 11.6 (Continued)

In this regard, lithium hexafluorophosphate (LiPF$_6$) was introduced into the solvent-free myoglobin melt and their electrochemical responses were investigated [23]. The blended melts showed no evidence of microphase separation, indicating that LiPF$_6$ was homogeneously dispersed in the solvent-free liquids. Diffuse UV–vis reflectance measurements demonstrated the structure of the protein was retained, without LiPF$_6$ interfering. The melts showed protein-mediated redox responses even at 150 °C, which was consistent with the high thermal stability of myoglobin in the solvent-free state. Significantly, the redox behaviors of the heme prosthetic group varied systematically and reversibly with temperature, which was consistent with hyperthermophilic unfolding/refolding of the protein structure. The diffusion coefficient was an order of magnitude larger in the presence of LiPF$_6$, suggesting that the presence of the ionic species facilitated the charge-transport properties. To illustrate the charge-transport rate in the liquids, an electrochemical field-effect transistor was constructed (Figure 11.6c). This device enabled control over the redox state of the heme by setting the potential difference between the interdigitated electrodes and the quasi-reference electrode (E_{gate}). The current arises from two transport mechanisms. One represents electron hopping between heme redox centers. The other one is due to ion movement within the protein liquids. It was found that the lateral conductivity of the protein melts increased with increasing temperature (Figure 11.6d). The addition of LiPF$_6$ to the melt resulted in an order of magnitude increase in the conductivity at 30 °C. The maximum conductivity value at 150 °C was on the order of 1×10^{-6} S cm^{-1}.

11.3.3 Catalysis of Solvent-Free Enzyme Liquids

Water or other solvent molecules are considered very important in enzyme catalysis since they are responsible for mass transfer of substrates and products,

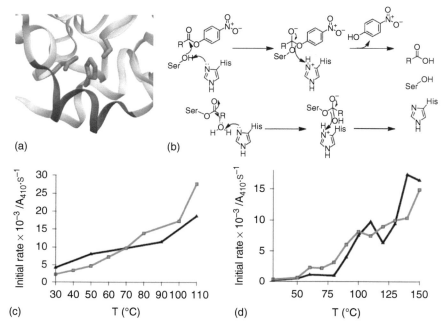

Figure 11.7 Hydrolysis of fatty acid esters in solvent-free lipase–surfactant liquids [22]. (a) Three-dimensional model showing the Ser144-His257-Asp203 catalytic triad of the lipase and the helical lid motif. (b) Two-step mechanism for lipase-based hydrolysis of p-nitrophenyl palmitate (pNPPal) and p-nitrophenyl butyrate (pNPB). Plot of initial rate of reactions of pNPB (c) and pNPPal (d) within solvent-free lipase–surfactant liquids as a function of temperature (Rhizomucor miehei-lipase catalysis, black curves; Thermomyces lanuginosus-lipase catalysis, gray curves). Source: Liu et al. 2017 [25]. Reproduced with permission from ACS.

nucleophilicity and proton transfer at the active site. In addition, they are required in solvent shell-mediated dynamics for accessing catalytically active conformations. Hence, enzyme catalysis in a solvent-free protein liquid represents a considerable challenge, as the prerequisites for such a process include the correct conformation of the active site and the presence of a medium for substrate and product mass transfer. In this regard, Mann and colleagues fabricated a series of solvent-free lipase–surfactant liquids and demonstrated that these biofluids can be directly used for the hydrolysis of fatty acid esters (Figure 11.7) [22]. Two types of lipases from the mesophile *Rhizomucor miehei* (RML) and thermophile *Thermomyces lanuginosus* (TLL) were used for the experiments. The solvent-free lipase liquids were thermally stable and underwent refolding at around 170 °C. Synchrotron radiation CD and attenuated total reflection FTIR spectra indicated the native structure of the lipases in the water-free enzyme liquids.

Regarding esterase activity, two substrates were investigated, i.e. p-nitrophenyl palmitate (pNPPal), a waxy solid, and p-nitrophenyl butyrate (pNPB), a liquid (Figure 11.7a,b). Upon addition of pNPB or pNPPal to the solvent-free lipase melts, an intense yellow color that was distributed throughout the entire enzyme–surfactant melt was detected, indicating the formation of the p-nitrophenol (pNP) product. UV–vis spectroscopy was used to analyze the lipase activity and revealed that the mesophile-derived RML–surfactant liquid was more effective in physiological temperatures. However, the initial reaction rates of the solvent-free lipase fluids were significantly lower than the catalysis in aqueous environment at 37 °C. This reduction in rate was attributed to low substrate and product mass transfer arising from the high viscosity of the melts. It was also found that enzyme activity in the solvent-free liquid state could be maintained over an extended temperature range. When the temperature was raised from 30 to 110 °C, the rate increased 4-fold and 12-fold for pNPB hydrolysis in the RML–surfactant and TLL–surfactant solvent-free reaction fluids, respectively (Figure 11.7c). Raising the temperature from 30 to 150 °C resulted in an 88-fold and 45-fold increase in the initial rate of pNPPal hydrolysis for the same set of surfactants as mentioned (Figure 11.7d). These observations indicated that the substrates in either liquid or solid form could be effectively dispersed and reacted within the solvent-free lipase melts. Mass transfer of the substrates to the active site was still possible in the viscous enzyme melt, although the enzyme molecules were embedded in a surfactant coronal layer. Therefore, the strategy to produce solvent-free enzymatic reactive liquids allows the investigation of biocatalysis at extreme temperatures and might provide new directions in industrial catalysis.

11.4 Solvent-Free Virus Liquids

In addition to the fabrication of solvent-free liquids based on nucleic acid and protein building blocks, liquids from bacteriophages and plant viruses are of special interest as they might act as novel storage and transport media, or they might be exploited for the development of nonaqueous processing routes in virus-based nanotechnology. According to a strategy for the fabrication of protein melts (Figure 11.8a), cowpea mosaic virus (CPMV) was engineered to produce a solvent-free liquid [16]. DSC characterization of the lyophilized CPMV–surfactant complex showed a melting transition at 28 °C. ATR-FTIR measurements indicated that surface modification, dehydration, and melting did not disturb the predominantly β-sheet secondary structure of the coat proteins and did not remove the RNA from the virus capsid interior. Interestingly, the virus melt was used as a highly concentrated agent for the infection of plants because the surfactant chains did not have any influence on host processing of the viral RNA. Direct application of the solvent-free CPMV liquid onto the leaf surface resulted in successful infection (Figure 11.8b,c), indicating that the viral RNA contained within the surface engineered capsids remained sufficiently intact and accessible to host cell processing. It was also found that the CPMV–surfactant complexes can be dissolved in a variety of organic solvents, including acetonitrile, isopropanol, chloroform, dichloromethane, and methyl ethyl ketone. Therefore, aerosol delivery of the virus in low-boiling-point organic solvents can be envisioned. Furthermore, a solvent-free liquid based on tobacco mosaic virus (TMV), which exhibits a rodlike shape, was produced exhibiting a melting temperature of 28 °C.

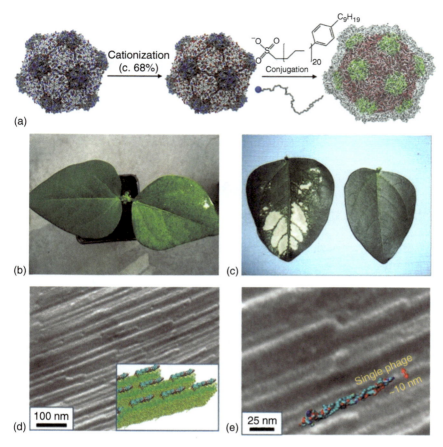

Figure 11.8 Fabrication of solvent-free virus–surfactant fluids. (a) Scheme showing the general route for the preparation of cowpea mosaic virus (CPMV) melt [16]. Optical images of symptomatic *Vigna unguiculata* plants inspected after infection with aqueous dispersions of wild-type CPMV (b) and solvent-free CPMV–surfactant droplet (c). In each case, pairs of leaves either treated or untreated with the infective agents are shown. (d, e) Anisotropic rodlike viruses (bacteriophage) were also used for fabrication of solvent-free virus liquid crystals and liquids by complexation with surfactant [17]. The magnifications of freeze-fracture transmission electron microscopy images of the phage–surfactant liquid crystals are shown. Long-range-ordered lamellar structure of the phage–surfactant mesophase, where individual phages were identified at the layer edges, indicating the orientational order of phages within the sublayer. The model of phage-based liquid crystals is sketched in the inset (side view). Source: Liu et al. 2017 [25]. Reproduced with permission from ACS.

Recently, even much larger and anisotropic virus particles were manipulated to form solvent-free liquids. For that purpose, engineered M13 bacteriophages were selected, which are monodisperse and anisotropic rodlike particles of 1 µm in length and ∼7 nm in width. The surface of the major coat protein of M13 contains negative charges, which can be complexed with cationic surfactants. After complexation of the phage with mixed surfactants of DOAB and DDAB, solvent-free M13–surfactant liquid crystals and liquids were produced [17]. DSC showed that the virus materials exhibit melting and clearing temperatures at 14 and 58 °C, respectively. SAXS and POM measurements indicated that a smectic mesophase with typical focal-conic birefringence was developed. Each repeating layer consists of tail-to-tail interacting surfactants that protrude from the phage particles. Long-range periodic layer structures in the mesophase were confirmed by FF-TEM studies (Figure 11.8d,e). The fractured plane revealed individual phages globally, along a preferred direction. Nematic orientational ordering has developed between different phages within the sublayer as a result of the rigidity and large length-to-diameter aspect ratio of the phage particles.

11.5 Mechanism for the Formation of Solvent-Free Bioliquids

Many small molecules can be present in all three physical states, i.e. solid, liquid, and gas. This is due to limited interactions between each other. With increase in molecular weight and presence of functional groups, intermolecular forces such as van der Waals and ionic interactions as well as hydrogen bonds restrict thermally induced motions, hence limiting adoption of different physical states under most pressure regimes [43–45]. An extreme case are biomacromolecules or large biological complexes with dimensions in the nanoscale, since they strongly interact in the absence of any solvent. Their phase behavior is largely limited since the size of the biomacromolecular architectures exceeds the range of their intermolecular force fields [46]. Hence, once a biopolymer powder obtained by freeze-drying is heated above a critical temperature, the material will degrade due to thermal scission of the biomacromolecular backbone. This is a general phenomenon of biomacromolecules; and only until very recently were their properties and functions almost exclusively investigated in solution, with water being the dominant solvent.

How can such folded biomacromolecules or even larger biopolymer complexes be rendered to show a richer phase behavior, and especially how can nucleic acid and protein liquids be induced? The answer lies in sterically shielding the strong intermolecular forces by introducing a surfactant or polymer surface layer as described in the previous paragraphs. It should be emphasized here that the surfactants or polymers do not act as a solvent or matrix but specifically interact via distinct electrostatic interactions to form well-defined biomacromolecule hybrid structures.

The artificial corona around the biomacromolecule surface lowers the intermolecular interaction forces by two mechanisms. On the one hand, repulsion between the biomacromolecule–surfactant hybrid is induced by the entropically driven osmotic force related to the compression of surfactant structures [46]. On the other hand, the reduction of the van der Waals forces is due to the similarities in the refractive indices between the biomacromolecule and the surfactant [46, 47]. In this way, upon supplying these biomacromolecular complexes with thermal energy, they can overcome their positional order of the solid state and transition into the liquid phase. This phase transition is characterized by an expansion in volume. During the formation of biomacromolecular liquid crystals, the surfactants serve a dual role. They sterically separate the biological moieties. Moreover, they contribute to achieve positional order in the solvent-free bioliquid crystals [48, 49].

11.6 Conclusions and Outlook

This chapter deals with new types of biomacromolecular physical states, i.e. solvent-free liquids and thermotropic liquid crystals. These states are enabled by enveloping nucleic acids, polypeptides, proteins, and multiprotein complexes in a well-defined surfactant shell. These materials are fabricated by very simple procedures relying on complexing a surfactant electrostatically with the biomacromolecule component, as the major step. The size of the individual biomacromolecule–surfactant hybrids ranges from very few nanometers to objects with an extension of more than one micrometer. Regarding the solvent-free nucleic acid liquids, they are thermally stable and can be processed in the absence of any solvent. Equipping the nucleic acid backbone with surfactants containing two alkyl chains also allows liquid crystal formation. Phase transitions in these fluidic materials can be controlled over an extremely broad temperature range by selection of the appropriate surfactants. Especially, the alkyl chain length plays a decisive role. The DNA melts providing a hydrophobic environment and lacking high dielectric water offer opportunities for the fabrication of DNA-based electrochemical devices. For instance, DNA liquids acted as a undiluted medium for charge transport in the absence of external electrolytes. Moreover, temperature-dependent electrochromism based on these new type of liquids was developed. This allows control of the volatility of the optoelectronic state simply by changing the phase of the DNA–surfactant materials. However, under ambient conditions, water molecules may diffuse into ionic DNA–surfactant materials and might negatively influence device performance. Therefore, restricting the amount of water molecules attached to the DNA chains appears to be a critical factor for obtaining more stable devices. In addition, the ability to control DNA sequence and secondary structure will allow creation of new classes of melts that undergo well-defined structural changes programmed by sequence and monitored by electrochemical signals. Since the solvent-free DNA fluids have negligible volatility, exhibit high DNA content, and low viscosity that depends on the surfactant alkyl chain length and molecular weight of the DNA, this type of soft DNA–surfactant liquids is suitable for a broad range of new studies and differ a lot from the current applications of DNA that are pursued in aqueous solutions. For example, they may act as a fluid depot for cargo storage when aptamer sequences are rendered as DNA liquids or molecules are intercalated into the DNA double helix, which is surrounded by surfactants. It is also expected that liquid bulk catalysts can be generated once ribozymes are converted into solvent-free nucleic acid liquids.

Electrostatic attachment of surfactants to the surface of proteins followed by removal of water and thermally induced melting was exploited for the preparation of protein-based fluids. Significantly, dehydration and subsequent melting of the protein–surfactant complexes had no considerable effect on the folded structure of proteins under investigation. The protein liquids exhibited hyperthermophilic behaviors and can be reversibly refolded by cooling from extreme thermal conditions. More importantly, the biological activity of proteins was retained in the solvent-free melts. For instance, the dehydrated myoglobin–surfactant liquids maintained their ability to reversibly bind dioxygen and other gaseous ligands as in their physiological environment. The lipase–surfactant fluids can directly solubilize substrates and catalyze fatty acid ester hydrolysis in the absence of any solvent. The maintained near-native structure of the protein at high temperatures beyond the boiling point of water enabled the increase of catalysis efficiency. The solvent-free liquids based on CPMV induced typical visual symptoms of infection when placed on leaves, indicating the remaining biological activity in the fluid state. These new findings suggest that retention of the structure and function of proteins in the absence of water was achieved effectively given the presence of appropriate intramolecular interactions and contacts in the molten liquid state. Therefore, water-free protein liquids present a significant challenge to existing theories on the role of water molecules in determining protein structure and function. In addition, other new behaviors and properties of the solvent-free protein melts have been found. The highly crowded protein melts made of myoglobin provided a new generation of biologically inspired charge transporting media, with tunable conductivities and chemical functionalities. Another type of water-free fluids based on SUPs exhibited high elasticity. The mechanical strength can be tuned conveniently by selecting the alkyl chain length or the molecular weight of the polypeptides as easily adjustable design parameters. In this context, fusion of other proteins to the unfolded SUPs might result in novel hybrid biofluids with smart functions and advanced properties or unexpected mechanical behavior.

The achievements discussed in this chapter will fuel further efforts of employing electrostatic complexation for the preparation of solvent-free biofluids based on a wide range of biomacromolecules, thus providing new directions in technological applications including biosensing, biocatalysis, biomedicine, and construction of bioelectronic devices.

References

1 Pettersson, L.G.M., Henchman, R.H., and Nilsson, A. (2016). Water determines the structure and dynamics of proteins. *Chem. Rev.* 13: 7673–7697.
2 Kim, W. and Conticello, V.P. (2007). Protein engineering methods for investigation of structure-function relationships in protein-based elastomeric materials. *Polym. Rev.* 47: 93–119.
3 Kyle, S., Aggeli, A., Ingham, E., and McPherson, M.J. (2009). Production of self-assembling biomaterials for tissue engineering. *Trends Biotechnol.* 27: 423–433.
4 Gordiichuk, P.I., Wetzelaer, G.H., Rimmerman, D. et al. (2014). Solid-state biophotovoltaic cells containing photosystem I. *Adv. Mater.* 26: 4863–4869.

5 Kwiat, M., Elnathan, R., Kwak, M. et al. (2012). Non-covalent monolayer-piercing anchoring of lipophilic nucleic acids: preparation, characterization, and sensing applications. *J. Am. Chem. Soc.* 134: 280–292.
6 Kwak, M., Gao, J., Prusty, D.K. et al. (2011). DNA block copolymer doing it all: from selection to self-assembly of semiconducting carbon nanotubes. *Angew. Chem. Int. Ed.* 50: 3206–3210.
7 Yuan, J.Y., Mecerreyes, D., and Antonietti, M. (2013). Poly(ionic liquid)s: an update. *Prog. Polym. Sci.* 38: 1009–1036.
8 Johnston, K.P., Maynard, J.A., Truskett, T.M. et al. (2012). Concentrated dispersions of equilibrium protein nanoclusters that reversibly dissociate into active monomers. *ACS Nano* 6: 1357–1369.
9 Ramundo, J. and Gray, M. (2008). Enzymatic wound debridement. *J. Wound Ostomy Continence Nurs.* 35: 273–280.
10 Leone, A.M., Weatherly, S.C., Williams, M.E. et al. (2001). An ionic liquid form of DNA: redox-active molten salts of nucleic acids. *J. Am. Chem. Soc.* 123: 218–222.
11 Bourlinos, A.B., Chowdhury, S.R., Herrera, R. et al. (2005). Functionalized nanostructures with liquid-like behavior: expanding the gallery of available nanostructures. *Adv. Funct. Mater.* 15: 1285–1290.
12 Liu, K., Shuai, M., Chen, D. et al. (2015). Solvent-free liquid crystals and liquids from DNA. *Chem. Eur. J.* 21: 4898–4903.
13 Perriman, A.W., Cölfen, H., Hughes, R.W. et al. (2009). Solvent-free protein liquids and liquid crystals. *Angew. Chem. Int. Ed.* 48: 6242–6246.
14 Perriman, A.W., Brogan, A.P.S., Cölfen, H. et al. (2010). Reversible dioxygen binding in solvent-free liquid myoglobin. *Nat. Chem.* 2: 622–626.
15 Liu, K., Pesce, D., Ma, C. et al. (2015). Solvent-free liquid crystals and liquids based on genetically engineered supercharged polypeptides with high elasticity. *Adv. Mater.* 27: 2459–2465.
16 Patil, A.J., McGrath, N., Barclay, J.E. et al. (2012). Liquid viruses by nanoscale engineering of capsid surfaces. *Adv. Mater.* 24: 4557–4563.
17 Liu, K., Chen, D., Marcozzi, A. et al. (2014). Thermotropic liquid crystals from biomacromolecules. *Proc. Natl. Acad. Sci. U.S.A.* 111: 18596–18600.
18 Faul, C.F.J. and Antonietti, M. (2003). Ionic self-assembly: facile synthesis of supramolecular materials. *Adv. Mater.* 15: 673–683.
19 Wenzel, A. and Antonietti, M. (1997). Superstructures of lipid bilayers by complexation with helical biopolymers. *Adv. Mater.* 9: 487–490.
20 Sascha, G. and Markus, A. (2002). Supramolecular organization of oligopeptides, through complexation with surfactants. *Angew. Chem. Int. Ed.* 41: 2957–2960.
21 Brogan, A.P.S., Siligardi, G., Hussain, R. et al. (2012). Hyper-thermal stability and unprecedented re-folding of solvent-free liquid myoglobin. *Chem. Sci.* 3: 1839–1846.
22 Brogan, A.P.S., Sharma, K.P., Perriman, A.W., and Mann, S. (2014). Enzyme activity in liquid lipase melts as a step towards solvent-free biology at 150 °C. *Nat. Commun.* 5: 5058.
23 Sharma, K.P., Risbridger, T., Bradley, K. et al. (2015). High-temperature electrochemistry of a solvent-free myoglobin melt. *ChemElectroChem* 2: 976–981.

24 Liu, K., Varghese, J., Gerasimov, J.Y. et al. (2016). Controlling the volatility of the written optical state in electrochromic DNA liquid crystals. *Nat. Commun.* 7: 11476.

25 Liu, K., Ma, C., Göstl, R. et al. (2017). Liquefaction of biopolymers: solvent-free liquids and liquid crystals from nucleic acids and proteins. *Accounts of Chemical Research* 50 (5): 1212–1221.

26 Leone, A.M., Tibodeau, J.D., Bull, S.H. et al. (2003). Ion atmosphere relaxation and percolative electron transfer in cobipyridine DNA molten salts. *J. Am. Chem. Soc.* 125: 6784–6790.

27 Szalai, V.A. and Thorp, H.H. (2000). Electron transfer in tetrads: adjacent guanines are not hole traps in G quartets. *J. Am. Chem. Soc.* 122: 4524–4525.

28 Candeias, L.P. and Steenken, S. (1989). Structure and acid-base properties of one-electron-oxidized deoxyguanosine, guanosine, and 1-methylguanosine. *J. Am. Chem. Soc.* 111: 1094–1099.

29 Rokhlenko, Y., Cadet, J., Geacintov, N.E., and Shafirovich, V. (2014). Mechanistic aspects of hydration of guanine radical cations in DNA. *J. Am. Chem. Soc.* 136: 5956–5962.

30 Bourlinos, A.B., Herrera, R., Chalkias, N. et al. (2005). Surface-functionalized nanoparticles with liquid-like behavior. *Adv. Mater.* 17: 234–237.

31 Rodriguez, R., Herrera, R., Archer, L.A., and Giannelis, E.P. (2008). Nanoscale ionic materials. *Adv. Mater.* 20: 4353–4358.

32 Cavallo, L., Kleinjung, J., and Fraternali, F. (2003). POPS: a fast algorithm for solvent accessible surface areas at atomic and residue level. *Nucleic Acids Res.* 31: 3364–3366.

33 Costantino, H.R., Curley, J.G., and Hsu, C.C. (1997). Determining the water sorption monolayer of lyophilized pharmaceutical proteins. *J. Pharm. Sci.* 86: 1390–1393.

34 Pauling, L. (1945). The adsorption of water by proteins. *J. Am. Chem. Soc.* 67: 555–557.

35 Rupley, J.A., Gratton, E., and Careri, G. (1983). The adsorption of water by proteins. *Trends Biochem. Sci.* 8: 18–22.

36 Gallat, F.X., Brogan, A.P.S., Fichou, Y. et al. (2012). A polymer surfactant corona dynamically replaces water in solvent-free protein liquids and ensures macromolecular flexibility and activity. *J. Am. Chem. Soc.* 134: 13168–13171.

37 Brogan, A.P.S., Sharma, K.P., Perriman, A.W., and Mann, S. (2013). Isolation of a highly reactive β-sheet-rich intermediate of lysozyme in a solvent-free liquid phase. *J. Phys. Chem. B* 117: 8400–8407.

38 Brogan, A.P.S., Sessions, R.B., Perriman, A.W., and Mann, S. (2014). Molecular dynamics simulations reveal a dielectric-responsive coronal structure in protein-polymer surfactant hybrid nanoconstructs. *J. Am. Chem. Soc.* 136: 16824–16831.

39 Sharma, K.P., Zhang, Y.X., Thomas, M.R. et al. (2014). Self-organization of glucose oxidase-polymer surfactant nanoconstructs in solvent-free soft solids and liquids. *J. Phys. Chem. B* 118: 11573–11580.

40 Hanski, S., Junnila, S., Almásy, L. et al. (2008). Structural and conformational transformations in self-assembled polypeptide-surfactant complexes. *Macromolecules* 41: 866–872.

41 Sharma, K.P., Bradley, K., Brogan, A.P.S. et al. (2013). Redox transitions in an electrolyte-free myoglobin fluid. *J. Am. Chem. Soc.* 135: 18311–18314.

42 Rusling, J.F. and Nassar, A.F. (1993). Enhanced electron transfer for myoglobin in surfactant films on electrodes. *J. Am. Chem. Soc.* 115: 11891–11897.

43 Doye, J.P.K. and Wales, D.J. (1996). The effect of the range of the potential on the structure and stability of simple liquids: from clusters to bulk, from sodium to C_{60}. *J. Phys. B: At. Mol. Opt. Phys.* 29: 4859–4894.

44 Min, Y., Akbulut, M., Kristiansen, K. et al. (2008). The role of interparticle and external forces in nanoparticle assembly. *Nat. Mater.* 7: 527–538.

45 Hagen, M.H.J., Meijer, E.J., Mooij, G.C.A.M. et al. (1993). Does C_{60} have a liquid-phase? *Nature* 365: 425–426.

46 Perriman, A.W. and Mann, S. (2011). Liquid proteins – a new frontier for biomolecule-based nanoscience. *ACS Nano* 5: 6085–6091.

47 Bishop, K.J.M., Wilmer, C.E., Soh, S., and Grzybowski, B.A. (2009). Nanoscale forces and their uses in self-assembly. *Small* 5: 1600–1630.

48 Faul, C.F.J. and Antonietti, M. (2003). Ionic self-assembly: facile synthesis of supramolecular materials. *Adv. Mater.* 15: 673–683.

49 Faul, C.F.J. (2014). Ionic self-assembly for functional hierarchical nanostructured materials. *Acc. Chem. Res.* 47: 3428–3438.

12

Ionic Liquids

Hiroyuki Ohno

Tokyo University of Agriculture and Technology, Graduate School of Engineering, 2-24 Nakacho, Koganei, Tokyo 184-8588, Japan

12.1 What Is Ionic Liquid?

Ionic liquids (ILs) are liquids composed only of ions. In other words, ILs are salts with very low melting point (T_m). Since ILs inherently contain no molecules such as organic solvents or water, they show quite unique properties [1]. ILs are composed of organic ions, and accordingly they have the advantages of both salts and organic compounds. Why are these salts liquid? The reason is because of both a weak electrostatic interaction force and an asymmetric ion structure. For example, inorganic salts such as sodium chloride are composed mainly of small ions. In the case of sodium chloride, there is a strong electrostatic force between Na^+ and Cl^-. Also, symmetric ions are easily crystallized. Against these inorganic salts, coupling of organic (and, accordingly, rather larger) ions makes the electrostatic force weaker due to longer distance between ions. The T_m of salts is lowered by weakening the electrostatic force between ions. Let us see some examples. As is well known, the T_m of NaCl is 803 °C. What happens if you change the cation species of this salt? The T_m of CsCl is 645 °C, and that of tetrapropylammonium chloride is 241 °C. Furthermore, 1-ethyl-3-methylimidazolium chloride shows its T_m at 87 °C. By changing from a Na^+ cation to an imidazolium cation, the T_m of chloride salt dropped for more than 700 °C! Then, what happens when you use organic anions instead of Cl^- to prepare imidazolium salts? 1-Ethyl-3-methylimidazolium (C2mim) nitrate and tetrafluoroborate showed their T_m at 38 and 15 °C, respectively. They are all ILs. As shown, there are many ILs prepared by coupling organic ions. It is therefore not so difficult to prepare (or design) ILs with organic ions.

As mentioned, use of large organic ions is quite effective to lower the T_m of the salts thus prepared. However, too large ions cannot be used for this IL preparation, because they behave like molecules, not ionic ones. Accordingly, there exists a suitable ion size to design ILs. There are, of course, other factors such as charge density and asymmetric feature of component ions for the design of ILs. These are mentioned later. At the beginning, it is important to understand how to prepare salts with very low T_m.

Functional Organic Liquids, First Edition. Edited by Takashi Nakanishi.
© 2019 Wiley-VCH Verlag GmbH & Co. KGaA. Published 2019 by Wiley-VCH Verlag GmbH & Co. KGaA.

12.2 Some Physicochemical Properties

Since ILs inherently contain no molecules, they show properties quite different from those of molecular liquids such as water and alcohols. For example, ILs show negligible vapor pressure even at high temperature. This attractive feature means that ILs are nonflammable organic fluids in a wide range of temperatures. From the viewpoint of industrial systems, these nonflammable fluids are quite attractive for the suppression of fire accidents. Since there are numerous papers and reviews that most chemical reactions can be carried out in many ILs, they are potential candidates as reaction solvents for industrial use. On the other hand, negligible vapor pressure means that these ILs cannot be purified by distillation even under reduced pressure. There are some techniques required to separate products from ILs used as solvents.

The advancement of ILs is their variation. Just as there are numerous kinds of ions reported to date, there are also numerous candidates of salt species prepared by the coupling of these ions. However, most of them should be solid with a very high T_m; a small part of the rest (but still many!) can be treated as ILs. There are also a variety of ILs with different properties, especially in density and viscosity. These variations can be found in the homepage of QUILL (http://quill.qub.ac.uk/), the first research group on ILs in the world. Taking into account the possible properties in a wide range, ILs can be considered to be new solvents, media, or matrices for quite diverse applications.

As mentioned, many ILs are nonflammable in a wide range of temperatures. However, they are flammable at very high temperatures. To make this clear, we should think about decomposition of ILs. ILs are nonflammable at any temperature as long as they are stable; but above a certain temperature, the so-called decomposition temperature (T_d), they are decomposed into fragments. Some of these fragments are not charged and, accordingly, easily catch fire. Therefore, T_d is a very useful parameter to estimate the upper limit temperature of the IL. The T_d of ILs deeply depends on the structure of component ions. Some ILs show T_d above 400 °C and are used as thermally stable organic fluid. Since the breakdown of the weakest bond in the ions govern T_d, it is quite important to design ion structures without weak bonds for the preparation of thermally stable ILs.

One of the characteristics of ILs is their very low glass transition temperature, T_g. There is a strong tendency that ILs having low T_m show low T_g. Kauzmann [2] first reported that T_g values of many materials were estimated by their $T_m \times 2/3$. Be careful to note that the SI unit of these temperatures is Kelvin. Angell et al. also found that this is effective for many ILs [3]. There are many ILs showing no T_m but only T_g. These ILs turned to amorphous solid by cooling without crystallization. Many mixtures of ILs also show only T_g. Since T_g is the temperature at which the molecules of the amorphous part start moving, the T_g strongly relates to ionic conductivity and viscosity of the ILs. The T_g is therefore a very important parameter of the ILs.

Since ILs are composed of only ions, they are believed to show very high ionic conductivity due to high ion density. However, formula weight of ILs is large (generally larger than 200), and the maximum ion density can be calculated to be around 5 $mol\,l^{-1}$. Furthermore, viscosity of ILs is very large (about

100–10000 times higher than that of water!). Considering the charge density and viscosity, we cannot expect extremely high ionic conductivity even in pure ILs.

ILs, composed of only ions, are generally believed to be very polar. However, this is not the case. Generally, the polarity of ILs is comparable to that of alcohols. Since ILs are conductive, it is not easy to determine their polarity. Instead of electrochemical methods, solvatochromic measurements are frequently used to estimate the polarity of ILs. The polarity of ILs has already been confirmed to deeply depend on the structure of their component ions by spectroscopic and chromatographic studies [4, 5]. This also suggests that it is possible to prepare ILs with polarity higher than that of polar molecules. Through these studies on the relation between polarity and component ion structure, chloride salts are confirmed to be one of the polar ILs. However, these chloride salts have a few drawbacks such as high T_m and high viscosity originating from a small chloride anion. There is a demand to prepare polar ILs with lower T_m and viscosity. For the design of such ILs with low T_m and low viscosity, it is necessary to prepare ions having suitable size and delocalized charges. It should be noted here that the polarity and some physicochemical properties such as T_m and viscosity are in a trade-off relation. In other words, it is quite difficult to realize both high polarity and low T_m (and low viscosity) in one IL. From this viewpoint, there were a few challenges to design such ILs satisfying these inconsistent requirements. In spite of the almost unlimited possibility of ion combination, there are not so many different ILs commercially available. We have synthesized several hundred different ILs to survey the relationship between component-ion structure and physicochemical properties of these ILs. This data set should be useful to make a protocol for design concept of functional ILs. Figure 12.1 shows some examples of cations and anions used as component ions for IL preparation.

As mentioned, polar ILs, especially ILs having strong proton-accepting ability, can be designed using suitable anions. There is a powerful method to

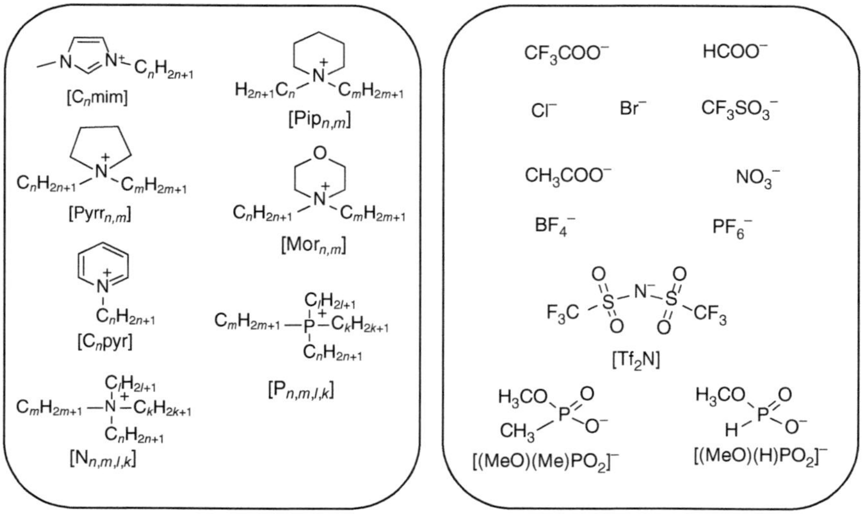

Figure 12.1 Typical ions used to prepare ILs.

evaluate the polarity of ILs with solvatochromic behavior of dye molecules in the IL [6–9]. According to literature, Kamlet–Taft parameters α and β related to proton-donating and proton-accepting ability of the ILs, can easily be determined spectroscopically. By comparing β values of several ILs having different anions, a series of carboxylate anions were confirmed to be effective to provide polar ILs comparable to chloride salts [10]. Furthermore, a series of carboxylate-anion-based ILs were found to be less viscous and to have a lower T_m than those of chloride salts.

One application of polar ILs can be found in the design of potential extraction solvents for polysaccharides such as cellulose from biomass under mild conditions. A series of carboxylate-based ILs were confirmed to dissolve cellulose [11]. For the design of more durable ILs having similar cellulose solubilization ability, phosphonates and analogs were examined as component anions. Among these, dimethylphosphonate-type ILs were found to be polar, as high as carboxylate-type ILs [12]. Further study on a series of phosphonate analogs opened up some new ILs with higher polarity than carboxylate-type ILs. These polar ILs dissolved cellulose without heating [12].

12.3 Preparation

There are a few methods of synthesizing ILs. The most popular is the quaternization of tertiary amines with nucleophilic reagents such as alkyl halides.

Another easy way is to neutralize tertiary amines with acids; this forms protonated amines as cations and acid residues as anions [13]. Thus, prepared salts have "mobile" protons and are generally called protic ILs.

Purification is sometimes easy depending on the properties of the prepared ILs. When the synthesized ILs are very hydrophobic, it is easy to wash unreacted materials and/or excess ions with a large amount of water; the washed salts should then be dried under reduced pressure. Purity of the prepared ILs can be confirmed by ion chromatography and mass spectra. ^{1}H NMR is not helpful for analysis of a trace amount of contaminants. For hydrophilic ILs, there are some other methods to purify them. Details should be found in the published books [14].

Whatever the drying process used, it should be noted here that it is extremely difficult to remove the quite small amount of water molecules tightly bound to ions. Presence of water molecules affects the physicochemical properties of ILs. For example, it is generally known that viscosity of the ILs decreased considerably by adding a small amount of water. The amount of water molecules should be considered when we discuss the properties of ILs. To measure the water content of ILs, the Carl–Fischer titration method is frequently used. It should also be noted that ^{1}H NMR is much less sensitive to measure the small amount of water contaminated in the ILs.

Regardless of the method used to prepare salts, we should think about their T_m. To lower the T_m of the salts, there are a few strategies such as to weaken the electrostatic interaction force between (among) cations and anions. To weaken the force, it is effective to lower the charge density of component ions using π-conjugated organic ions and/or to introduce electron-pulling residues to anions and electron-pushing residues to cations.

Suppression of crystallization of salts is another effective method to prepare liquid salts at lower temperature. Asymmetric component ions and multicomponent ions are frequently used to make the salts amorphous. Amorphous materials generally show no T_m but only T_g. The amorphous materials are in the liquid state until the temperature reaches the T_g. Evaluation of amorphous property can be carried out with thermal analysis such as differential scanning calorimetry (DSC).

12.4 IL Derivatives

ILs, composed of cations and anions, are liquid state and show unique properties. It is possible to put functions in ordinary ILs. Common way to prepare such functional ILs is carried out by introducing functional groups onto component ions *via* covalent bond. Here, a few examples such as zwitterions, assembled ions, and polymers of ILs are briefly mentioned.

12.4.1 Zwitterions

Ionic liquid can generally be expressed as a pair of cation and anion. Among the derivatives with distinctive properties, for example, zwitterion is one of the most unique. Zwitterion is a general name for a tethered ion pair with a covalent bond. There are a few different ways to synthesize zwitterions, but shown here is an easy chemistry of zwitterion synthesis. As shown in Figure 12.2, many different zwitterions can be prepared by the one-step coupling of tertiary amines and sultone. An equimolar amount of sultone is reacted with an amine such as ethylimidazole at room temperature to form onium cation having an anionic tail. This can be carried out under mild conditions. A spacer length between cationic site and anionic site can easily controlled by the use of sultone. Seen at the top of Figure 12.2 is sultone with a different number of methylene groups directly related to the spacer length of the prepared zwitterions.

Also, polymerizable zwitterion was obtained from vinylimidazole; see derivative **2** in Figure 12.2. Polymerized IL was first reported in 1998 based on the same strategy as that to introduce polymerizable unit onto component ions [15]. The preparation is the same as mentioned for polymerizable zwitterions. Starting from vinylimidazole, monomeric IL can be prepared after quaternization. Polymerized ILs are mentioned later. It is also possible to synthesize quaternized amine with functional tail first, then the anionic charge is introduced, as shown at the bottom of Figure 12.2. This method is effective to prepare zwitterions having other anionic sites.

12.4.2 Self-Assembled ILs

To construct a unique environment with ILs, the self-assembly technique has frequently been used. The strategy to design a self-assembled IL is the same as

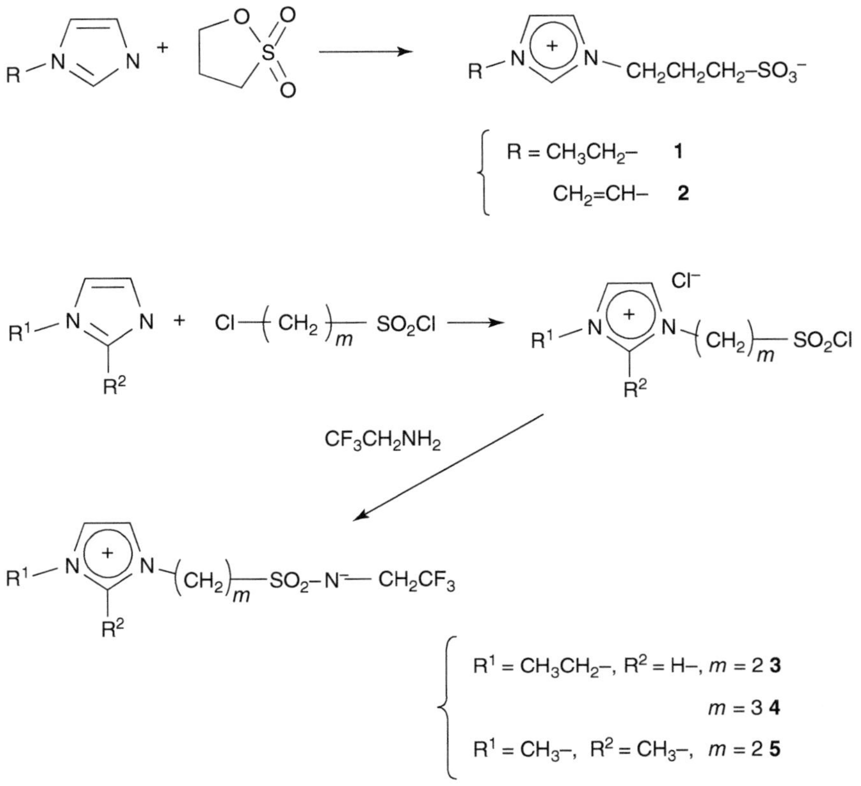

Figure 12.2 Examples of zwitterions easily synthesized.

that for ordinary amphiphilic molecules. The point is the segmental design of the ionophilic part and ionophobic part within one ion or one ion pair of ILs [16]. For example, imidazole with a long alkyl chain was neutralized with HBF_4 to form protic ILs [17]. This salt showed liquid crystalline phase between 91 and 112 °C. Furthermore, the mixture of this imidazole and imidazolium BF_4 salt showed high ion conductivity including proton conduction due to the orientated structure of imidazole and imidazolium cation. When imidazolium cation with dodecyl chain was coupled with dodecyl sulfonate (**3** in Figure 12.3), the mixture showed a smectic A phase [18]. A quite interesting point is that the ionic conductivity increased with heating. This was also found in the smectic phase, but further heating induced a small drop in the ionic conductivity. This means that the ion conduction in the liquid crystalline phase is faster than that in the isotropic liquid phase at higher temperature. This can be understood as the effect of ordered

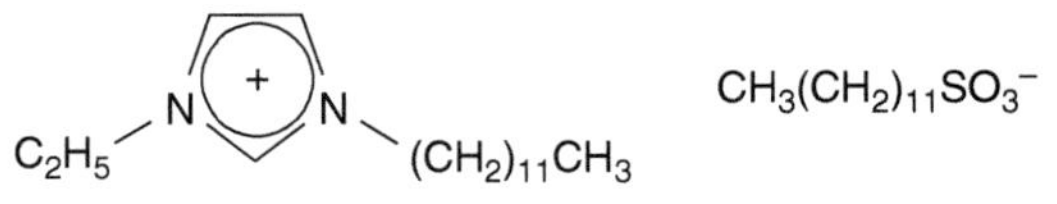

Figure 12.3 Structure of 1-ethyl-3-dodecylimidazolium dodecylsulfonate.

(or suitable) structure for ion conduction. Also, anisotropic ion conduction was observed in this system, strongly suggesting the contribution of the assembled structure on the ion conduction.

Ordered structure of ILs is quite important for a few aspects including ion transport. Not only the smectic phase but the columnar liquid crystalline structure was also constructed by introducing a fan-shape structure onto imidazolium cation [19]. Its BF_4 salt formed a columnar structure to show clear anisotropic ion conduction. Orientation of these columns can be carried out by outer stimuli such as rubbing. These assembled ILs provide an anisotropic environment, and these are expected to be useful for electronics (ionics), optics, and other areas.

12.4.3 Polymers

As mentioned in Section 12.4.1 on zwitterions, polymerized ILs can be prepared from ILs having polymerizable groups. In 1998, we first reported polymerizable ILs based on vinylimidazolium salts [15]. The polymerized ILs are characterized as fully charged polymers with very low T_g. Most of the charged polymers, called polyelectrolytes, are all solid because of their very high T_g. We were preparing many different ion conductive polymers, and trying to design them with high ionic conductivity. For high ionic conductivity, polymers with high ion density and low T_g are generally required. Then, we were attracted by the very low T_g of ILs. This is the reason why we started to polymerize ILs.

There are many different ways to get IL polymers. Almost all of the polymerization methods were used for nonionic monomers such as anion polymerization, cation polymerization, and polycondensation. It should be noted here that the polymerized ILs are all solid, and ionic conductivity decreased dramatically from that for monomers due to the decrease in the number of mobile ions and increase in the local viscosity (which leads to a decrease in the diffusion coefficient of free ions). There were, however, a few trials to improve ionic conductivity of polymerized ILs such as introduction of a spacer between the polymerizable unit and IL unit [20].

These advanced polymerized ILs are examined as potential components of electrolyte layers for electrochemical devices such as batteries, capacitors, and actuators [21]. Detailed discussion should be found in the book [21] and references therein. There are many further polymerized ILs for different purposes. The recent progress in these polymerized ILs can be seen in our review [22].

12.5 IL/Water Functional Mixture

For the functionalization of ILs, there are several approaches to introduce functional groups into component ions of ILs. There are many excellent reviews on the design and properties of functional ILs [23, 24]. Introduced here is a different approach to prepare a unique function for ILs. The strategy is quite simple; just mix ILs with water. There are mainly two groups of ILs such as water-miscible ILs and water-immiscible ILs. Water-immiscible ILs are especially of great interest, because they can be used as reaction matrix, separation liquid systems, and so on as well as from the point of view of thermodynamics. Hydrophobic ILs are water-immiscible and they are easily synthesized from hydrophobic ions such as fluorine-containing ions and/or ions with hydrophobic groups. For example, ILs containing bis(trifluoromethylsulfonyl)amide (Tf_2N) anion or hexafluorophosphate (PF_6) anion are classified into typical hydrophobic ILs [25, 26]. Ions containing benzene unit or other hydrophobic groups are also used to prepare hydrophobic ILs [27]. During the detailed analysis of hydrophobicity of a series of ILs with many different component ions, we found very unique ILs. They were miscible with water at low temperature, but phase separated by heating the mixture [28]. This unique phase change is called lower critical solution temperature (LCST)-type phase transition. As seen in Figure 12.4, homogeneously mixed IL and water changed to a phase-separated state by increasing the temperature for only a few degree Celsius. This is quite a temperature-sensitive transition, and such ILs can be found in the group between hydrophilic ILs and hydrophobic ILs.

This means that these ILs, showing such LCST-type phase change after mixing with water, can be designed by coupling cation and anion considering the total hydrophobicity of the ion pair. We then measured the saturated water content of these ILs after mixing them with water; and the ILs, which never show such a type of temperature-sensitive phase transition, contain less than seven water molecules per ion pair. They are always immiscible with water. However, relatively hydrophobic ILs with more than seven water molecules per ion pair show the LCST-type phase change. From these data, we also confirmed that the

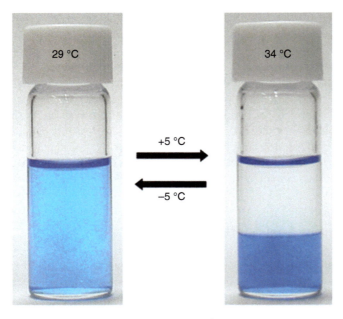

Figure 12.4 LCST-type phase transition of IL/water mixture.

LCST-type phase change could be seen in the mixture of relatively hydrophobic IL and water only when the maximum water content was more than seven water molecules per ion pair. It is also possible to design such IL/water mixtures showing the LCST-type phase separation by mixing ILs with different hydrophobicity. For example, a mixture of hydrophilic and hydrophobic IL can behave like the abovementioned IL with suitable hydrophobicity to show the LCST-type phase transition only when the total hydrophobicity of the mixture is almost the same as that for the IL. However, it is difficult to express the hydrophobicity of ion pairs quantitatively. So we use the maximum number of water molecules per ion pair instead. When the mixed ILs having average number of water molecules per ion pair after phase separation with water exceeded seven, they are strongly expected to show the LCST-type phase transition with water [28]. Of course, this concept cannot be applied for hydrophilic ILs because they are freely miscible with water.

Control of total hydrophobicity of ILs may lead to novel use of the ILs as functional materials. There are also several applications waiting after the progress of IL/water mixtures showing the LCST-type phase transition.

12.6 Application

As mentioned, there are many interesting ILs having unique physicochemical properties. Even ordinary ILs have properties quite different from those of organic molecular liquids, and some functional ILs should be expected to open new fields in basic science and applications. This is the major advantage of the science of ILs. Here, only a few applications are introduced. Wider aspects should be found in other books [29–31].

12.6.1 Reaction Solvents

Since ILs are nonvolatile liquids, they are perfect solvents from the viewpoint of safety. From this point of view, numerous attempts have been carried out to examine chemical reactions in many ILs. Detailed results are summarized in the review papers and books [1, 32–35]. Almost all of the organic (and some inorganic) reactions were found to be proceeded in ILs. Here, it should be noted that although most chemical reactions can be carried out in ILs, there are still a few advantages and disadvantages. One of the advantages should be that a unique charged environment (high ion density) of ILs may be helpful for certain reactions to suppress electrostatic interactions. Against this, examples of disadvantages are the inherent properties of ILs such as negligible vapor pressure (that means distillation cannot help the separation and purification processes for ILs). With the aid of other techniques, ILs should be used as solvents, especially at the industrial scale.

12.6.2 Electrolyte Solution

Since ILs are liquids composed of only ions, they are expected to be highly ion-conductive liquids. In spite of the high concentration of ions, ILs are generally very viscous and, accordingly, the mobility of ions is not always high enough to drive electrochemical devices. Similarly, since the formula weight of general ILs is quite large comparing inorganic salts, it is not possible to make extremely high ion concentration even without molecular solvents. Since ionic conductivity is the product of number of ions and their mobility, very viscous liquid is not so effective for high ionic conductivity. Taking these properties into account, it is not possible to make super ion-conductive liquids with ionic conductivity higher than that of aqueous salt solutions. However, due to a few advantageous (unique) properties of ILs such as nonvolatility, there are potentials for ILs to be used as electrolyte solution substituents [21]. Since the viscosity of ILs decreased considerably by adding small amount of organic solvent, it is quite effective to add small amount of polar but less viscous liquids to the ILs to enhance the ionic conductivity.

Selective ion transport in ILs is another subject to be solved. Since there are, of course, quite many ions in ILs, it is quite difficult to transport only selected ions such as lithium cations for lithium-ion batteries. Some electrochemical energy devices such as batteries require transportation of specific (active) ions, ILs used as electrolyte solutions for these devices require additional properties, the so-called selective ion transport. It is not easy because all ions of ILs migrate under potential gradient. To solve this, there are a few ideas to inhibit migration of specific ions in ILs. Use of zwitterions is one of the trials to inhibit migration of component ions of ILs. Since zwitterions are tethered ion pairs, they cannot migrate under potential gradient. Only added ions can migrate in the zwitterions. The remaining problem is that the T_m of zwitterions is much higher than that of ordinary ILs without tethering. However, almost all zwitterions are solid at room temperature, but mixtures of zwitterions and certain salts were found to be liquid [36]. For example, when imidazolium-type zwitterion was mixed with equimolar amount of lithium Tf_2N salt, the mixture was obtained as a liquid (or it should be better to say that the mixture showed no T_m and it was defined as amorphous state). Such mixtures showed quite high ionic conductivity than mixtures of zwitterions and other salts such as LiCl or LiBr, because these anions cannot form a low T_g environment. This means that when a suitable anion was selected as a partner of target ions (in this case, Li^+), the mixture of zwitterions and the salt should show relatively high ionic conductivity with a high Li^+ transport number.

In case of capacitor and supercap, there is almost no restriction of ion species, and these devices can be functionalized with ILs. However, the selection of ILs is still important; and ILs with lower viscosity, high ion density, and wide potential window are more favorable. For further information, application of ILs in the electrochemical fields has been reviewed [21, 37, 38].

12.6.3 Biomass Treatment

ILs are considered to be very polar, but they are generally not so. However, it is possible to control their polarity. Some ILs are more polar than ordinary molecular solvents, and they are expected to be potential solvents for hardly soluble materials which cannot be dissolved by polar molecular solvents. A typical example is the dissolution of cellulose. Cellulose was first dissolved by imidazolium chloride under heating [39]. After this report, there have been increasing numbers of papers concerning dissolution of cellulose using some polar ILs. Simultaneously, required factors to dissolve cellulose have also been proposed.

Through the study on the relationship between polarity of ILs and solubility of cellulose, the proton-accepting property was found to be important to dissolve cellulose. Since ILs are conductive materials, it is rather difficult to measure the polarity as mentioned. Instead of polarity value, Kamlet–Taft parameters are quite useful parameters in these days. Especially, α and β values related to proton-donating and proton-accepting abilities were used to discuss the polarity of ILs. There was a positive relation between β value and solubility of cellulose, and ILs having β value higher than 0.8 were strongly expected to dissolve cellulose [12]. According to this, some polar ILs were synthesized and their ability to dissolve cellulose was examined [40].

There is a challenging trial to extract components from biomass using polar ILs. Bran was added to imidazolium methylphosphonate at 50 °C; this was then filtered to remove the undissolved part, which was analyzed to be mainly lignin. Poor solvents such as water or methanol were added to filtrate to precipitate polysaccharides such as cellulose. After addition of such poor solvents, white precipitates were obtained. This precipitate was confirmed to be the mixture of cellulose and hemicellulose by NMR and infrared (IR) measurements. The filtrate after removing polysaccharides can be separated into IL and poor solvent, and these are reused to extract polysaccharides and precipitants, respectively. This is an energy-saving process and a closed system, and polysaccharides can be extracted from biomass under quite mild conditions.

There are other valuable components in biomass such as lignin. Lignin is known to be a valuable starting material to produce aromatic compounds and other useful starting materials for industrial use. Generally, lignin was fragmented by treating it in concentrated sulfuric acid at 200 °C. This is a very energy-consuming process, and there are strong requests to reduce the treatment energy, in other words, fragmentation and dissolution of lignin under mild condition are most requested. However, it is not easy to degrade a highly cross-linked network structure of lignin in nature without using agitation at high temperature. Recently, a mixture of IL and hydrogen peroxide aqueous solution was found to be effective to partly break the network structure of lignin [41]. Another potential solvent for woody biomass is the series of organic onium hydroxide aqueous solutions [42, 43]. Actually, organic onium hydroxides such as tetrabutylphosphonium hydroxide are not categorized as ILs, but they are composed of onium cations and hydroxide anions (and water). In the broad sense of the term "ionic liquids," these organic onium hydroxide aqueous solutions can be considered to be the mixture of IL and water. Anyway, tetrabutylphosphonium hydroxide aqueous solution shows power to dissolve biomass. Wood chips were almost completely dissolved in this tetrabutylphosphonium hydroxide aqueous solution at 60 °C [43]. There is a water content dependence on the dissolution ability, and the solution containing excess amount of water has no power to dissolve biomass [43].

There should be a few systems to extract valuable chemicals from biomass using ILs (and their derivatives) in the near future.

12.6.4 Solvents for Proteins and Biofuel Cell

There is a challenge to use ILs as solvents for proteins. Ordinary ILs can solubilize many biomaterials including proteins, but most proteins lose their higher ordered structure in these ILs. That is, it is essentially important to dissolve proteins without changing their higher ordered structure when certain activity of these proteins was expected in the ILs. This is a crucial aspect to keep the protein activity, but it is quite difficult to design such ILs having equivalent performance to aqueous buffer solution. A few proteins such as lipase are exceptionally stable in some ILs, and many papers have been published to realize their activity in ILs. These are exceptions and we have to consider totally new ILs to satisfy the following two contradictory requirements such as dissolve proteins monomolecularly and keep their higher-ordered structure. There are, however, a few efforts to find candidates to dissolve proteins without denaturation. One of the potential candidates is the mixture of polar ILs with a small amount of water, the so-called "hydrated ILs." It is quite interesting that no free water exists in these hydrated ILs because all water molecules strongly interacted (hydrated) with ions, and accordingly these hydrated ILs show properties quite similar to those of typical ILs. Cholinium dihydrogen phosphate was mixed with a small amount of water (to reach three water molecules per ion pair) to use it as solvent for proteins [44–46]. In this hydrated IL, electron transfer of redox-active proteins was detected and strongly suggested that this liquid is effective in solubilizing such proteins without losing their functions.

These trials easily pushed us to apply some ILs as reaction matrix and electrolyte solution of biofuel cells. Among a few different classes of biofuel cells, the most interesting and widely applicable one should be enzyme-aided fuel cells. These use enzymes as catalysts to oxidize substrates to get electrons. There are many studies on these devices using aqueous electrolyte solutions because aqueous buffer solutions are known to be the most comfortable environment for enzymes. Recently, there are some challenges to substitute aqueous buffer solutions with ILs for such cells to improve them more durable ones. The key point to apply ILs for these biofuel cells is the stability of catalysts, i.e. enzymes in them. It is however known that most enzymes are denatured in ILs due to different charged environment of the ILs from aqueous buffer solutions. Some hydrated ILs are reported to be excellent media for many proteins [46]. Hydrated cholinium dihydrogenphosphate was used as matrix to examine oxidation of cellobiose with cellobiose dehydrogenase. Oxidation of cellobiose was confirmed spectroscopically in the hydrated IL, and this strongly suggested that this hydrated IL worked as an electrolyte solution for biofuel cells. Enzymes are known to be very delicate molecules and easily denatured by changing the physical and chemical environment such as temperature. However, when proteins were dissolved in certain ILs, they were found to be active in the very wide temperature range. For example, cytochrome c, a kind of heme protein, was reported to be active at a wide temperature range up to surprisingly 140 °C in some ILs [47]. Taking these results into account, ILs and hydrated ILs may open a new field of biofuel cells that allows using the cells in a wider condition than that with aqueous media. In the near future, there will be new biofuel cells to get electricity when biomass such as used clothes, papers, and even grass or woodchip were added into the cell composed of enzyme-fixed electrode and IL electrolyte solution. In the cell, cellulose was dissolved from biomass and decomposed into oligomers and/or glucose. Then these were oxidized with the aid of enzymes to collect electrons under mild conditions [42]. These studies should open a new field of biofuel cells based on ILs as electrolyte solutions in the near future after suitable design of enzyme-friendly hydrated ILs.

12.7 Summary

ILs, composed of cations and anions, are discussed in this chapter. ILs show both properties of salts and liquids at moderate temperature, which cannot be realized by ordinary materials. Among many functional fluids mentioned in this book, the history of ILs is not so long, but there is increasing attention on these ILs because they are quite interesting liquids and have an unlimited future. There are many research fields and chances of applications for ILs. Of course, there are still some points to be solved to apply ILs in various fields. For that, continuous study is needed on ILs. This chapter should be helpful to welcome many readers to the world of ILs.

Acknowledgments

Most of the works introduced in this chapter have been done in our laboratory with the aid of financial support (KAKENHI) from the Japan Society for the Promotion of Science.

References

1 Welton, T. (1999). Room-temperature ionic liquids. solvents for synthesis and catalysis. *Chem. Rev.* 99: 2071–2084.
2 Kauzmann, W. (1948). The nature of the glassy state and the behavior of liquids at low temperatures. *Chem. Rev.* 43: 219–256.
3 Alba, C., Fan, J., and Angell, C.A. (1999). Thermodynamic aspects of the glass transition phenomenon. II. Molecular liquids with variable interactions. *J. Chem. Phys.* 110: 5262–5272.
4 Crowhurst, L., Mawdsley, P.R., Perez-Arlandis, J.M. et al. (2003). Solvent–solute interactions in ionic liquids. *Phys. Chem. Chem. Phys.* 5: 2790–2794.
5 Anderson, J.L., Ding, J., Welton, T., and Armstrong, D.W. (2002). Characterizing ionic liquids on the basis of multiple solvation interactions. *J. Am. Chem. Soc.* 124: 14247–14254.
6 Kamlet, M.J. and Taft, R.W. (1976). The solvatochromic comparison method. 1. The β-scale of solvent hydrogen-bond acceptor (HBA) basicities. *J. Am. Chem. Soc.* 98: 377–383.

7 Taft, R.W. and Kamlet, M.J. (1976). The solvatochromic comparison method. 2. The α-scale of solvent hydrogen-bond donor (HBD) acidities. *J. Am. Chem. Soc.* 98: 2886–2894.
8 Kamlet, M.J., Abboud, J.L., and Taft, R.W. (1977). The solvatochromic comparison method. 6. The π^* scale of solvent polarities. *J. Am. Chem. Soc.* 99: 6027–6038.
9 Yokoyama, T., Taft, R.W., and Kamlet, M.J. (1976). The solvatochromic comparison method. 3. Hydrogen bonding by some 2-nitroaniline derivatives. *J. Am. Chem. Soc.* 98: 3233–3237.
10 Fukaya, Y., Sugimoto, A., and Ohno, H. (2006). Superior solubility of polysaccharides in low viscosity, polar, and halogen-free 1,3-dialkylimidazolium formates. *Biomacromolecules* 7: 3295–3297.
11 Sun, N., Rahman, M., Qin, Y. et al. (2009). Complete dissolution and partial delignification of wood in the ionic liquid 1-ethyl-3-methylimidazolium acetate. *Green Chem.* 11: 646–655.
12 Fukaya, Y., Hayashi, K., Wada, M., and Ohno, H. (2008). Cellulose dissolution with polar ionic liquids under mild conditions: required factors for anions. *Green Chem.* 10: 44–46.
13 Hirao, M., Sugimoto, H., and Ohno, H. (2000). Preparation of novel room temperature molten salts by neutralization of amines. *J. Electrochem. Soc.* 147: 4168–4172.
14 Freemantle, M. (2010). *An Introduction to Ionic Liquids*. London: RSC Publishing.
15 Ohno, H. and Ito, K. (1998). Room-temperature molten salt polymers as a matrix for fast ion conduction. *Chem. Lett.* 751–752.
16 Gordon, C.M., Holbrey, J.D., Kennedy, A.R., and Seddon, K.R. (1998). Ionic liquid crystals: hexafluorophosphate salts. *J. Mater. Chem.* 8: 2627–2636.
17 Mukai, T., Yoshio, M., Kato, T. et al. (2005). Self-organization of protonated 2-heptadecylimidazole as an effective ion conductive matrix. *Electrochemistry* 73: 623–626.
18 Mukai, T., Yoshio, M., Kato, T. et al. (2005). Anisotropic ion conduction in a unique smectic phase of self-assembled amphiphilic ionic liquids. *Chem. Commun.* 1333–1335.
19 Yoshio, M., Mukai, T., Ohno, H., and Kato, T. (2004). One-dimensional ion transport in self-organized columnar ionic liquids. *J. Am. Chem. Soc.* 126: 994–995.
20 Yoshizawa, M. and Ohno, H. (2001). Synthesis of molten salt-type polymer brush and effect of brush structure on the ionic conductivity. *Electrochim. Acta* 46: 1723–1728.
21 Ohno, H.E. (2011). *Electrochemical Aspects of Ionic Liquids*, 2e. New York: Wiley Interscience.
22 Nishimura, N. and Ohno, H. (2014). 15th Anniversary of polymerised ionic liquids. *Polymer* 55: 3289–3297.
23 Ohno, H. (2006). Functional design of ionic liquids. *Bull. Chem. Soc. Jpn.* 79: 1665–1680.
24 Davis, J.H. (2004). Task-specific ionic liquids. *Chem. Lett.* 33: 1072–1077.

25 Huddleston, J.G., Willauer, H.D., Swatloski, R.P. et al. (1998). Room temperature ionic liquids as novel media for 'clean' liquid–liquid extraction. *Chem. Commun.* 1765–1766.
26 Bonhôte, P., Dias, A.P., Papageorgiou, N. et al. (1996). Hydrophobic, highly conductive ambient-temperature molten salts. *Inorg. Chem.* 35: 1168–1178.
27 Kohno, Y., Arai, H., Saita, S., and Ohno, H. (2011). Material design of ionic liquids to show temperature-sensitive LCST-type phase transition after mixing with water. *Aust. J. Chem.* 64: 1560–1567.
28 Kohno, Y. and Ohno, H. (2012). Temperature-responsive ionic liquid/water interfaces: relation between hydrophilicity of ions and dynamic phase change. *Phys. Chem. Chem. Phys.* 14: 5063–5070.
29 Ohno, H.E. (2003). *Ionic Liquids-I, The Front and Future of Material Development*. Tokyo: CMC Press.
30 Ohno, H.E. (2006). *Ionic Liquids-II, Marvelous Developments and Colourful Near Future*. Tokyo: CMC Press.
31 Ohno, H.E. (2010). *Ionic Liquids-III, Challenges to Nano-bio Sciences*. Tokyo: CMC Press.
32 Plechkova, N.V. and Seddon, K.R. (2008). Applications of ionic liquids in the chemical industry. *Chem. Soc. Rev.* 37: 123–150.
33 Dupont, J., de Souza, R.F., and Suarez, P.A. (2002). Ionic liquid (molten salt) phase organometallic catalysis. *Chem. Rev.* 102: 3667–3692.
34 Endres, F., MacFarlane, D.R., and Abbott, A.D.E. (2017). *Electrodeposition in Ionic Liquids*, 2e. New York: Wiley.
35 Wasserscheid, P. and Welton, T.E. (2008). *Ionic Liquids in Synthesis*, 2e. New York: Wiley.
36 Narita, A., Shibayama, W., and Ohno, H. (2006). Structural factors to improve physico-chemical properties of zwitterions as ion conductive matrices. *J. Mater. Chem.* 16: 1475–1482.
37 Fujita, K., Murata, K., Masuda, M. et al. (2012). Ionic liquids designed for advanced applications in bioelectrochemistry. *RSC Adv.* 2: 4018–4030.
38 Armand, M., Endres, F., MacFarlane, D.R. et al. (2009). Ionic-liquid materials for the electrochemical challenges of the future. *Nat. Mater.* 8: 621–629.
39 Swatloski, R.P., Spear, K., Holbrey, J.D., and Rogers, R.D. (2002). Dissolution of cellulose with ionic liquids. *J. Am. Chem. Soc.* 124: 4974–4975.
40 Earle, M.J., Esperança, J.M.S.S., Gilea, M.A. et al. (2006). The distillation and volatility of ionic liquids. *Nature* 439: 831–834.
41 Yamanaka, S., Yoshioka, K., Miyafuji, H., and Ohno, H. (2017). Effect of hydrogen peroxide on the extraction of components of cedar powder with tetra-n-butylphosphonium hydroxide aqueous solution at 60 °C. *Australian J. Chem.* 70: 322–327.
42 Abe, M., Fukaya, Y., and Ohno, H. (2012). Fast and facile dissolution of cellulose with tetrabutylphosphonium hydroxide containing 40 wt% water. *Chem. Commun.* 48: 1808–1810.
43 Abe, M., Yamanaka, S., Yamada, H. et al. (2015). Almost complete dissolution of woody biomass with tetra-n-butylphosphonium hydroxide aqueous solution at 60 °C. *Green Chem.* 17: 4432–1438.

44 Fujita, K., MacFarlane, D.R., and Forsyth, M. (2005). Protein solubilising and stabilising ionic liquids. *Chem. Commun.* 4804–4806.
45 Fujita, K., MacFarlane, D.R., Forsyth, M. et al. (2007). Solubility and stability of cytochrome c in hydrated ionic liquids: effect of oxo acid residues and kosmotropicity. *Biomacromolecules* 8: 2080–2086.
46 Fujita, K., Nakamura, N., Igarashi, K. et al. (2009). Biocatalytic oxidation of cellobiose in a hydrated ionic liquid. *Green Chem.* 11: 351–354.
47 Tamura, K., Nakamura, N., and Ohno, H. (2012). Cytochrome c dissolved in 1-allyl-3-methylimidazolium chloride type ionic liquids undergoes a quasi-reversible redox reaction up to 140 °C. *Biotech. Bioeng.* 109: 729–735.

13

Room-Temperature Liquid Metals as Functional Liquids

Minyung Song and Michael D. Dickey

North Carolina State University, Department of Chemical and Biomolecular Engineering, 911 Partners Way, Raleigh, NC 27606, USA

13.1 Introduction: Room-temperature Liquid Metals

This chapter discusses the properties and applications of liquid metals. We focus on metals that derive their utility and function from being simultaneously liquids and metals at or near room temperature.

Compared to other liquids (e.g. water, ionic liquids, organic liquids), liquid metals have outstanding thermal and electrical conductivity, as well as distinct optical properties (i.e. reflectivity). Liquid metals generally have low viscosity (similar to water), yet many form surface oxides that alter the rheological properties in a manner that enables them to be patterned into nonspherical shapes (as discussed herein). The resulting shapes maintain metallic electrical conductivity while being deformed significantly and can therefore be used as stretchable wires, interconnects, and antennas. In fact, liquid metals have the best combination of conductivity and deformability relative to any other known material. Taken together, these interesting and unique properties enable a number of applications [1, 2], including elastomeric circuits [3], stretchable electronics [4], electronic skins [5], components for soft robotics [6, 7], and self-healing electronic components [8, 9].

Besides mercury, francium, cesium, gallium, and rubidium, most metals are solid at or near room temperature. However, since francium is radioactive and cesium and rubidium are both explosively reactive, these materials are typically avoided for practical applications. Thus, the remaining metals, mercury, gallium, and alloys containing these metals, are the most popular liquid metals.

13.1.1 Mercury

Mercury (Hg) is the most well-known liquid metal (melting point −38.8 °C). It is readily available and is slow to oxidize. Its high electrical conductivity ($\sigma = 1.04 \times 10^6$ S m^{-1}) enables it to be utilized for electrodes [10, 11], microfluidics [12, 13] thermostat switches, and microelectromechanical system (MEMS) devices [14–16]. Nevertheless, its toxicity [17] and its vapor pressure, along

Functional Organic Liquids, First Edition. Edited by Takashi Nakanishi.
© 2019 Wiley-VCH Verlag GmbH & Co. KGaA. Published 2019 by Wiley-VCH Verlag GmbH & Co. KGaA.

with the potential danger of absorption via skin and vapor inhalation, limit its utility. In addition, its large surface tension (>400 mN m^{-1}) causes it to adopt a spherical shape and therefore it is difficult to pattern into useful shapes like wires and antennas.

13.1.2 Gallium-Based Alloys

Gallium, discovered in 1875, has low toxicity [18, 19] and an extremely low vapor pressure (it is effectively zero at room temperature) [20]. Even at an extremely high temperature of 1037 °C, the vapor pressure of gallium is only 1 kPa, which implies that gallium-based alloys will not evaporate readily [21] and therefore do not pose an inhalation danger at room temperature. Although gallium has a melting point above room temperature, some of its binary/ternary eutectic alloys have melting points below room temperature. Two commercial popular liquid metals are Galinstan (an alloy of 68.5 wt% of Ga, 21.5 wt% of In, 10 wt% of Sn) [22] and EGaIn (75.5 wt% of Ga, 24.5 wt% of In) [23, 24]. Galinstan and EGaIn have a melting point of −19 and 15.5 °C, respectively, making them room-temperature liquid metals. In the absence of surface species (e.g. oxides), the surface tension of liquid metals is significantly larger than the surface tension of other familiar fluids like water (e.g. they are typically >400 mN m^{-1} [23, 25]).

As opposed to mercury, gallium-based alloys rapidly form an oxide layer (~1–3 nm) [23, 26–30] that passivate the surface in the presence of oxygen [31]. The presence of the oxide is an important consideration for fluidic applications because it can cause the metal to stick to surfaces, including the walls of capillaries and tubing. It can also interfere with electrochemical measurements [32]. Although these considerations have often limited the utility of gallium alloys, the oxide does provide some unique benefits and opportunities. For example, the oxide layer also allows the metal to be manipulated into stable nonspherical shapes such as cones (Figure 13.1a) [24] or a stack of drops (Figure 13.1b) [33]. The oxide also prevents the metal from contacting directly its surroundings.

13.1.3 Oxide Skin on Ga Alloys

The surface oxide of EGaIn is composed primarily of gallium oxide, despite the presence of indium in the bulk [23].

The oxide is passivating, which means it does not get thicker with time (although in wet environments, the oxide changes properties and becomes less passivating [34]). The oxide-coated metal adheres to many surfaces [35] (see also Figure 13.1). The presence of the solid oxide "skin" around the droplet complicates the wetting behavior [36].

13.2 Removal of Oxide Skin

The oxide layer can be removed easily using acid, base, or electrochemistry [37]. The oxide is "amphoteric" since it can react with both acid and base. The use of

Figure 13.1 (a) Series of photographs showing that an EGaIn droplet featuring a native oxide "skin" sticks to surfaces and deforms into a stable cone shape as the syringe tip pulls away from the substrate. Formation of the oxide layer enables these nonspherical structures despite the large surface tension of the liquid. Source: Chiechi et al. 2008 [24]. Reproduced with permission from John Wiley & Sons. (b) Stacks of liquid metal drops that are dispensed on the head of an insect (left) and form freestanding structures (right). Source: Ladd et al. 2013 [33]. Reproduced with permission from John Wiley & Sons.

HCl solution converts Ga_2O_3 to gallium trichloride ($GaCl_3$) and NaOH solution forms sodium tetrahydroxogallate (III) ($Na[Ga(OH)_4]$).

$$Ga_2O_3 + 6\,HCl \rightarrow 2\,GaCl_3 + 3\,H_2O$$

$$Ga_2O_3 + 2\,NaOH + 3\,H_2O \rightarrow 2\,Na[Ga(OH)_4]$$

Upon removing the oxide, liquid gallium alloys bead up due to their large surface tension, as shown in Figure 13.2.

13.3 Patterning Techniques for Liquid Metals

Gallium-based alloys have some unique properties that have been harnessed for patterning. First, they can be injected into channels and capillaries due to their low viscosity. Second, they can hold their shape (both within microchannels and on surfaces) due to the rapid formation of the native oxide skin. Lastly, their freezing/melting temperatures are near room temperature; consequently, they can be manipulated as both solids and liquids at reasonable temperatures.

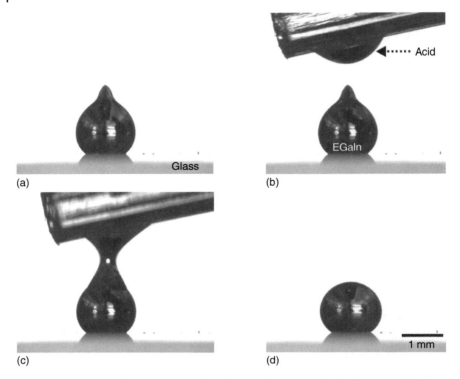

Figure 13.2 (a) Liquid metal in a nonspherical, cone shape in air. (b–c) After removal of the native oxide layer by a drop of acid (HCl), (d) the droplet beads up due to its large surface tension. Source: Dickey 2014 [1]. Reproduced with permission from John Wiley & Sons.

There are several methods in patterning techniques of liquid metal. Although there are likely other methods to pattern liquid metals, a recent review [38] organized existing methods into four categories: (i) lithographic processes, (ii) injection, (iii) subtractive, and (iv) additive. We briefly review those methods here.

13.3.1 Lithography-enabled Processes

In general, photolithography is the most commonly used lithographic process for patterning materials [39]. Photolithography is typically done on rigid, planar substrates and is therefore difficult to utilize directly for patterning liquid metal. It has, however, been used to "lift off" films of liquid metal spread over lithographic features [40]. Lithography can create molds or stencils for patterning liquid metal (Figure 13.3A). Spreading liquid metal across a stencil placed on a substrate can achieve features as small as 100 μm [43, 44].

Thin films of liquid metal can be imprinted using topographically patterned elastomeric mold, such as polydimethylsiloxane (PDMS) (Figure 13.3B). Gallium oxide adheres to the cavities in the PDMS mold to produce features as small as 2-μm-wide lines [42].

Patterning of liquid metal is also possible by selective surface wetting [45, 46]. For example, Galinstan will reactively wet tin patterned on a surface [43].

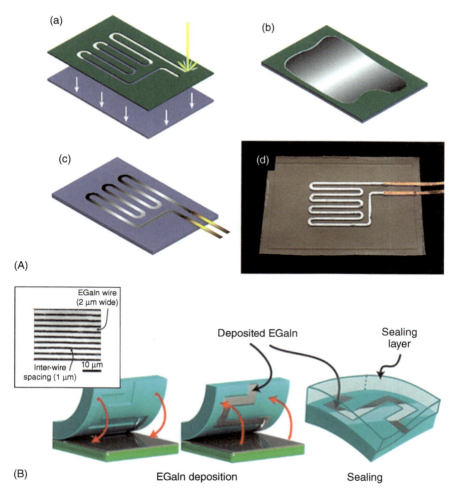

Figure 13.3 Patterning of liquid metals using lithography-enabled patterning. (A) Stencil printing method using masks. Source: Tabatabai et al. 2013 [41]. Reproduced with permission from ACS. (B) Imprint lithography. Source: Gozen et al. 2014 [42]. Reproduced with permission from John Wiley & Sons.

13.3.2 Injection

Injecting a gallium-based alloy into microfluidic channels is a common approach for creating soft and stretchable liquid metal devices due to its simplicity and tendency to faithfully duplicate the dimension of channels [23, 47, 48]. By applying a sufficient pressure (or vacuum [49]), gallium oxide breaks and the low viscosity metal flows readily into microchannels. The skin adheres to the wall of most microchannels to create stable structures [23]. Precoating the microchannels with Au facilitates filling with liquid metal [48].

In some cases, it is preferable to keep the liquid metal from sticking to the walls of the channel (e.g. for moving plugs of the metal back and forth). In this case, gallium oxide can be removed using hydrochloric acid or sodium hydroxide

[23, 50–53]. Alternatively, prefilling channels with water (or other fluids) creates a slip layer between the oxide and microchannel wall upon injection of the liquid metal [34, 54].

Liquid metals can be injected into holes [55], hollow fibers [4], and channels with complex geometries [56–58]. EGaIn can also be injected into capillaries with diameters as small as 150 nm by employing large pressures [59]. Mercury (Hg) can be injected in microchannels; however, it will adopt its shape to minimize surface energy causing formation of voids inside the channels. It is therefore limited to plugs of metal in microchannels.

13.3.3 Subtractive

Electrochemical reactions can locally remove the oxide skin of liquid metal, which causes the metal to flow within microchannel via capillary action (Figure 13.4a) [56]. This technique is called "recapillarity" because it operates with reductive potential to induce capillary behavior. Direct laser patterning is another facile method to pattern liquid metals (Figure 13.4b) embedded in elastomer with features as small as 100 μm [60].

13.3.4 Additive

Additive techniques deposit liquid metals in desired locations. For example, three-dimensional (3D) printing builds objects by depositing materials in a

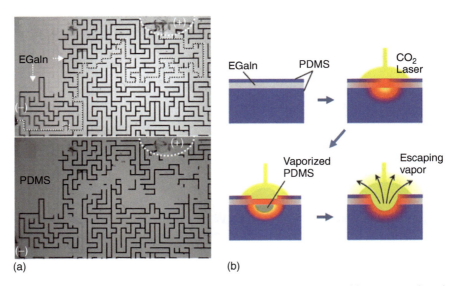

Figure 13.4 Subtractive methods for liquid metal printing. (a) It is possible to remove liquid metal selectively from complex microchannels using induced capillarity action via electrochemical reduction of the oxide layer. Source: Khan et al. 2015 [56]. Reproduced with permission from John Wiley & Sons. (b) Using CO_2 laser to ablate material from undesired regions, liquid metal can be easily patterned. Source: Lu et al. 2014 [60]. Reproduced with permission from John Wiley & Sons.

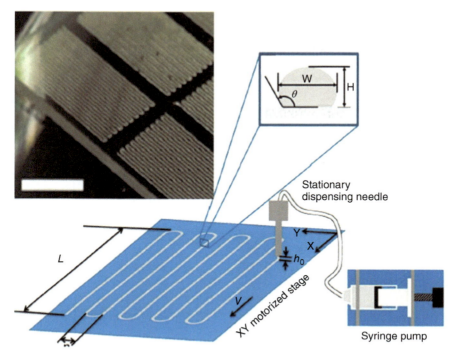

Figure 13.5 Liquid metal pattern produced by additive pattering techniques. It is stabilized by the oxide skin on the liquid metal. Scale bar is 5 mm. Source: Boley et al. 2014 [61]. Reproduced with permission from John Wiley & Sons.

layer-by-layer manner (Figure 13.5). Due to the oxide skin, liquid metal can be 3D printed and maintain its shape despite being a fluid [33], which may find applications in 3D printing electronics [62]. This same concept can be used for direct writing [61]. After encasing printed liquid metal structures, the metal can be removed as a "fugitive ink" to produce microvasculature (i.e. hollow channels within a monolithic structure) [63].

Other processes like tape transfer printing [64], microcontact printing [41], and roller-ball dispensing [65] can produce liquid metal patterns in two dimensions.

There are challenges with patterning by inkjet printing because of the oxide skin and the high surface tension of the metal. However, it is possible to inkjet print suspensions of small droplets of liquid metal dispersed in a solvent [66]. Normally, printed metal particles require sintering at high temperatures to be rendered conductive, but the liquid metal particles can be sintered by simply applying pressure (i.e. "mechanical sintering" at room temperature [66–68].

13.4 Controlling Interfacial Tension

Liquid metals have enormous surface tension, which is a dominant force at sub-millimeter length scales. The ability to control the surface tension offers a way to manipulate the shape, flow, and position of liquid metals in microsystems.

A recent review article summarizes ways to electrically control and manipulate the interfacial tension of liquid metals [69]. Thus, we only briefly mention the techniques here and point the reader to the review for more information.

Electrocapillarity is the change in interfacial tension of a fluid (e.g. Hg in water) caused by charges across the metal/electrolyte interface (i.e. electrical double layer) [70]. This capacitive effect lowers interfacial tension proportionally to the square of the potential across the interface. Only modest potentials can be applied to the metal (relative to the electrolyte) before Faradaic reactions occur across the interface (e.g. bubble formation via electrolysis of water), which limits the ability to bring about large changes in tension.

Electrowetting on dielectric (EWOD) is similar to electrocapillarity, but avoids Faradaic reactions by placing a dielectric between the liquid metal and counter electrode (i.e. a capacitor-like geometry). The large surface tension of liquid metal and the dielectric barrier necessitate large voltages to achieve modest changes in contact angle. Ultimately, dielectric breakdown across the dielectric limits the voltages that can be applied [70, 71].

Continuous electrowetting (CEW) utilizes gradients in surface tension to move plugs or droplets of liquid metal. Displacement of ions on the surface of the metal in response to an electric field creates gradients in surface tension via electrocapillarity. CEW is an effective way to move plugs of liquid metal within capillaries [71–74].

Taken in sum, the large interfacial tension of liquid metals has generally limited the utility of voltage-driven methods to manipulate liquid metals, which motivated a new approach.

13.4.1 Surface Activity of the Oxide on Liquid Metal Droplets

In 2009, Tsai et al. [75] observed that liquid metals spread dramatically in response to voltage, but attributed the spreading to electrocapillarity. As noted in the previous section, electrocapillarity only brings about modest changes in surface tension. In the following years, our group and others repeated the work of Tsai, but it was not until five years later that the mechanism became clearer.

Our group measured the interfacial tension of EGaIn versus voltage in electrolyte and showed that the formation of the surface oxide significantly lowers the interfacial tension [76, 77]. Figure 13.6a shows the "electrocapillary" curve of EGaIn measured in 1.0 M NaOH. Between −2.0 to −1.4 V (relative to a reference electrode), in the absence of oxide, the EGaIn droplet behaves according to classical electrocapillarity; that is, it voltage modulates the interfacial tension by controlling the electrical double layer. At potentials less than −2.0 V, water reduces and creates hydrogen bubbles. Therefore, −2.0 V is the lower limit. In the absence of the oxide, the metal assumes a large interfacial tension and therefore beads up (similar to that shown in Figure 13.2).

The oxide layer begins to form on the surface at a voltage over −1.3 V; at this value, interfacial tension drops significantly. As confirmed by the cyclic voltammogram (Figure 13.6b), the oxide layer forms at the potential near −1.3 V in 1.0 M NaOH. The implication is that the oxide layer behaves as a surfactant; but unlike regular surfactants, this "surfactant" brings about enormous changes in

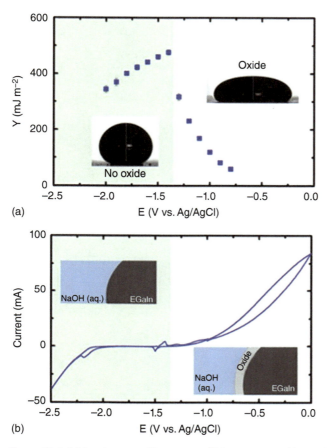

Figure 13.6 (a) An electrocapillary curve of EGaIn measured by analyzing the profile of a sessile drop (insert) in 1.0 M NaOH. (b) A cyclic voltammogram of EGaIn shows that the oxide forms exactly at the potential where the interfacial tension lowers the most (−1.3 V). Source: Khan et al. 2014 [77]. Reproduced with permission from PNAS.

interfacial tension and is completely reversible (that is, it can be deposited or removed rapidly using voltage). Normally, an oxide would prevent metals from flowing, but the presence of NaOH continuously removes the oxide, allowing it to flow. According to Figure 13.6b, the interfacial tension can be modulated from ∼500 mN m^{-1} to near zero using approximately one volt of applied potential (relative to the open circuit potential of −1.5 V). Understanding the details of this mechanism is an ongoing area of research.

Electrochemical control of surface oxidation requires minimal energy, and provides rapid and reversible control of interfacial tension over an enormous range compared to conventional molecular surfactants. Moreover, it is not dependent on the properties of the substrate. Spreading occurs in a variety of electrolytes, over a wide range of pH, with and without dissolved oxygen. Figure 13.7 illustrates how the liquid metal EGaIn spreads in response to a change of applied voltage [77].

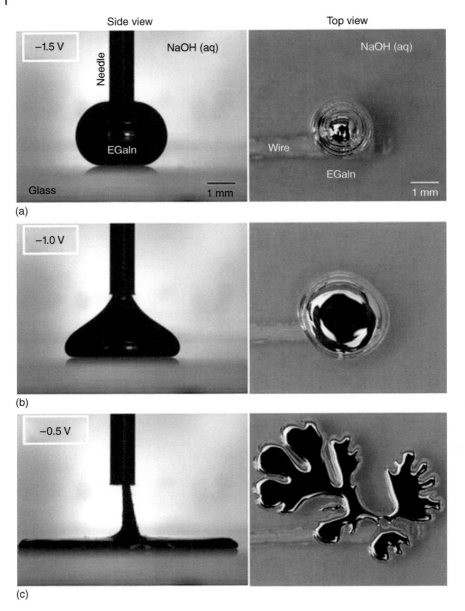

Figure 13.7 Oxidative spreading of a bead of liquid metal in 1.0 M NaOH solution (left column). A needle serves as a bottom electrical contact to the droplet (right column). (a) Without applied potential (i.e. at the open circuit potential of −1.5 V), the metal is oxide free and beads up due to the large tension. The NaOH keeps the metal oxide free. (b) Electrochemical oxidation deposits a surface oxide. The oxide acts like a surfactant and lowers the interfacial tension. (c) The metal spreads dramatically since the interfacial tension approaches zero at this potential. Source: Khan et al. 2014 [77]. Reproduced with permission from PNAS.

13.5 Applications of Liquid Metals

Liquid metals have a wide range of potential applications. Because liquid metal can maintain conductivity during deformation, they offer the possibility of creating soft and stretchable analogs of rigid conductors. They can also be incorporated readily into microsystems as electrodes, pumps, and other components. We provide several illustrative examples, although the following is not intended to be exhaustive due to space constraints.

Liquid metals may be used in microsystems and microfluidic channels for pumping [13, 78–80], thermal cooling [81], Joule heating [82], energy harvesting [83], microvalving [84], mixing [85], and as electrodes [58]. It is possible to form liquid metal droplets [86–90], steer them [91], and self-propel them [92–96].

Gallium-based alloys can be utilized in conformal electrodes for various applications. They have shown excellent performance in contacting and characterizing self-assembled monolayer [97], nanoparticles and quantum dots [98, 99], carbon nanotubes [100], graphene [101, 102], and organic solar cells [103, 104]. This general principle can be extended to other applications by making electrical contacts; stretchable organic solar cells [105], dielectric elastomers [106, 107], and e-skin [108], for example. Liquid metals can also be utilized as electrodes for electrochemical reactions [109] and the surfaces can be used catalytically [110].

Either printed or patterned structures (cf. Figure 13.8A) can be utilized as wires, antennas, interconnects, or sensors in electronic skins, stretchable electronics, and active conformable devices.

Liquid metals can form self-healing circuits. When a liquid metal wire is cut, the oxide layer stabilizes the metal and prevents it from leaking. Once the interfaces of the cut wires are brought back together, the circuit regains conductivity. By encasing the metal in a self-healing polymer [114], it is possible for the circuit to also heal mechanically [8] (Figure 13.8B).

Liquid metal can play an important role in functional circuit elements such as memristors, sensors [115], and diodes [116]. A memristor refers to a "memory resistor" device based on a two-terminal electrical component that has a binary output of 1's or 0's based on whether the devices are in the resistive or conductive state [117]. It is possible to build an entirely soft and liquid-like memristor using two liquid metal electrodes separated by hydrogel (Figure 13.8C) [111]. The soft device records 1's (conductive) and 0's (resistive) using electrochemical reactions to control the oxide layer thickness.

Liquid metal enables devises that are soft and stretchable by injecting liquid metal within elastomeric microchannels (Figure 13.8D, E). Figure 13.8D shows a sensor composed of multilayers of elastomeric microchannels filled with liquid metal. These sensors can detect changes in resistance or capacitance due to physical deformation of the channels (e.g. by touch) [112, 118–120].

The spectral properties of antennas, for example, such as frequency, depend on shape and therefore stretchable antennas can be wireless sensors of strain [113]. It is possible to form patch antennas [57, 121–125], coils [126–129], radiofrequency antennas [130, 131], loops [132], spherical caps [133], phase-shifting

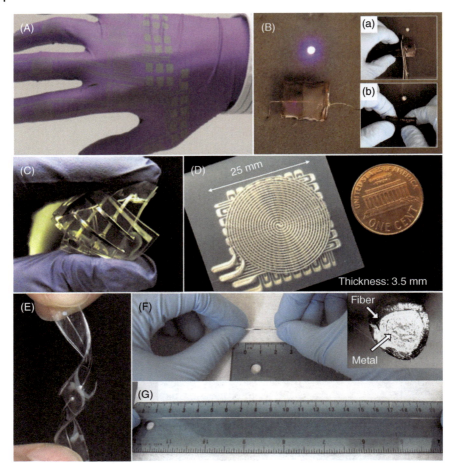

Figure 13.8 Applications of liquid metals. (A) Printed colloidal particles of liquid metal dispersed in a solvent using inkjet printing. Source: Boley et al. 2015 [66]. Reproduced with permission from John Wiley & Sons. (B) Self-healing wire with liquid metal in the core. Source: Palleau et al. 2013 [8]. Reproduced with permission from John Wiley & Sons. (C) A resistive memory device composed of hydrogel and liquid metals. Source: Koo et al. 2011 [111]. Reproduced with permission from John Wiley & Sons. (D) Soft artificial skin sensor embedded with liquid metals that is sensitive to pressure and touch by change of electrical resistivity. Source: Park et al. 2012 [112]. Reproduced with permission from AAAS. (E) Flexible and stretchable dipole antennas formed by injecting liquid metals. Source: So et al. 2009 [113]. Reproduced with permission from John Wiley & Sons. (F, G) Ultra-stretchable metallic wires composed of an elastomeric shell filled with liquid metals in the core of hollow fiber. Source: Zhu et al. 2013 [4]. Reproduced with permission from John Wiley & Sons.

coaxial transmission lines [134], monopoles [135], beam-steering antennas [136], and reconfigurable antennas [54, 123, 137–144].

Hollow elastomeric tubes infiltrated with liquid metal can stretch up to ∼700% (Figure 13.8F, G). Normally, conductivity and stretchability are conflicting material demands; this composite breaks this trade-off and offers the best combination of strain and conductivity reported to date [4]. These fibers may find uses in electronic textiles or stretchable wiring. One of the earliest examples of stretchable electronic devices used Hg filled in an elastomeric tube as a strain sensor [145].

13.6 Conclusions and Outlook

Liquid metals are uniquely both liquids and metals. Gallium is considered to have low toxicity and therefore offers an alternative to Hg. The formation of an oxide skin on the surface of gallium allows it to be patterned into useful shapes, such as wires, antennas, and sensors. These structures are soft and offer an unrivaled combination of conductivity and deformation (strain). Liquid metal can be patterned using a variety of methods, including injection into microchannels and 3D printing. In addition, the ability to deposit or remove the oxide electrochemically enables unprecedented control over interfacial activity, which provides a means to reconfigure and manipulate the shape and position of liquid metal (for reconfigurable devices). We believe that as more researchers learn of the potential of liquid metal, more applications will emerge since it is the only material that combines metallic and fluidic properties at room temperature.

References

1 Dickey, M.D. (2014). Emerging applications of liquid metals featuring surface oxides. *ACS Appl. Mater. Interfaces* 6: 18369–18379.
2 Rogers, J.A., Ghaffari, R., and Kim, D.-H. (eds.) (2016). *Stretchable Bioelectronics for Medical Devices and Systems*. Cham: Springer International Publishing.
3 Zhang, B., Dong, Q., Korman, C.E. et al. (2013). Flexible packaging of solid-state integrated circuit chips with elastomeric microfluidics. *Sci. Rep.* 3: 1098.
4 Zhu, S., So, J.-H., Mays, R. et al. (2013). Ultrastretchable fibers with metallic conductivity using a liquid metal alloy core. *Adv. Funct. Mater.* 23: 2308–2314.
5 Hammock, M.L., Chortos, A., Tee, B.C.-K. et al. (2013). 25th anniversary article: the evolution of electronic skin (e-skin): a brief history, design considerations, and recent progress. *Adv. Mater.* 25: 5997–6038.
6 Majidi, C. (2013). Soft robotics: a perspective – current trends and prospects for the future. *Soft Rob.* 1: 5–11.
7 Kim, S., Laschi, C., and Trimmer, B. (2013). Soft robotics: a bioinspired evolution in robotics. *Trends Biotechnol.* 31: 287–294.
8 Palleau, E., Reece, S., Desai, S.C. et al. (2013). Self-healing stretchable wires for reconfigurable circuit wiring and 3D microfluidics. *Adv. Mater.* 25: 1589–1592.
9 Blaiszik, B.J., Jones, A.R., Sottos, N.R., and White, S.R. (2014). Microencapsulation of gallium–indium (Ga—in) liquid metal for self-healing applications. *J. Microencapsulation* 31: 350–354.

10 Holmlin, R.E., Haag, R., Chabinyc, M.L. et al. (2001). Electron transport through thin organic films in metal–insulator–metal junctions based on self-assembled monolayers. *J. Am. Chem. Soc.* 123: 5075–5085.

11 Grahame, D.C. (1949). Measurement of the capacity of the electrical double layer at a mercury electrode. *J. Am. Chem. Soc.* 71: 2975–2978.

12 Lee, J. and Kim, C.J. (2000). Surface-tension-driven microactuation based on continuous electrowetting. *J. Microelectromech. Syst.* 9: 171–180.

13 Yun, K.-S., Cho, I.-J., Bu, J.-U. et al. (2002). A surface-tension driven micropump for low-voltage and low-power operations. *J. Microelectromech. Syst.* 11: 454–461.

14 Simon, J., Saffer, S., and Kim, C.J. (1997). A liquid-filled microrelay with a moving mercury microdrop. *J. Microelectromech. Syst.* 6: 208–216.

15 Sen, P. and Kim, C.-J. (2009). Microscale liquid–metal switches a review. *IEEE Trans. Ind. Electron.* 56: 1314–1330.

16 Andrews, J.R. (1973). Random sampling oscilloscope for the observation of mercury switch closure transition times. *IEEE Trans. Instrum. Meas.* 22: 375–381.

17 Clarkson, T.W. and Magos, L. (2006). The toxicology of mercury and its chemical compounds. *Crit. Rev. Toxicol.* 36: 609–662.

18 Chandler, J.E., Messer, H.H., and Ellender, G. (1994). Cytotoxicity of gallium and indium ions compared with mercuric ion. *J. Dent. Res.* 73: 1554–1559.

19 Lu, Y., Hu, Q., Lin, Y. et al. (2015). Transformable liquid–metal nanomedicine. *Nat. Commun.* 6: 10066.

20 CRC (2012). *Handbook of Chemistry and Physics*, 92e. CRC Press Book.

21 Geiger, F., Busse, C.A., and Loehrke, R.I. (1987). The vapor pressure of indium, silver, gallium, copper, tin, and gold between 0.1 and 3.0 bar. *Int. J. Thermophys.* 8: 425–436.

22 Morley, N.B., Burris, J., Cadwallader, L.C., and Nornberg, M.D. (2008). GaInSn usage in the research laboratory. *Rev. Sci. Instrum.* 79: 056107.

23 Dickey, M.D., Chiechi, R.C., Larsen, R.J. et al. (2008). Eutectic gallium–indium (EGaIn): a liquid metal alloy for the formation of stable structures in microchannels at room temperature. *Adv. Funct. Mater.* 18: 1097–1104.

24 Chiechi, R.C., Weiss, E.A., Dickey, M.D., and Whitesides, G.M. (2008). Eutectic gallium–indium (EGaIn): a moldable liquid metal for electrical characterization of self-assembled monolayers. *Angew. Chem. Int. Ed.* 47: 142–144.

25 Liu, T., Sen, P., and Kim, C.J. (2012). Characterization of nontoxic liquid–metal alloy Galinstan for applications in microdevices. *J. Microelectromech. Syst.* 21: 443–450.

26 Cademartiri, L., Thuo, M.M., Nijhuis, C.A. et al. (2012). Electrical resistance of Ag-TS-S(CH_2)(n-1)CH_3//Ga_2O_3/EGaIn tunneling junctions. *J. Phys. Chem. C* 116: 10848–10860.

27 Regan, M.J., Tostmann, H., Pershan, P.S. et al. (1997). X-ray study of the oxidation of liquid–gallium surfaces. *Phys. Rev. B: Condens. Matter* 55: 10786–10790.

28 Scharmann, F., Cherkashinin, G., Breternitz, V. et al. (2004). Viscosity effect on GaInSn studied by XPS. *Surf. Interface Anal.* 36: 981–985.
29 Plech, A., Klemradt, U., Metzger, H., and Peisl, J. (1998). In situ X-ray reflectivity study of the oxidation kinetics of liquid gallium and the liquid alloy. *J. Phys. Condens. Matter* 10: 971.
30 Chabala, J.M. (1992). Oxide-growth kinetics and fractal-like patterning across liquid gallium surfaces. *Phys. Rev. B* 46: 11346–11357.
31 Tostmann, H., DiMasi, E., Pershan, P.S. et al. (1999). Surface structure of liquid metals and the effect of capillary waves: X-ray studies on liquid indium. *Phys. Rev. B: Condens. Matter Mater. Phys.* 59: 783–791.
32 Giguère, P.A. and Lamontagne, D. (1954). Polarography with a dropping gallium electrode. *Science* 120: 390–391.
33 Ladd, C., So, J.-H., Muth, J., and Dickey, M.D. (2013). 3D printing of free standing liquid metal microstructures. *Adv. Mater.* 25: 5081–5085.
34 Khan, M.R., Trlica, C., So, J.-H. et al. (2014). Influence of water on the interfacial behavior of gallium liquid metal alloys. *ACS Appl. Mater. Interfaces* 6: 22467–22473.
35 Ye, Z., Lum, G.Z., Song, S. et al. (2016). Phase change of gallium enables highly reversible and switchable adhesion. *Adv. Mater.* 28: 5088–5092.
36 Doudrick, K., Liu, S., Mutunga, E.M. et al. (2014). Different shades of oxide: from nanoscale wetting mechanisms to contact printing of gallium-based liquid metals. *Langmuir* 30: 6867–6877.
37 Pourbaix, M. (1974). *Atlas of Electrochemical Equilibria in Aqueous Solutions*. Houston, TX: National Association of Corrosion.
38 Joshipura, I.D., Ayers, H.R., Majidi, C., and Dickey, M.D. (2015). Methods to pattern liquid metals. *J. Mater. Chem. C* 3: 3834–3841.
39 Madou, M.J. (2002). *Fundamentals of Microfabrication: The Science of Miniaturization*. CRC Press.
40 Park, C.W., Moon, Y.G., Seong, H. et al. (2016). Photolithography-based patterning of liquid metal interconnects for monolithically integrated stretchable circuits. *ACS Appl. Mater. Interfaces* 8: 15459–15465.
41 Tabatabai, A., Fassler, A., Usiak, C., and Majidi, C. (2013). Liquid-phase gallium–indium alloy electronics with microcontact printing. *Langmuir* 29: 6194–6200.
42 Gozen, B.A., Tabatabai, A., Ozdoganlar, O.B., and Majidi, C. (2014). High-density soft-matter electronics with micron-scale line width. *Adv. Mater.* 26: 5211–5216.
43 Kramer, R.K., Majidi, C., and Wood, R.J. (2013). Masked deposition of gallium–indium alloys for liquid-embedded elastomer conductors. *Adv. Funct. Mater.* 23: 5292–5296.
44 Jeong, S.H., Hagman, A., Hjort, K. et al. (2012). Liquid alloy printing of microfluidic stretchable electronics. *Lab Chip* 12: 4657–4664.
45 Kramer, R.K., Boley, J.W., Stone, H.A. et al. (2014). Effect of microtextured surface topography on the wetting behavior of eutectic gallium–indium alloys. *Langmuir* 30: 533–539.
46 Li, G., Wu, X., and Lee, D.-W. (2015). Selectively plated stretchable liquid metal wires for transparent electronics. *Sens. Actuators, B* 221: 1114–1119.

47 Cheng, S. and Wu, Z. (2012). Microfluidic electronics. *Lab Chip* 12: 2782–2791.
48 Kim, H.-J., Son, C., and Ziaie, B. (2008). A multiaxial stretchable interconnect using liquid-alloy-filled elastomeric microchannels. *Appl. Phys. Lett.* 92: 011904.
49 Fassler, A. and Majidi, C. (2013). 3D structures of liquid-phase GaIn alloy embedded in PDMS with freeze casting. *Lab Chip* 13: 4442–4450.
50 Wang, J., Liu, S., Vardeny, Z.V., and Nahata, A. (2012). Liquid metal-based plasmonics. *Opt. Express* 20: 2346–2353.
51 Wang, J., Liu, S., Guruswamy, S., and Nahata, A. (2013). Reconfigurable liquid metal based terahertz metamaterials via selective erasure and refilling to the unit cell level. *Appl. Phys. Lett.* 103: 221116.
52 Kim, D., Thissen, P., Viner, G. et al. (2013). Recovery of nonwetting characteristics by surface modification of gallium-based liquid metal droplets using hydrochloric acid vapor. *ACS Appl. Mater. Interfaces* 5: 179–185.
53 Kim, D., Pierce, R.G., Henderson, R. et al. (2014). Liquid metal actuation-based reversible frequency tunable monopole antenna. *Appl. Phys. Lett.* 105: 234104.
54 Koo, C., LeBlanc, B.E., Kelley, M. et al. (2015). Manipulating liquid metal droplets in microfluidic channels with minimized skin residues toward tunable RF applications. *J. Microelectromech. Syst.* 24: 1069–1076.
55 Kim, H.-J., Maleki, T., Wei, P., and Ziaie, B. (2009). A biaxial stretchable interconnect with liquid-alloy-covered joints on elastomeric substrate. *J. Microelectromech. Syst.* 18: 138–146.
56 Khan, M.R., Trlica, C., and Dickey, M.D. (2015). Recapillarity: electrochemically controlled capillary withdrawal of a liquid metal alloy from microchannels. *Adv. Funct. Mater.* 25: 671–678.
57 Hayes, G.J., So, J.-H., Qusba, A. et al. (2012). Flexible liquid metal alloy (EGaIn) microstrip patch antenna. *IEEE Trans. Antennas Propag.* 60: 2151–2156.
58 So, J.-H. and Dickey, M.D. (2011). Inherently aligned microfluidic electrodes composed of liquid metal. *Lab Chip* 11: 905–911.
59 Zhao, W., Bischof, J.L., Hutasoit, J. et al. (2015). Single-fluxon controlled resistance switching in centimeter-long superconducting gallium–indium eutectic nanowires. *Nano Lett.* 15: 153–158.
60 Lu, T., Finkenauer, L., Wissman, J., and Majidi, C. (2014). Rapid prototyping for soft-matter electronics. *Adv. Funct. Mater.* 24: 3351–3356.
61 Boley, J.W., White, E.L., Chiu, G.T.-C., and Kramer, R.K. (2014). Direct writing of gallium–indium alloy for stretchable electronics. *Adv. Funct. Mater.* 24: 3501–3507.
62 Parekh, D., Cormier, D., and Dickey, M. (2015). Multifunctional printing: incorporating electronics into 3D parts made by additive manufacturing. In: *Additive Manufacturing* (ed. A. Bandyopadhyay and S. Bose), 215–258. CRC Press.
63 Parekh, D.P., Ladd, C., Panich, L. et al. (2016). 3D printing of liquid metals as fugitive inks for fabrication of 3D microfluidic channels. *Lab Chip* 16: 1812–1820.

64 Jeong, S.H., Hjort, K., and Wu, Z. (2014). Tape transfer printing of a liquid metal alloy for stretchable RF electronics. *Sensors* 14: 16311–16321.
65 Zheng, Y., Zhang, Q., and Liu, J. (2013). Pervasive liquid metal based direct writing electronics with roller-ball pen. *AIP Adv.* 3: 112117.
66 Boley, J.W., White, E.L., and Kramer, R.K. (2015). Mechanically sintered gallium–indium nanoparticles. *Adv. Mater.* 27: 2355–2360.
67 Fassler, A. and Majidi, C. (2015). Liquid-phase metal inclusions for a conductive polymer composite. *Adv. Mater.* 27: 1928–1932.
68 Lin, Y., Cooper, C., Wang, M. et al. (2015). Handwritten, soft circuit boards and antennas using liquid metal nanoparticles. *Small* 11: 6397–6403.
69 Eaker, C.B. and Dickey, M.D. (2015). Liquid metals as ultra-stretchable, soft, and shape reconfigurable conductors. Proceedings Volume 9467, Micro- and Nanotechnology Sensors, Systems, and Applications VII; 946708.
70 Mugele, F. and Baret, J.-C. (2005). Electrowetting: from basics to applications. *J. Phys. Condens. Matter* 17: R705.
71 Moon, H., Cho, S.K., Garrell, R.L., and Kim, C.-J. (2002). Low voltage electrowetting-on-dielectric. *J. Appl. Phys.* 92: 4080.
72 Gough, R.C., Morishita, A.M., Dang, J.H. et al. (2015). Rapid electrocapillary deformation of liquid metal with reversible shape retention. *Micro Nano Syst. Lett.* 3: 1–9.
73 Gough, R.C., Morishita, A.M., Dang, J.H. et al. (2014). Continuous electrowetting of non-toxic liquid metal for RF applications. *IEEE Access* 2: 874–882.
74 Beni, G., Hackwood, S., and Jackel, J.L. (1982). Continuous electrowetting effect. *Appl. Phys. Lett.* 40: 912–914.
75 Tsai, J.T.H., Ho, C.-M., Wang, F.-C., and Liang, C.-T. (2009). Ultrahigh contrast light valve driven by electrocapillarity of liquid gallium. *Appl. Phys. Lett.* 95: 251110.
76 Eaker, C.B., Khan, M.R., and Dickey, M.D. (2016). A method to manipulate surface tension of a liquid metal via surface oxidation and reduction. *J. Vis. Exp.* (107): e53567. https://doi.org/10.3791/53567.
77 Khan, M.R., Eaker, C.B., Bowden, E.F., and Dickey, M.D. (2014). Giant and switchable surface activity of liquid metal via surface oxidation. *Proc. Natl. Acad. Sci. U.S.A.* 111: 14047–14051.
78 Tang, S.-Y., Khoshmanesh, K., Sivan, V. et al. (2014). Liquid metal enabled pump. *Proc. Natl. Acad. Sci. U.S.A.* 111: 3304–3309.
79 Gao, M. and Gui, L. (2014). A handy liquid metal based electroosmotic flow pump. *Lab Chip* 14: 1866–1872.
80 Knoblauch, M., Hibberd, J.M., Gray, J.C., and van Bel, A.J. (1999). A galinstan expansion femtosyringe for microinjection of eukaryotic organelles and prokaryotes. *Nat. Biotechnol.* 17: 906–909.
81 Hodes, M., Zhang, R., Lam, L.S. et al. (2014). On the potential of Galinstan-based minichannel and minigap cooling. *IEEE Trans. Compon. Packag. Manuf. Technol.* 4: 46–56.
82 Je, J. and Lee, J. (2014). Design, fabrication, and characterization of liquid metal microheaters. *J. Microelectromech. Syst.* 23: 1156–1163.

83 Krupenkin, T. and Taylor, J.A. (2011). Reverse electrowetting as a new approach to high-power energy harvesting. *Nat. Commun.* 2: 448.

84 Pekas, N., Zhang, Q., and Juncker, D. (2012). Electrostatic actuator with liquid metal–elastomer compliant electrodes used for on-chip microvalving. *J. Micromech. Microeng.* 22: 097001.

85 Tang, S.-Y., Sivan, V., Petersen, P. et al. (2014). Liquid metal actuator for inducing chaotic advection. *Adv. Funct. Mater.* 24: 5851–5858.

86 Friedman, H., Reich, S., Popovitz-Biro, R. et al. (2013). Micro- and nano-spheres of low melting point metals and alloys, formed by ultrasonic cavitation. *Ultrason. Sonochem.* 20: 432–444.

87 Hohman, J.N., Kim, M., Wadsworth, G.A. et al. (2011). Directing substrate morphology via self-assembly: ligand-mediated scission of gallium–indium microspheres to the nanoscale. *Nano Lett.* 11: 5104–5110.

88 Han, Z.H., Yang, B., Qi, Y., and Cumings, J. (2011). Synthesis of low-melting-point metallic nanoparticles with an ultrasonic nanoemulsion method. *Ultrasonics* 51: 485–488.

89 Thelen, J., Dickey, M.D., and Ward, T. (2012). A study of the production and reversible stability of EGaIn liquid metal microspheres using flow focusing. *Lab Chip* 12: 3961–3967.

90 Hutter, T., Bauer, W.-A.C., Elliott, S.R., and Huck, W.T.S. (2012). Formation of spherical and non-spherical eutectic gallium–indium liquid–metal microdroplets in microfluidic channels at room temperature. *Adv. Funct. Mater.* 22: 2624–2631.

91 Tang, S.-Y., Lin, Y., Joshipura, I.D. et al. (2015). Steering liquid metal flow in microchannels using low voltages. *Lab Chip* 15: 3905–3911.

92 Tersoff, J., Jesson, D.E., and Tang, W.X. (2009). Running droplets of gallium from evaporation of gallium arsenide. *Science* 324: 236–238.

93 Zavabeti, A., Daeneke, T., Chrimes, A.F. et al. (2016). Ionic imbalance induced self-propulsion of liquid metals. *Nat. Commun.* 7: 12402.

94 Zhang, J., Yao, Y., Sheng, L., and Liu, J. (2015). Self-fueled biomimetic liquid metal mollusk. *Adv. Mater.* 27: 2648–2655.

95 Tang, S.-Y., Ayan, B., Nama, N. et al. (2016). On-chip production of size-controllable liquid metal microdroplets using acoustic waves. *Small* 12: 3861–3869.

96 Mohammed, M., Sundaresan, R., and Dickey, M.D. (2015). Self-running liquid metal drops that delaminate metal films at record velocities. *ACS Appl. Mater. Interfaces* 7: 23163–23171.

97 Zhang, Y., Zhao, Z., Fracasso, D., and Chiechi, R.C. (2014). Bottom-up molecular tunneling junctions formed by self-assembly. *Isr. J. Chem.* 54: 513–533.

98 Du, K., Glogowski, E., Tuominen, M.T. et al. (2013). Self-assembly of gold nanoparticles on gallium droplets: controlling charge transport through microscopic devices. *Langmuir* 29: 13640–13646.

99 Weiss, E.A., Chiechi, R.C., Geyer, S.M. et al. (2008). Size-dependent charge collection in junctions containing single-size and multi-size arrays of colloidal CdSe quantum dots. *J. Am. Chem. Soc.* 130: 74–82.

100 Yazdanpanah, M.M., Chakraborty, S., Harfenist, S.A. et al. (2004). Formation of highly transmissive liquid metal contacts to carbon nanotubes. *Appl. Phys. Lett.* 85: 3564–3566.
101 Ordonez, R.C., Hayashi, C.K., Torres, C.M. et al. (2016). Conformal liquid–metal electrodes for flexible Graphene device interconnects. *IEEE Trans. Electron Devices* 63: 4018–4023.
102 Jiao, Y., Young, C., Yang, S. et al. (2016). Wearable Graphene sensors with microfluidic liquid metal wiring for structural health monitoring and human body motion sensing. *IEEE Sens. J.* 16: 7870–7875.
103 Du Pasquier, A., Miller, S., and Chhowalla, M. (2006). On the use of Ga—in eutectic and halogen light source for testing P3HT-PCBM organic solar cells. *Sol. Energy Mater. Sol. Cells* 90: 1828–1839.
104 Ongul, F., Yuksel, S.A., Bozar, S. et al. (2015). Vacuum-free processed bulk heterojunction solar cells with E-GaIn cathode as an alternative to Al electrode. *J. Phys. Appl. Phys.* 48: 175102.
105 Lipomi, D.J., Tee, B.C.-K., Vosgueritchian, M., and Bao, Z. (2011). Stretchable organic solar cells. *Adv. Mater.* 23: 1771–1775.
106 Anderson, I.A., Gisby, T.A., McKay, T.G. et al. (2012). Multi-functional dielectric elastomer artificial muscles for soft and smart machines. *J. Appl. Phys.* 112: 041101.
107 Liu, Y., Gao, M., Mei, S. et al. (2013). Ultra-compliant liquid metal electrodes with in-plane self-healing capability for dielectric elastomer actuators. *Appl. Phys. Lett.* 103: 064101.
108 Lipomi, D.J., Vosgueritchian, M., Tee, B.C.-K. et al. (2011). Skin-like pressure and strain sensors based on transparent elastic films of carbon nanotubes. *Nat. Nano* 6: 788–792.
109 Fahrenkrug, E., Gu, J., and Maldonado, S. (2013). Electrodeposition of crystalline GaAs on liquid gallium electrodes in aqueous electrolytes. *J. Am. Chem. Soc.* 135: 330–339.
110 Zhang, W., Ou, J.Z., Tang, S.-Y. et al. (2014). Liquid metal/metal oxide frameworks. *Adv. Funct. Mater.* 24: 3799–3807.
111 Koo, H., So, J., Dickey, M.D., and Velev, O.D. (2011). Toward all-soft matter circuits: prototypes of quasi-liquid devices with memristor characteristics. *Adv. Mater.* 23: 3559–3564.
112 Park, Y.-L., Chen, B.-R., and Wood, R.J. (2012). Design and fabrication of soft artificial skin using embedded microchannels and liquid conductors. *IEEE Sens. J.* 12: 2711–2718.
113 So, J.-H., Thelen, J., Qusba, A. et al. (2009). Reversibly deformable and mechanically tunable fluidic antennas. *Adv. Funct. Mater.* 19: 3632–3637.
114 Cordier, P., Tournilhac, F., Soulié-Ziakovic, C., and Leibler, L. (2008). Self-healing and thermoreversible rubber from supramolecular assembly. *Nature* 451: 977–980.
115 Ota, H., Chen, K., Lin, Y. et al. (2014). Highly deformable liquid-state heterojunction sensors. *Nat. Commun.* 5: 5032.
116 So, J., Koo, H., Dickey, M.D., and Velev, O.D. (2012). Ionic current rectification in soft-matter diodes with liquid–metal electrodes. *Adv. Funct. Mater.* 22: 625–631.

117 Strukov, D.B., Snider, G.S., Stewart, D.R., and Williams, R.S. (2008). The missing memristor found. *Nature* 453: 80–83.
118 Menguc, Y., Park, Y.-L., Pei, H. et al. (2014). Wearable soft sensing suit for human gait measurement. *Int. J. Rob. Res.* 33: 1748–1764.
119 Chossat, J.-B., Park, Y.-L., Wood, R.J., and Duchaine, V. (2013). A soft strain sensor based on ionic and metal liquids. *IEEE Sens. J.* 13: 3405–3414.
120 Shin, H.-S., Ryu, J., Majidi, C., and Park, Y.-L. (2016). Enhanced performance of microfluidic soft pressure sensors with embedded solid microspheres. *J. Micromech. Microeng.* 26: 025011.
121 Aïssa, B., Nedil, M., Habib, M.A. et al. (2013). Fluidic patch antenna based on liquid metal alloy/single-wall carbon-nanotubes operating at the S-band frequency. *Appl. Phys. Lett.* 103: 063101.
122 Dey, A., Guldiken, R., and Mumcu, G. (2016). Microfluidically reconfigured wideband frequency-tunable liquid–metal monopole antenna. *IEEE Trans. Antennas Propag.* 64: 2572–2576.
123 Wang, M., Trlica, C., Khan, M.R. et al. (2015). A reconfigurable liquid metal antenna driven by electrochemically controlled capillarity. *J. Appl. Phys.* 117: 194901.
124 Reig, C. and Avila-Navarro, E. (2014). Printed antennas for sensor applications: a review. *IEEE Sens. J.* 14: 2406–2418.
125 Entesari, K. and Saghati, A.P. (2016). Fluidics in microwave components. *IEEE Microwave Mag.* 17: 50–75.
126 Qusba, A., RamRakhyani, A.K., So, J.-H. et al. (2014). On the design of microfluidic implant coil for flexible telemetry system. *IEEE Sens. J.* 14: 1074–1080.
127 Lazarus, N., Meyer, C.D., Bedair, S.S. et al. (2014). Multilayer liquid metal stretchable inductors. *Smart Mater. Struct.* 23: 085036.
128 Fassler, A. and Majidi, C. (2013). Soft-matter capacitors and inductors for hyperelastic strain sensing and stretchable electronics. *Smart Mater. Struct.* 22: 055023.
129 Kong, T.F. and Nguyen, N.-T. (2015). Liquid metal microcoils for sensing and actuation in lab-on-a-chip applications. *Microsyst. Technol.* 21: 519–526.
130 Kubo, M., Li, X., Kim, C. et al. (2010). Stretchable microfluidic radiofrequency antennas. *Adv. Mater.* 22: 2749–2752.
131 Cheng, S. and Wu, Z. (2010). Microfluidic stretchable RF electronics. *Lab Chip* 10: 3227–3234.
132 Cheng, S., Rydberg, A., Hjort, K., and Wu, Z. (2009). Liquid metal stretchable unbalanced loop antenna. *Appl. Phys. Lett.* 94 (14): 144103.
133 Jobs, M., Hjort, K., Rydberg, A., and Wu, Z. (2013). A tunable spherical cap microfluidic electrically small antenna. *Small* 9: 3230–3234.
134 Hayes, G.J., Desai, S.C., Liu, Y. et al. (2014). Microfluidic coaxial transmission line and phase shifter. *Microwave Opt. Technol. Lett.* 56: 1459–1462.
135 Morishita, A.M., Kitamura, C.K.Y., Shiroma, W.A., and Ohta, A.T. (2014). Two-octave tunable liquid-metal monopole antenna. *Electron. Lett* 50: 19–20.
136 Rodrigo, D., Jofre, L., and Cetiner, B.A. (2012). Circular beam-steering reconfigurable antenna with liquid metal parasitics. *IEEE Trans. Antennas Propag.* 60: 1796–1802.

137 Liyakath, R.A., Takshi, A., and Mumcu, G. (2013). Multilayer stretchable conductors on polymer substrates for conformal and reconfigurable antennas. *IEEE Antennas Wirel. Propag. Lett.* 12: 603–606.
138 Saghati, A.P., Singh Batra, J., Kameoka, J., and Entesari, K. (2016). A microfluidically-reconfigurable dual-band slot antenna with a frequency coverage ratio of 3:1. *IEEE Antennas Wirel. Propag. Lett.* 15: 122–125.
139 Damgaci, Y. and Cetiner, B.A. (2013). A frequency reconfigurable antenna based on digital microfluidics. *Lab Chip* 13: 2883–2887.
140 King, A.J., Patrick, J.F., Sottos, N.R. et al. (2013). Microfluidically switched frequency-reconfigurable slot antennas. *IEEE Antennas Wirel. Propag. Lett.* 12: 828–831.
141 Li, M., Yu, B., and Behdad, N. (2010). Liquid-tunable frequency selective surfaces. *IEEE Microw. Wirel. Compon. Lett.* 20: 423–425.
142 Bhattacharjee, T., Jiang, H., and Behdad, N. (2016). A fluidically-tunable, dual-band patch antenna with closely-spaced bands of operation. *IEEE Antennas Wirel. Propag. Lett.* 15: 118–121.
143 Saghati, A.P., Batra, J.S., Kameoka, J., and Entesari, K. (2015). A miniaturized microfluidically reconfigurable coplanar waveguide bandpass filter with maximum power handling of 10 W. *IEEE Trans. Microw. Theory Tech.* 63: 2515–2525.
144 Wu, B., Okoniewski, M., and Hayden, C. (2014). A pneumatically controlled reconfigurable antenna with three states of polarization. *IEEE Trans. Antennas Propag.* 62: 5474–5484.
145 Whitney, R.J. (1953). The measurement of volume changes in human limbs. *J. Physiol.* 121: 1–27.

Index

a

additive techniques   256–257
aggregation-induced emission (AIE) phenomenon   95, 96
air-cured TDI coupling with amino nanofluid   195–196
alkylated liquid porphyrins   13
alkyl chain engineering   2–3
α-cyclodextrin (α-CD)   162
amplified stimulated emission (ASE) thresholds   15
anthracenes   6–8
arylethynyl porphyrins   28, 31
A_4-tetraarylporphyrins   30
atomic force microscopy (AFM)   75, 189, 190
attenuated total reflectance-Fourier transform infrared (ATR–FTIR)   219, 226
azobenzene
   alkyl chain length of   80–82
   isomerization   77
   3-methyl-4,4′-dialkoxyazobenzenes   76
   photo-melting of   75
   polymer film via AFM   75
   sugar alcohol derivatives
      chemical structures of   76
      mixed arms   82–83
      number of   79–80
      structure of   83–84
azobenzene chromophores   5, 102–105, 107, 108, 112, 121
azobenzenes   5–6, 76, 78, 79, 81, 102–104

b

benzothiadiazoles (BTDs)   12
BF_3OEt_2 54, 55
biomacromolecules
   solvent-free bio-liquids   217–226
   solvent-free liquids   212–217
biomass treatment   244–245
bis(1-bromopropoxy)benzene   58
bis-porphyrins   31–35
butadiynyl-bridged bis-porphyrins   32

c

calix[*n*]arenes   53, 55, 60–61, 70
carbazole   3–5, 130–132, 134
carbon nanotubes (CNT)   12, 103, 107, 151, 154–156, 171, 175, 202, 261
circular dichroism (CD)   212
continuous electrowetting (CEW)   258
conventional mechanochromism   88, 91, 95
conventional solid-state OLED   128–130, 137, 143, 147
conventional solution-processing techniques   2
conventional spectroscopic tools, FML   17
core-free corona–cap nanofluid   186
corroles   34
covalent organic frameworks (COFs)   40

Functional Organic Liquids, First Edition. Edited by Takashi Nakanishi.
© 2019 Wiley-VCH Verlag GmbH & Co. KGaA. Published 2019 by Wiley-VCH Verlag GmbH & Co. KGaA.

cowpea mosaic virus (CPMV) 226, 227, 230
crown ethers 53, 60, 70
Cry-Iso phase transition 22, 26, 27, 31, 34
CuAAC reaction 54, 65, 69, 70
cucurbit[n]urils 53, 60–61
cyclodextrins 42, 53, 58–60, 70

d
decaimidazolium bromide 58
decomposition temperature 17, 236
Dianin's compound 41
1,12-diazidedodecane 65
1,12-dibromododecane 62–64
differential scanning calorimetry (DSC) 82
   organic supercooled liquids 89
   porphyrins 22, 23, 26
diketopyrrolopyrrole (DPP) derivative 91–95
1,4-dimethoxybenzene units 54
direct laser patterning 256
D-mannitol 82, 84
DNA–surfactant melts 213, 215
D-sorbitol 84

e
EGaIn droplet 253, 258
elastomeric microchannels 261
electrocapillarity 258, 259
electrochemical methods 237
electrochromic DNA-surfactant liquid devices 217
electroluminescence quantum yield (EQE) 5, 10
electroluminescent organic materials 129
electrowetting on dielectric (EWOD) 258
epoxy coupling with amino nanofluid 198–199
esterase activity 225
ethylene glycol dimethacrylate (EGDMA) 191
Eutectic gallium–indium (EGaIn) 252, 253, 256, 258, 259
excimeric-fluorescent liquid pyrene 10

f
ferritin 176, 179, 218–220
fluidic functional quantum dots 161–162
fluidic graphene/graphene oxide 156–158
Förster-type energy transfer 132, 140, 144
functional colloidal fluids 160–161
functional molecular liquids (FMLs)
   alkylated π-molecular liquids
      anthracenes 6–8
      azobenzenes 5–6
      carbazoles 3–5
      fullerene 12–13
      naphthalenes 6
      π-conjugated oligomers 10–12
      porphyrins 12
      pyrenes 8–10
   alkyl chain engineering 2–3
   alkylsilane chain appended π-molecular liquids
      oligofluorene liquids 15
      phthalocyanines 15
      triarylamines 14
   analytical tools
      calorimetric analysis 17
      fluorescence lifetime 17
      fluorescence measurements 17
      FT-IR 17
      microscopy techniques 16
      rheology 16
      structural analysis 16
      UV-Vis analysis 17
   organic soft materials 1, 2
functional organic liquids
   advantages 101
   molecular design 101
   photoresponsive π-liquids for molecular solar thermal fuels 102–107

g

Galinstan 252, 254
gallium-based alloys 252, 253, 255, 261
graphene oxide epoxy nanocomposites 201

h

hexamethylene diisocyanate (HMDI) 193, 196–198
highest occupied molecular orbital (HOMO) 129, 132, 134, 144
highly emissive supercooled liquids 95–97
hydrated cholinium dihydrogenphosphate 246
hydrated ILs 246, 247
hydrophobic amphiphiles 13, 152
hydroxyl-nanofluid 198

i

incoherent neutron scattering (INS) 219
inorganic nanoparticles
  fluidic functional quantum dots 161–162
  functional colloidal fluids 160–161
  silica nanoparticles 159–160
ionic liquid (IL)
  application of
    biomass treatment 244–245
    electrolyte solution 243–244
    proteins and biofuel cell solvents 246–247
    reaction solvents 243
  derivatives
    polymers 241
    self-assembly technique 239–241
    zwitterions 239
  description of 235
  physico-chemical properties 236, 237
  preparation of 238–239
  water functional mixture 241–243
isothiocyanate nanofluids 192–195

l

large area flexible microfluidic OLEDs 129, 137–140
lignin 245
liquid anthracenes 7, 8
liquid carbazoles 3, 5, 130–132, 134
liquid-crystalline (LC) materials 1
liquid crystalline $trans$-A_2B_2-arylethynyl Porphyrins 31
liquid fullerenes 13, 22, 152–154
liquid metals 251–263
liquid nanocarbons
  CNTs 154–156
  graphene 156–158
  liquid fullerenes 152–154
  π-surface of 151
  solvent-free fluid 151
  synthetic strategies 151–152
  van der Waals (vdW) forces 151
lithium hexafluorophosphate (LiPF$_6$) 224
lithography-enabled processes 254–255
lower critical solution temperature (LCST)-type phase transition 242, 243
lowest unoccupied molecular orbital (LUMO) 129, 135, 144

m

matrix assisted laser desorption/ionization time-of-flight mass spectroscopy (MALDI-TOF-MS) 83
mechanically interlocked molecules (MIMs)
  catenanes 70
  polycatenanes 70
  polyrotaxanes 66
  [2]rotaxane 64
mechanochromism 88, 91, 95
memristor 261
mercaptoacetic acid 162, 175
mercury 6, 108, 251–252, 256
metal organic frameworks (MOFs) 40, 42, 48, 49

methyl tetrahydrophthalic anhydride (MeTHPA)   201
micro-contact printing   257
microfluidic OLEDs
  fabrication method   138
  large area flexible   137–140
  multicolor   140–143
  refreshable liquid electroluminescent devices   135–137
  white emission   143–147
MnSn(OH)$_6$ thread epoxy nanocomposites   201
multiwall carbon nanotubes (MWCNTs)   155, 156, 175
MWCNT polyamide nanocomposites   200
myoglobin   219, 220, 222–224, 230
myoglobin liquid   219, 222

n

nanofillers
  graphene oxide epoxy nanocomposites   201
  MnSn(OH)$_6$ thread epoxy nanocomposites   201
  MWCNT polyamide nanocomposites   200
  nanosilica polyacrylate nanocomposites   199–200
nanoscale ionic materials (NIMs)   152, 158, 159, 162, 171
6-(1)-naphthalene-2-carboxylic acid hexyl ester (BAPTNCE)   3
naphthalenes   6
N-methylfulleropyrrolidine derivatives   153
non-flammable organic fluids   236
nuclear magnetic resonance (NMR)   62, 82, 83, 134, 180, 238, 245
nucleic acid liquids   212–217

o

oligo(ethylene oxide) chains   58, 60
oligofluorene liquids   15
Oligo-(p-phenyleneethylene)s (OPEs)   11–12
Oligo-(p-phenylenevinylene)s (OPVs)   10–11
onium hydroxides   245
organic light emitting diodes (OLEDs)   3
  microfluidic
    fabrication method   138
    large area flexible   137–140
    multicolor   140–143
    refreshable liquid electroluminescent devices   135–137
    white emission   143–147
  organic solid thin films   128
  solvent-free liquid organic light-emitting layer
    carbazole   130–131
    conventional solid-state OLED   129–130
    electrolyte   132–134
    liquid material issues   134–135
organic liquid dyes   3
organic soft materials   1, 2
organic supercooled liquids
  AIE   95, 96
  DPP8   91, 92, 95
  kinetically trapped metastable state   87, 88
  rapid temperature quenching   87
  research of   87
  stimuli responsive optical properties
    shear-triggered crystallization   88–89
    TFMAQ derivatives   89
    TFMAQ derivatives   90

p

perchloropropene   47
per-hydroxylated pillar[n]arenes   56, 58
photo-isomerizable compound   76, 94
photolithography   76, 130, 137, 140, 144, 254
phthalocyanines   15
Piers–Rubinsztajn reaction   14
pillar[n]arenes

host-guest complexation 63
host properties of 61–62
molecular design
　bis(1-bromopropoxy)benzene 58
　calix[n]arenes possess 60–61
　crown ethers 60
　cucurbit[n]urils 61
　cyclodextrins 58–60
　melting points 56–61
　oligo(ethylene oxide) chains 58
　POSS 58
polyrotaxanes 65
pseudopolyrotaxanes 67
pseudo[2]rotaxane 69
[2]rotaxanes 64, 67
synthesis and structure of 54–55
versatile functionality of 55–56
π-conjugated backbones 2
π-conjugated oligomers
　BTDs 12
　OPEs 11–12
　OPV 10–11
π-expanded porphyrins 21
polarized optical microscopy (POM) 16, 22, 108, 158, 213
poly(ethylene glycol) (PEG) 45, 46, 155, 212, 213
poly(L-lysine) (PLL) 220, 221
polydimethylsiloxane (PDMS) 175, 197, 254
polyethylene glycol (PEG) chains 45, 162
polyhedral oligomeric silsesquioxane (POSS) 58, 160, 170, 175, 179, 181, 202
polymerizable zwitterion 239
polymerized ILs 239, 241
polymers of intrinsic microporosity (PIMs) 41, 49
polyoxometalates (POM) 158
polyrotaxanes 54, 65, 67, 70
polytetrahydrofuran (PolyTHF) 65
polyurethane and polyurea coupling, nanofluids
　air-cured, isothiocyanate 192–195

air-cured TDI coupling, amino nanofluid 195–196
hydroxyl-nanofluid 198
PDMS-amino nanofluids coupling, HMDI 197–198
shape memory materials 196–197
porous coordination polymers (PCPs) 40
porous liquids
　application for 49
　porosity
　　in liquids 41–43
　　in solids 40–41
　solvent-free liquid phases 48–49
　type 1 43–46
　type 2 46–48
　type 3 48
porous organic cages (POCs) 41, 44
porous organic polymers (POPs) 40–41
porphyrins 12
　bis-porphyrins 31–34
　complexes vs. free bases 27
　corroles 34
　DSC 22, 23, 26
　isotropisation temperatures 22, 28
　liquid-crystalline
　　trans-A_2B_2-arylethynyl Porphyrins 31
　low-melting trans-A_2B_2-arylethynyl porphyrins 28–31
　meso-substituted A4-porphyrins possessing 22, 24
　POM 22
　tetrakis(pentafluorophenyl)borate anion 27
PPG-di-IPDI 192–194
pseudopolyrotaxanes 67
pseudo[2]rotaxane 67, 69
pyrenes 8–10

q

quantum dots (QDs) 160–162, 171, 175, 261

r

reactive solvent-free nanofluids 169, 183
recapillarity 256
regioisomeric alkylated naphthalenes 6
Rhizomucor miehei (RML) 225, 226
roller-ball dispensing 257
room temperature liquid metals
  applications of 261–263
  controlling interfacial tension
    CEW 258
    electrocapillarity 258–259
    EWOD 258
  gallium-based alloys 252
  low viscosity 251
  mercury 251
  oxide skin, Ga-alloys 252
  oxide skin removal 252–253
  patterning techniques
    additive 256–257
    injection 255–256
    lithography-enabled processes 254–255
    subtractive methods 256
  thermal and electrical conductivity 251
[2]rotaxanes 54, 63–65, 67, 70

s

self-assembled ILs 239–241
shear-triggered crystallization 88–89, 91–95
silica nanoparticles 159–160, 181
SiPNMDD-NPEOPS 185–188, 191, 199
small and wide angle X-ray scattering (SWAXS) 16
small angle neutron scattering (SANS) 10, 16, 159
small-angle X-ray scattering (SAXS) measurements 213
solvent-free bio-liquids 228–229
solvent-free fluidic nanocarbons 151
solvent-free liquids
  nucleic acid
    electrical applications 215–217
    fabrication of 212–215
  protein
    electrochemical applications 222–224
    enzyme catalysis 224–226
    fabrication of 217–222
    viruses 226–228
solvent-free lysozyme-surfactant liquids 220
solvent-free nanofluids
  corona-cap type 176, 181
  metal oxides 175
  molecular dynamics (MD) 180, 181
  nanoparticles surface modification 171
  NMR analyses of 180
  reactive 183
  solid-solid transitions 180
  structures for 172, 175
  surface chemical functionality 169, 170
  syntheses of
    core-corona-cap nanofluid 184–186
    core-coronan nanofluid 186–187
    core-free-corona-cap nanofluid 186
  types and classes of 170
  UV reactive nanofluids
    alkyl glycidyl ether functionalized fullerene 187–188
    atomic force microscopy (AFM) 189
    fullerene epoxy materials 187
    THPETA 188, 189
    2D layered material 190
    UV protective coatings 191
  zero-dimensional systems 175
sugar alcohol derivatives
  chemical structures of 76
  mixed arms 82–83
  number of azobenzene units 79–80
  structure of 83–84
supercharged polypeptides (SUPs) 221, 222, 230

superconducting quantum interference device (SQUID) measurement   161
supramolecular ionic liquids (SIL)   170, 171
surface relief grating (SRG)   75

t

tape transfer printing   257
tertiary amines   238, 239
tetrabutylammonium hexafluorophosphate (TBAHFP)   3, 132
tetrahydroxyethylpentaerythritol tetraacrylate (THPETA)   188, 189
tetraphenylethene   95, 96
thermogravimetric analysis (TGA)   17, 27, 218
thermophile *Thermomyces lanuginosus*   225
thiol containing ionic liquid (IL)   158
thiol-functionalized imidazolium-type IL (TFI-IL)   160
tobacco mosaic virus (TMV)   226
tolyldiisocyanate (TDI)   195–196
trans-A_2B_2-diarylethynyl porphyrins   28
triarylamines   14
trifluoromethylquinoline (TFMAQ) derivatives   89–91

tris[(1-benzyl-1H-1,2,3-triazol-4-yl) methyl]amine (TBTA)   65
two-photon absorption (2PA)   28, 30–32, 35
type 1 porous liquids   42, 44, 45
type 2 porous liquids   42, 46
type 3 porous liquids   42, 48, 49

u

UV protective coatings   191
UV reactive nanofluids
  alkyl glycidyl ether functionalized fullerene   187–188
  atomic force microscopy (AFM)   189
  fullerene epoxy materials   187
  THPETA   188, 189
  2D layered material   190
  UV protective coatings   191

v

van der Waals (vdW) forces   151
van der Waals (vdW) interactions   3, 26, 108, 120, 228

w

white OLEDs   144, 147

z

zwitterions   239–241, 244